Mastering Solid-State Amplifiers

Other books in the series
Mastering IC Electronics

Mastering Solid-State Amplifiers

Joseph J. Carr

FIRST EDITION
FIRST PRINTING

© 1993 by **TAB Books**.
TAB Books is a division of McGraw-Hill, Inc.

Library of Congress Cataloging-in-Publication Data

Carr, Joseph J.
    Mastering solid-state amplifiers / by Joseph J. Carr.
      p.   cm.
    Includes index.
    ISBN 0-8306-3081-1   ISBN 0-8306-3081-3 (pbk.)
    1. Transistor amplifiers. I. Title.
  TK7871.2.C343 1992
  621.381′535—dc20   92-10488
                                          CIP

TAB Books offers software for sale. For information and a catalog, please
contact TAB Software Department, Blue Ridge Summit, PA 17294-0850.

Acquisitions Editor: Roland S. Phelps
Editor: Suzanne L. Cheatle
Managing Editor: Sandra Johnson
Director of Production: Katherine G. Brown
Book Design: Jaclyn J. Boone
Cover Design: Holberg Design, York, Pa.   EL1

Contents

Introduction

ALTHOUGH ELECTRONICS IS A VERY BROAD AND VARIED FIELD, ONE OF THE central themes is *amplifiers.* Indeed, when radio kicked off the electronics industry prior to World War I, the milestone event that separated earlier efforts from the modern era was the invention of the amplifying vacuum tube in 1903 by Lee DeForest. Prior to that time, electronics progress was limited because of the insensitivity of radio receivers (radio sets were the only electronic devices in those days). Today, even though we are quite sophisticated in all areas of electronics, the old-fashioned amplifier — no longer a clunky vacuum tube, but rather a sleek integrated circuit that packs more punch in a tiny fraction of the space — is still the defining circuit in electronics.

This book is designed to teach you about amplifiers from the most basic to some reasonably sophisticated designs. If you diligently follow the text, perform the experiments, and build some of the projects, you will come to an understanding of how amplifiers work.

No time is spent on archaic vacuum tubes, so this book is essentially a book on solid-state amplifiers; e.g., transistors and integrated circuits. Following some introductory material electronics (chapter 1), electronic circuit construction techniques (chapter 2), and basic electrical theory (chapter 3), you will find a chapter on dc power supplies (chapter 4). This chapter contains a project for a dc power supply that can be used to run the experiments and operate the projects found later in the book. If you already own a suitable bench supply, intend to buy one, or intend to use batteries (usually a good idea for newcomers), then you might want to skip this chapter. However, the material is of general interest to anyone delving into electronics, so it is recommended that you at least read chapter 4, even though you don't intend to build the project.

Semiconductor materials are introduced in chapter 3 during the discussion of basic electrical principles. The use of semiconductors in building a simple pn junction diode is discussed in chapter 5. The pn junction diode isn't an amplifier, but understanding this device is central to understanding transistor operation. Amplifiers are introduced generically in chapter 6, followed by a discussion of bipolar npn and pnp transistors in chapter 7 and transistor amplifiers in chapter 8. The field-effect transistor (JFETs and MOSFETs) is introduced in chapter 9.

No invention, save those of the transistor and the vacuum tube, was more important to the advance of electronics than the integrated circuit (IC). These amplifier devices are covered starting in chapter 10. The operational amplifier is, perhaps, the most widely used linear IC amplifier, so it has an entire chapter of its own (chapter 11), followed by a general discussion (chapter 12) on other forms of linear IC amplifiers (CDAs and OTAs).

Applications are discussed, beginning with audio amplifiers in chapter 13 and small-signal RF amplifiers in chapter 14. If you are a ham radio operator or shortwave listener, then chapter 14 will hold special interest for you.

Finally, the book is rounded out with chapters on testing solid-state amplifiers (chapter 15) and substituting transistors in circuits (chapter 16).

The appendix discusses electrical safety. I recommend that you read the appendix first so that you are more familiar with some of the very real hazards found in electronics.

1
Electronics
is a verb!

ELECTRONICS IS FOR DOING, AND THIS BOOK IS INTENDED FOR DOERS. IN IT you will learn the basics of solid-state amplifier circuits. When you finish reading this book, you will be able to design some forms of amplifier, and either have the basis for a long-time hobby, or have a leg up on further study in the field.

Electronics is based historically in radio, and also on the primitive electrical industry that existed at the turn of the century. The first radio receivers were not terribly sensitive because they used simple circuits and devices, such as Branly coherers and galena crystals, for the detection of signals. A crucial need at that time was for a better detector and, above all, an amplifier device. The device that provided both of these functions was the vacuum tube — the first true electronic component.

For many years, the electronics industry was wedded to radio, but that situation changed as World War II approached. Many of the components, circuits, and techniques of radio were found to be equally useful for other purposes as well. For example, the amplifier could be used for consumer audio, motion picture, and other entertainment products, and also for electronic instruments and recording devices. The first scientific and medical instruments appeared circa World War I, and by World War II were an established fact. Indeed, they represented a new industry separate from radio.

World War II presented the electronics industry with an unprecedented opportunity for advancement. Many new military applications of electronics were invented, and many translated into civilian technology after the war: communications, radar, control systems, and many others. During the war, the first solid-state signal diodes were created for the VHF through microwave regions of the electromagnetic spectrum. These advances led directly to the invention of the transistor in 1948.

The transistor was invented by a team of three physicists at Bell Laboratories, a research arm of American Telephone and Telegraph's Bell System. One form of the device was actually postulated in scientific circles prior to the war by one of the three inventors, but it took the mighty research effort of the war years to produce the purity of semiconductor materials needed to invent these new devices.

Perhaps one of the most momentous decisions of the electronics era was the decision by AT&T to license the transistor to other companies. Although arguably in its own best interest, for it spurred on new technology at an unprecedented pace, this decision opened the solid-state era at a much earlier date than would have been otherwise possible. Within ten years, vacuum tubes had disappeared from large areas of electronics technology.

The first transistor radios appeared in 1954, with most of the first models being portable AM radios. The low power consumption and low voltage requirements of transistors made them ideal for portable radios. Motorola produced a universal car radio using transistor technology, but it took until 1962 for the first transistor car radio to be produced for a major automobile manufacturer. That year, the Delco Radio division of General Motors produced all-transistor radios for GM cars. From 1957 to 1962 (or 1963 for other brands), the design of car radios used low-voltage vacuum tubes for all stages except the detector and audio stages, where diodes and transistors were used. By 1963, however, all car radios were transistorized, including AM/FM models.

When I first became interested in electronics as a high school student in the late 1950s, one of my teachers predicted that semiconductor manufacturers would develop technology that would permit putting an entire audio amplifier chain, or even an entire radio, on a single crystal chip not larger than a thumbnail. Although that visionary prediction seemed a little on the wild side in 1959, by 1969 it was a fact, and devices were in production in home, portable, and car radio receivers.

The prediction that my first electronics teacher made was the invention of the *integrated circuit*, or *chip*. These devices integrate onto a single monolithic crystal of semiconductor material all of the transistors, diodes, and resistors needed to make a circuit. Some modern circuits also include inductances and capacitances, but only at very high frequencies. These components, being built onto a small crystal, are very dense, so the size of electronic devices can be reduced considerably.

Digital electronics also progressed at a tremendous pace. (The modern computer would not be practical for most users were it not for the semiconductor industry.) Digital circuits are those that obey the rules of the binary, or *base-2*, arithmetic system. In these devices, the numerals "0" and "1" are presented by two different voltage levels, all other voltage levels being prohibited. The first digital devices in large-scale production were the *resistor transistor logic* (RTL) and *diode transistor logic* (DTL). In the 1960s, the now ubiquitous *transistor transistor logic* (TTL) devices were produced. The TTL devices were followed soon with the *complementary metal oxide semiconductor* (CMOS) devices. The latter required such low power that many portable devices previously impossible now became easy.

One of the startling aspects of integrated electronics that has been repeated endless times is the well-known "learning curve" phenomenon. The price of IC devices typically starts out very high when the technology is new, and then drops rapidly. For example, the μA-709 operational amplifier started out over \$120 in the early 1960s (when dollars were bigger!), but now cost less than 50¢; the μA-703 RF/IF amplifier cost \$15 to replace in the middle to late 1960s, when I was servicing electronic equipment, but now cost two bits, when you can find them. Similarly, I paid \$10 for a 741 operational amplifier in the mid-1960s, but now they cost 20¢ in commercial plastic packages. The low cost of these devices makes it reasonable for hobbyists, amateurs, and students to experiment with ICs, and use them in actual, practical projects.

Today, electronics includes a bewildering variety of functions from radio broadcasting, communications of many types, control systems, instrumentation, and many thousands of other applications. Even your automobile is highly dependent on electronics. While older cars had only a radio as an electronics device, modern cars are run on central computers, and have dozens of other electronics devices on board. Even the ignition system is now an electronics circuit.

This book is about amplifiers, and more specifically, solid-state amplifiers. You will first be introduced to electronics components and electronic construction practices. You will also be introduced to electronic instruments you'll need to work the experiments, build the projects, or enjoy electronics in any deeper sense. Indeed, I recommend strongly that the newcomer to electronics build or buy a number of electronic instruments for his or her own use.

This book is intended to be open-ended, for I hope that you will pursue electronics in ever deeper circles. This publisher has an outstanding collection of electronics textbooks, most of them quite suitable for either self-study or classroom use, from the most elementary levels through the intermediate and most advanced levels. Once you master a level of difficulty on any given topic in electronics, then you might want to either go deeper into the same topic or explore another topic. Practical books will help you in gaining this deeper understanding.

If I ever have the pleasure of visiting your lab or workshop, then I hope to find a copy of this book on the workbench, bent, dirtied, and well marked as you pursue the material presented herein. It is no dishonor to the author of a "how-to" self-instructional book to find that readers have been using the book "where the solder meets the hot tip . . . on the bench."

Now that I've laid some foundation by telling you where we came from, and where we are going, let's start the journey and actually start where we intend to go! The first topic that I'll discuss is electronic construction methods. The discussion is not exhaustive, but rather deals with the principal methods needed for building amplifier circuits and performing the experiments in this book.

2
Electronic workbench skills

ALTHOUGH ELECTRONICS IS CLEARLY A VERY PRACTICAL FIELD, IF THE theory is not somehow converted to practice — hardware on the bench — then the whole thing is a bit of a waste of time. Much of this book is about electronic theory, although written from a practical perspective. In this chapter I'll take a look at the real "down and dirty" workbench aspects of electronics.

Power supplies for laboratory experiments

Many readers perform laboratory experiments both independently and as part of a formal class. Guidelines for experiments are given at various points in this book. A power supply must be selected (or built) to work these experiments. All of them can be worked with batteries, although you might prefer to buy or build your own power supply.

Unless otherwise specified, the experiments in this book are designed to use +12 volt (±12 volt) dc from a regulated power supply. The power supply selected should offer either a single bipolar power supply or two independent 12-volt dc (Vdc) supplies that are not ground referenced. The nongrounded feature allows you to create a bipolar supply by connecting the positive output terminal of one supply to the negative output terminal of the other (Fig. 2-1).

Desirable features to have on a bench power supply include the following:

- Output voltages fixed at ±12 Vdc (or ±15 Vdc), or
- Output voltages adjustable from zero to greater than 12 Vdc
- Current available from each polarity not less than 100 milliamperes
- Metered outputs (I and V)
- Voltage regulation
- Current limiting for output short circuit protection
- Overvoltage protection

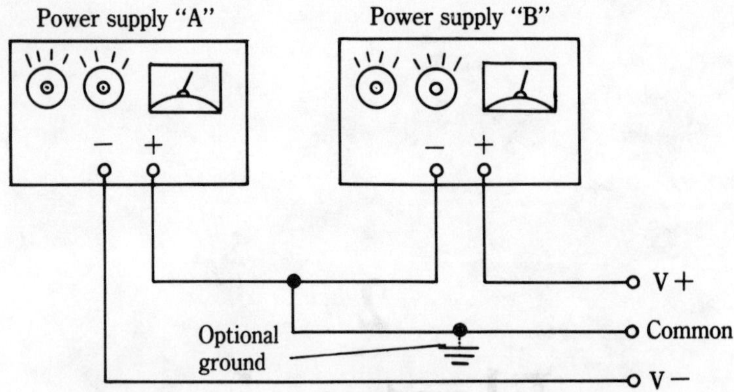

2-1 dc power supply connection to make op amps and other linear ICs work.

Although not all good dc power supplies have all of these features, those that do are clearly superior for most laboratory applications. All of the experiments in this book can be run on a power supply that substantially meets these requirements.

Electronic construction media

Electronic projects must be constructed on some sort of base or media. Although novices tend to build their first few projects "rat's nest style" (i.e., in sort of a three-dimensional "wire sculpture" form), more experienced people use a chassis, printed circuit board, perforated board, or some other media.

Successfully performing experiments and designing or debugging circuits on a professional level requires certain skills and a knowledge of construction practices. In this section I'll take a look at some of these practical matters.

There are various levels of construction practice, with four major identifiable categories:

1. partial breadboard
2. complete breadboard
3. brassboard
4. full-scale development model

The two "breadboard" categories involve using special construction methods to check circuit validity and make certain preliminary measurements. The breadboard is in no way usable in actual equipment, but rather is a bench or laboratory device.

The difference between partial and complete breadboards is merely one of scale. The partial breadboard is used to check out a partial circuit, or a small group of circuits. The complete breadboard will be the full-up circuit, and is much more extensive than the partial breadboard. There are fundamental differences in the type of breadboarding hardware used for both types, especially in large circuits.

A commonly used type of breadboard hardware is shown in Fig. 2-2. The device is an "IC pin socket" breadboard. Some pin socket breadboards offer two or three built-in dc power supplies. Typical specifications are: +12 V at 100 mA, −12 V at 100 mA and +5 V

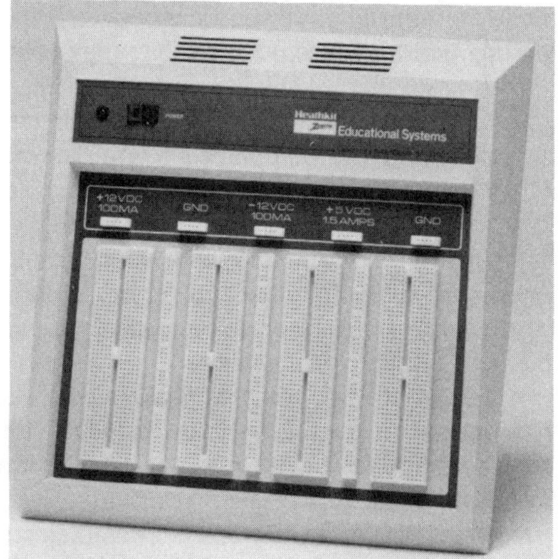

2-2   Electronic construction breadboard.

at 1.5 A (for TTL digital logic IC devices). Located between the large multipin IC sockets are bus sockets in which all pins in a given row are strapped together. These sockets are generally used for power distribution and common ("ground") buses. Interconnections between sockets, power sources, and components are made using #22 insulated hook-up wire. The bared ends of the wires are pressed into the socket holes.

The other method of breadboarding is wire-wrapping (Fig. 2-3). Sockets with special

2-3   Wire-wrap connections to printed board terminals.

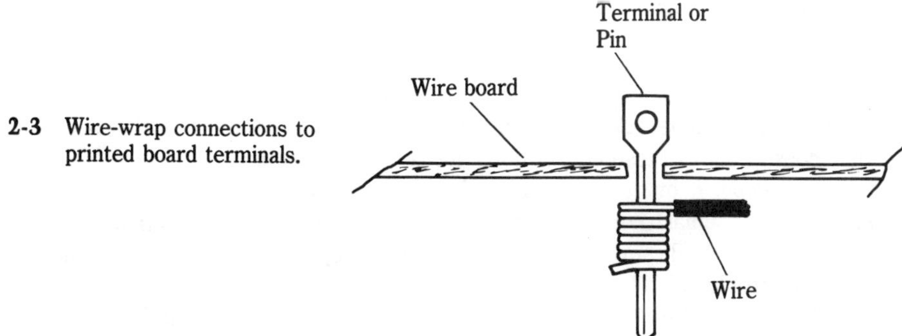

square or rectangular wirewrap pins are mounted on a piece of "universal" printed circuit board. These boards are perforated to accept wirewrap IC sockets and usually have "printed" power distribution and ground buses. Using a special tool, the wire interconnects are strapped from point to point as needed by the circuit. The wire is wrapped around each contact tightly enough to break through the insulation to make the electrical connection.

In general, socketed breadboards are typically used for student applications, for partial breadboarding in professional laboratories, and for small design projects that don't justify the cost of the wirewrap board. Wirewrapping is used for larger, more extensive, projects. Although the distinctions between types of breadboard, and the respective supporting hardware, are quite flexible, they hold true for a wide range of situations.

Brassboards are full-up model shop versions of the circuit that will plug into the actual equipment cabinet in which the final circuit will reside. Some brassboards are wire-wrapped, but most are printed circuits. The important consideration is that the brassboard is near its final configuration and can be tested in situ in the actual equipment. In contrast, the breadboard is purely a laboratory model. The brassboard is usually made using model shop handwork methods, not routine automated production methods. The brassboard differs from the final version in that it might contain "dead bug" and "kluge card" modifications in which components are informally mounted on the board for purposes of testing circuit changes.

The full-scale development (FSD) model is the highest preproduction model. It might look very much like the first article production model that is eventually made and is intended for field tests of the final product. For example, an FSD model of a two-way land-mobile radio transceiver might be mounted in an actual vehicle and used for its intended purpose in tests or field trials. For aircraft electronics, the FSD model must be built and tested according to flight-worthiness criteria. The FSD model is built as near to regular production methods as possible.

Some points to remember

There are several principles to remember when breadboarding electronic circuits. Although it is possible to get away with ignoring these rules, they constitute good practice and ignoring them is risky. The rules are:

- Insert and remove components only with the power turned off.
- Wire and make changes to wiring with the power turned off.
- Check all wiring prior to applying power for the first time.

Power distribution and ground wiring is always done first, and then checked before any other wiring is done. (*Note:* in many schematics the V− and V+ wiring is omitted, but that doesn't mean these connections are not to be made.)

Use *single-point*, or *star*, grounding wherever possible. Do not use a ground plane or ground bus unless signal frequencies are low, signal voltage levels are relatively large, and current drain from the dc power supply is low. Otherwise, ground noise and "ground loop" voltage drop problems will exist.

Applying a signal to IC input pins when dc is not applied can cause the substrate "diode" to be forward biased, with the potential for damage to the device being very high. Therefore:

- Do not apply signal until the dc power is turned on
- Turn off the signal source prior to turning off the dc power supply
- Never apply a signal with a positive peak that exceeds $V+$ or a negative peak that exceeds $V-$.

All measurements are to be made with respect to common or ground, unless special test equipment is provided. For example, in Fig. 2-4 differential voltage V_d is composed of two ground-referenced voltages, V_1 and V_2. Measure these voltages separately and then take the difference between them (i.e., $V_2 - V_1$).

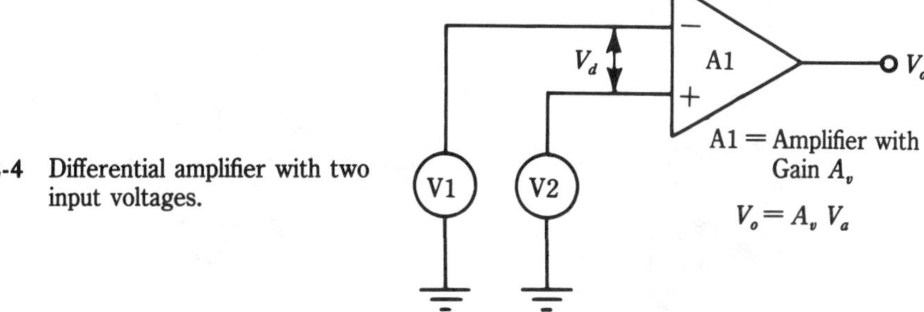

2-4   Differential amplifier with two input voltages.

Although it is relatively easy today to make a small printed circuit board using "homebrew" techniques, most small projects are constructed, at least initially, on perforated wiring board (usually just called *perf board*). Figure 2-5 shows both RF (Fig. 2-5A)

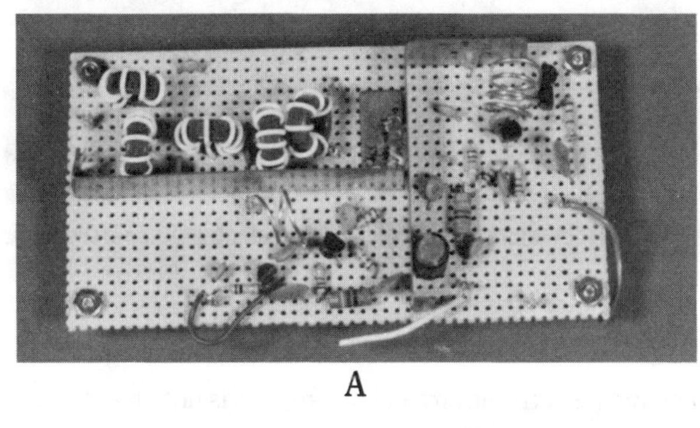

A

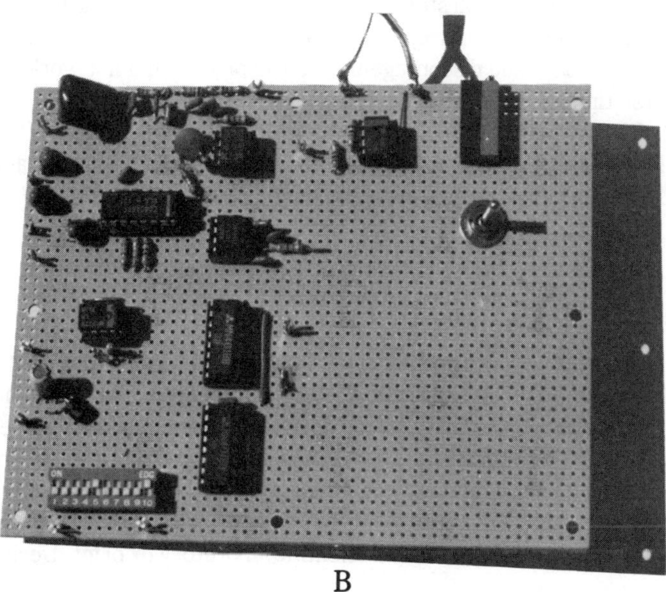

B

2-5   A. RF project on perfboard; B. digital project on perfboard.

and subaudio (Fig. 2-5B) projects built on perf board. RF projects should be housed in metal shielded chassis, such as those shown in Fig. 2-6.

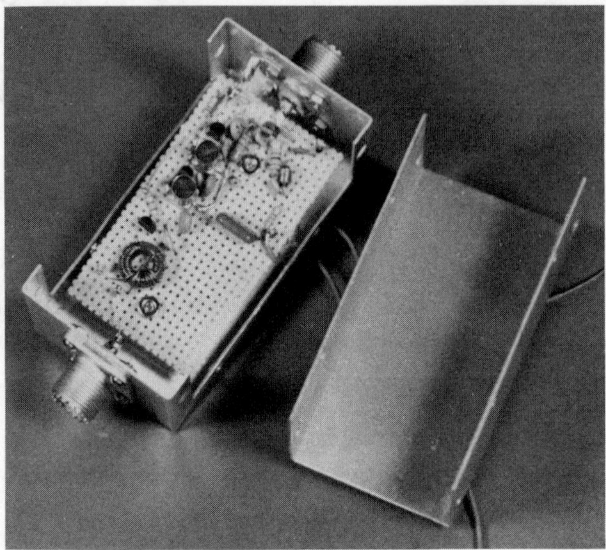

2-6   Shielded aluminum box used for electronic construction.

Designing electronic projects

The process of designing electronic circuits or projects is not an arcane art, open only to a few initiates. Rather, it is a logical step-by-step process that can be learned. Like any skill, design skill is improved with practice, so you are cautioned against both excessive expectations "first time at bat" and discouragement if the process did not work out exactly as planned the first time.

Some of the material in this chapter might be called "philosophy" (by which the speaker usually means not true philosophy but the other person's opinion), and that might be a fair label. Although many technical people claim to disdain "philosophy," we all have it and use it.

A design procedure

The procedure that you adopt to designing might well be different from what is presented here, and that's all right. The purpose of offering a procedure is to systematize the process. Although it is conceivable that you can design an instrument by a process similar to Brownian motion (as, theoretically, a Tom Clancy novel can be written by an infinite collection of monkeys pounding randomly on an infinite number of PC keyboards), it is the systematic approach that most often yields success. This procedure assumes a product that is a one-of-a-kind instrument, as in a scientific laboratory or plant. Designing a product for production and sale follows a similar procedure, but involves marketing and production problems as well. The steps in the procedure, some of which are iterative with respect to each other, are offered here:

1. Define and tentatively solve the problem.
2. Determine the crucial attributes required of the final product and incorporate them into a specification and a test plan that determines objective criteria for acceptance or rejection.
3. Determine the crucial parameters and requirements.
4. Attempt a "block diagram" solution.
5. Apportion requirements (e.g., gain, frequency response, etc.) to the various blocks.
6. Perform analysis and do simulations on the block diagram to test the validity of the approach.
7. Design specific circuits to fill in the blocks.
8. Build and test the circuits.
9. Combine the circuits with each other on a breadboard.
10. Test the breadboarded circuit according to a fixed test plan.
11. Build brassboard that incorporates all changes made in the previous steps.
12. Test the brassboard and correct problems.
13. Design and construct the final configuration.
14. Test the final configuration.
15. Ship the product.

Solving the (right) problem   The purpose of the designer is to solve some problem or another using analog circuits, digital circuits, a computer, or whatever else is available in the armamentarium. Two related problems are often seen in the efforts of novices.

First, often the designer has a tentative favorite approach in mind before the problem is properly understood. Decisions are made based on what the designer is most comfortable doing. For example, many younger designers are likely to select the digital solution in a knee-jerk manner that excludes any consideration of the analog solution. Both should be evaluated, and the one that best fits the need, selected.

Second, be sure you understand the problem being solved. While this advice seems trivial, it is also true that failure to understand the problem at hand often sinks designs before they have a chance to manifest themselves. There are several facets to this problem. For example, a natural tendency of engineers is to think that an elegant solution is complex and large scale. If this mistake is made, it is likely that the product will be over-designed and have too many whistles and bells. It was, after all, designed to solve a much harder problem than was actually presented.

Another aspect to understanding the problem is to understand the final customer's use for the product. It is all too easy to get caught up in the specification, or our own ideas about the job, and overlook altogether what the user needs to accomplish with the product. An example is derived from biomedical instrumentation. A physiologist requested a pressure amplifier that would measure blood pressures over a range of 0 to 300 mmHg (Torr). What the salesman never told the plant was that it would be used on humans (safety and regulatory issues); that blood would come in contact with the diaphragm (cleaning and/or liquid isolation issues); and that it would occasionally be used for measuring 1 to 5 mmHg central venous pressures (low-end linearity issues).

Part of the problem in determining the level of complexity, or the specific design's function, is miscommunication between the end user and the designer. Although miscommunications occur frequently between in-house designers and their "clients," it is probably most common between distant customers and engineers in the plant. Of course, marketing people might never let the engineer and customer get together (either from ignorance or a fear that the salesman's little lies will surface).

The proper role of the designer is to scope out the problem at hand, to understand what the circuit or instrument is supposed to do, how and where the user is going to use it, and exactly what the user wants and expects from the product.

Determine crucial attributes Determining crucial attributes is basically the fruit of understanding and solving the correct problem. From the solution of the problem you can determine, and write down, a set of attributes, characteristics, parameters, and other indices of the product's final nature.

It is at this point that you must write a specification that documents what the final product is supposed to do. The specification must be written clearly so that others can understand it. A concept or idea does not really exist, except in the mind of the originator. One must, according to W. Edwards Deming,* create *operational definitions* for the attributes of the product. One cannot simply say "it must measure pressure to a linearity of 1 percent over a range of 0 to 100 psi." Rather, it might be necessary to specify a test method under which this requirement can be met. There might be, after all, more than one standard for pressure and measurement, and there is certainly more than one definition of linearity. The operational definition serves the powerful function of providing everyone with the same set of rules—basically it levels the playing field.

Part of this step, and of making an operational definition, is to write a test plan for the final product. It is here that you determine (and often contractually agree) exactly what the final product will do and define the objective criteria of goodness or badness that will be used to judge the product.

Determine crucial parameters and requirements Once the product is properly scoped out, it is time to determine the crucial technical parameters that are needed to meet the test requirements (and hopefully the user's needs—if the test requirements are properly written). Parameters such as frequency response and gain tend to vary in multimode instruments, so you must determine the worst case for each specification item and design for it.

**Out of the Crisis,* W. Edwards Deming, MIT Press.

Attempt a block diagram solution   The block diagram is a signals flow or function diagram that represents stages, or collections of stages, in the final instrument. In large instruments there might be several indexed levels of block diagram, each one going finer in detail.

Apportion requirements to the blocks   Once the block diagram solution is on paper, tentatively apportion system requirements to each block. Distribute gain, frequency response, and other attributes to each block. Keep in mind that factors such as gain distribution can have a profound effect on dynamic range. Also, the noise factor and drift of any one amplifier can have a tremendous effect on the final performance. It's in these types of parameters (where crucial placement of one high-quality stage might be sufficient) that added cost and complexity often arise.

Analyze and simulate   Once the block diagram is set and the requirements apportioned to the various stages, it is time to analyze the circuit and run simulations to see if it actually works. A little "desk checking" goes a long way toward eliminating problems later when the thing is first prototyped. Plug in typical input values and see what happens on a stage-by-stage basis. Check for the reasonableness of outputs at each stage. For example, if the input signal should drive an output signal to, say, 17 volts, and the operational amplifiers are operated only from 5-volt power supplies, then something is wrong and will have to be corrected.

Design specific circuits for each block   On this point the remainder of this book is most useful to you, for it is on specific circuits that I later dwell. In this step, fill in those blocks with real circuit diagrams.

Build and test the circuits   At this point you must actually construct the individual circuits and test them to make sure they work as advertised (unless, of course, the circuit is so familiar that no testing is needed). Keep in mind that some of your best ideas for simplified circuits might not actually work — and now is the time to find out. Use a benchtop breadboard that allows circuit construction using plug-in, stripped-end wires.

Combine the circuits in a formal breadboard   Once the validity of the individual circuits is determined, combine them together in a formal breadboard. Whether you build on a benchtop breadboard or on a prototyping board, make sure that the layout is similar to that expected in the final form.

Test the breadboard   Test the overall circuit according to formally established objective criteria. This test plan should be developed earlier (in step two). As problems arise and are solved, make changes or corrections — and document the results. It is, perhaps, the main failing of inexperienced designers that they do not properly document their work, even in an engineer's or scientist's notebook.

Build and test a brassboard version   Although breadboarding techniques can be a little sloppy, the brassboard should be a proper printed circuit board. The test criteria should be the same as before, updated only for changes that occurred. If problems turn up, correct them prior to proceeding further. Keep in mind that the most common problems that occur in leaping from breadboard to brassboard are layout (e.g., coupling between stages), power distribution, and ground noise. (These are the areas of difference between the two configurations.)

Design, build, and test the final version   Once all of the problems are known and solved and changes are incorporated, it is time to build the final product as it will be given to the end user. It is at this point that the reputation of the designer is made or broken because it is now that the product is finally evaluated by the client.

Electronic instruments and their use

Some particularly skilled electronics enthusiasts seem to be able to divine the problem in a circuit, almost if by magic. Although it seems like magic to the noninitiate, it is really only a reflection of the fact that a large amount of experience is at work. However, even the most skilled, seemingly magical, electronics technologist will agree that a good selection of test equipment is the key to wringing out circuit problems and repairing faults.

The nature of the test equipment you select depends on the job you want to do. I assume that the readers of this book are interested in solid-state circuits, integrated circuits, and dc and relatively low frequency (< 1 MHz) ac circuits. The test equipment discussed in this chapter is aimed at those readers.

For integrated circuit work you need several different types of instruments: multimeter, signal generators, oscilloscope, power supplies, and certain accessories, which I'll deal with shortly. Let's take a look at these classes, and some practical examples, each in their turn.

Kit-built instruments are a good idea, especially for newcomers. In addition to saving you a little bit of money, the kit-built instrument provides you with project construction experience under controlled circumstances. I have built scores of *Heathkits*, made by the Heath Co., Benton Harbor, MI 49022.

Multimeters

A *multimeter* is an instrument that combines into one package a multirange voltmeter, milliammeter (or ammeter), ohmmeter, and possibly some other functions. The single instrument measures all of the basic electrical parameters that the technician needs to understand to troubleshoot a receiver.

VOMs   The earliest multimeters were instruments called *volt-ohm-milliammeters* (VOMs). These instruments are passive devices, and the only power used in them is a battery for the ohmmeter function. Because they are passive instruments, they are still preferred for troubleshooting medium- to high-power radio transmitters. They are included here, even though not recommended, because they are low cost. They are available at places like Radio Shack®.

The sensitivity of the VOM is a function of the full-scale deflection of the meter movement used in the instrument. Sensitivity is rated in terms of *ohms per volt fullscale*. Three common standard sensitivities were found: 1,000 ohms/volt; 20,000 ohms/volt; and 100,000 ohms/volt, representing full-scale meter sensitivities of 1 mA, 100 μA, and 10 μA, respectively.

Voltage is read on the VOM by virtue of a multiplier resistor in series with the current meter, which forms the basic instrument movement. The front panel switch selects which multiplier resistor, hence which voltage scale, is used.

VTVMs   The earliest form of active multimeter was the *vacuum tube voltmeter* (VTVM). These instruments used a differential balanced amplifier circuit based on dual triode tubes (e.g., 12AX7) to provide amplification for the instrument. As a result, the VTVM was a lot more sensitive than the VOM. It typically had an input impedance of 1 megohm, with an additional 10 megohms in the probe, for a total input impedance of 11 megohms. Compare this feature with the VOM, which had a different impedance for each voltage scale.

FETVMs   The next improvement in meters was the solid-state meter. These instruments used a field-effect transistor in the front end, and other transistors in the rest of the circuit. As a result, these instruments were called *field-effect transistor voltmeters* (FETVM). The FETVM has a very high input impedance, typically 10 megohms or more.

DMMs   Finally, the most modern forms of instrument are various digital multimeters (DMM), both hand-held and bench (Fig. 2-7). These instruments are the meter of choice today if you are going to purchase a new model. In fact, it might prove a little difficult to find a nondigital model anywhere except Radio Shack.

The DMM is preferred today because it provides the same high-input impedance as the FETVM. However, because the display is numeric instead of analog, the readout is less ambiguous.

There is one bit of ambiguity, however, and that is *last-digit bobble.* The digital meter allows only certain discrete states, so if a minimum value is between two of the allowed states, the reading might switch back and forth between the two (especially if noise is present). For example, a voltage might be 1.56456 volts, but the instrument is only capable of reading to three decimal places. Thus, the actual reading might bobble back and forth between 1.564 and 1.565 volts.

Last-digit bobble is not really a problem, however, unless you bought an instrument that is not good enough. The number of digits on the meter should be one more than that required for the actual service work. The way that the digits are specified in DMMs is in terms of the number places that the meter can read. For example, some instruments are called 2½-digit models. The "½" digit refers to the fact that the most significant digit (on the left side) can only be 1 or 0 (in which case it is blanked off). Thus, the full scale of this meter is always "199." Where the decimal point is placed is a function of the range setting.

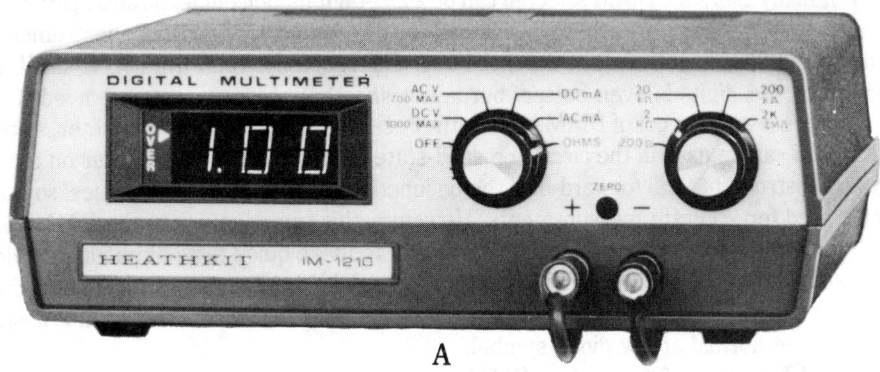

A

2-7   A. Bench type digital multimeter (DMM); B. handheld DMM.

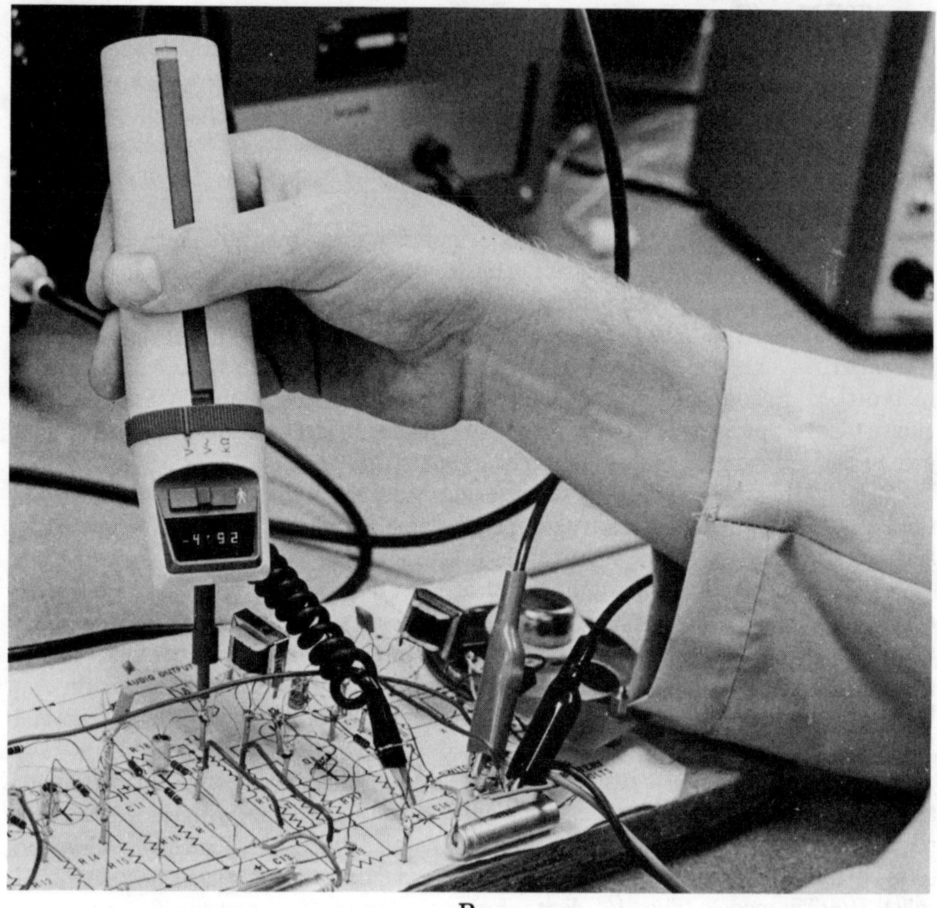

B

2-7   Continued

For many users the DMM selected can be a 2½-digit model, but if you anticipate a lot of detailed solid-state work, then opt instead for a 3½-digit model. These instruments read to "1999" on each scale. Oddly, the cost difference between the two is small. A model with 4½-digits is even better, but is not strictly necessary for most needs.

An interesting aspect of DMMs is that they use low voltage for the ohmmeter, so the instrument can be used in the circuit on solid-state equipment. The ohmmeter on other forms of instruments will forward-bias the pn junctions of the solid-state devices, so cannot be used for accurate measurements. However, this same feature of the DMM also means that the normal ohmmeter cannot be used to make quick tests of pn junction devices, such as transistors and diodes. However, some DMMs have a special switch setting that allows this operation. In some it is a "high-power" setting, while in others it is designated by the normal arrow diode symbol.

Another feature of the modern DMM is an aural continuity tester. It sounds a beep anytime the resistance between the probes is low, indicating a short circuit exists. This feature is especially useful for testing multiconductor cables for continuity when the actual resistance is not a factor.

Older instruments   As is true with other equipment, the multimeter shows up on the used market (especially hamfests) quite often. It is easy to obtain workable, if older, instruments at cheap prices, especially now that everyone seems to need to dump old analog instruments in favor of the new digital varieties. However, there are some pitfalls.

First, make sure that the instrument isn't so old that it uses a 22.5-volt (or higher) battery in the ohmmeter section. Those instruments were made prior to about 1956 and will blow most transistors or integrated circuits that you measure with them. Look out especially for the oversized RCA and Hickcok instruments that were once the mainstay of radio service shops.

Second, make sure that there is nothing wrong with the instrument that cannot be repaired easily. VTVMs tend to be in good shape, or at least repairable condition. Unfortunately, a lot of VOMs are in terrible shape. Part of the problem comes from the fact that the meter measures current easily enough, but if the operator left the meter on the current setting and then measured a voltage, *pooff*! The instrument burned up. When examining an instrument, try to make a measurement on each scale. Alternatively, remove the case and look for charred resistors and burnt switch contacts.

In any event, when you obtain a used meter, be sure to take it out of the case and clean the switch contacts and generally spruce it up. The usual forms of contact and tuner cleaner work wonders on used multimeters as well. A simple cleaning of the switch(es) might stop intermittent or erratic operation.

Signal generators

The purpose of a signal generator is to produce a signal that can be used to troubleshoot, align, adjust, or simply prove the performance of a piece of electronic equipment. Fortunately, only a few basic forms of signal generator are needed. Signal generators for this market come in several varieties, including audio generators, function generators, and RF signal generators.

Audio generators produce at least sinewaves on frequencies within the range of human hearing (20 Hz to 20 kHz). Some models produce frequencies over a greater range, while others produce squarewaves as well as sinewaves. The standard audio signal generator has a variable amplitude or level control, and a fixed output impedance of 600 ohms. Audio generators might or might not have an output level meter.

A *function generator* is much like an audio generator but also outputs a triangle waveform in addition to the sinewave and squarewave signals. Some function generators also produce pulses, sawtooths, and other waveforms. Typical function generators operate from less than 1 Hz to more than 100 kHz. Many models have a maximum output frequency of 500 kHz, 1 MHz, 2 MHz, 5 MHz, and in at least one case 11 MHz. Like the audio generator, the standard output impedance is 600 ohms. However, some instruments also offer 50-ohm outputs (standard for RF circuits) and a TTL-compatible output. The latter is a digital output that is compatible with the ubiquitous transistor transistor logic (TTL or

T²L) family of digital logic devices (these are the ones with either 74xx or 54xx type numbers).

The *sweep function generator* allows the output frequency to sweep back and forth across a range around the set center frequency. As a result, you can do frequency response evaluations using the sweep function generator and an oscilloscope.

The *RF signal generator* outputs signals generally in the range above 20 kHz and typically has an output impedance of 50 Ω. This value of impedance is common for RF circuits except in the TV industry (where 75 Ω is the standard). Various types of RF signal generators are available, but they can be grouped into two general categories: service grade and laboratory grade. For service work, either type is usable.

My own bench has several different signal generator instruments: the Heath IG-18, a sweep function generator, a Model 80 (2-400 MHz) RF signal generator, a Japanese signal generator that is sold under different brand names, and an elderly Precision Model E-200C that I refurbished after buying it at a hamfest in nonworking condition.

Oscilloscopes

Probably no other instrument is as useful as the oscilloscope in working on IC circuits. You can look at signals and waveforms instead of just averaged dc voltages (as are measured on the DMM). Figure 2-8 shows a simple service-grade instrument that can be used for most service jobs on AM, FM, and shortwave receivers, as well as most CB and amateur radio transmitters. The oscilloscope is formally called the *cathode ray oscillograph*, or CRO. It displays the input signal on a viewing screen provided by a cathode ray tube (CRT).

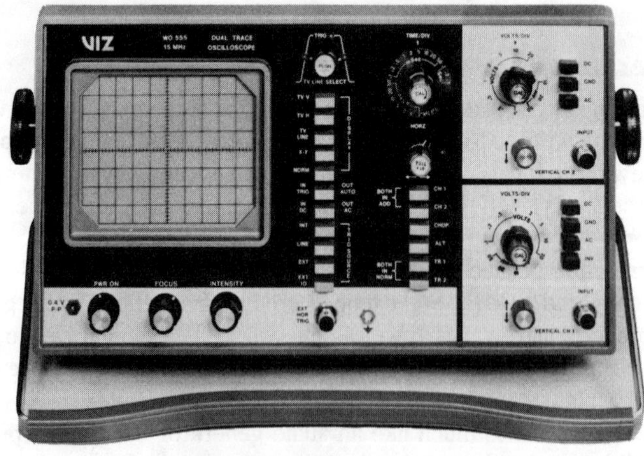

2-8   Dual-trace oscilloscope.

The light on the viewing screen is produced by the CRT by deflecting an electron beam vertically (*Y*-direction) and horizontally (*X*-direction). When two signals are viewed with respect to each other on an X-Y oscilloscope, the result is a *Lissajous figure*.

The quality of oscilloscopes is often measured in terms of the *vertical bandwidth* of the instrument. This specification refers to the −3 dB frequency of the vertical amplifier (or amplifiers if it is a dual-beam 'scope). The higher the bandwidth, the higher the frequency that can be displayed and the sharper the risetime pulse that will be faithfully reproduced. For ordinary service work, a 5 MHz 'scope suffices for most applications. However, there will be times when you need a high-frequency 'scope, so buy as much bandwidth as you can afford. Although once in the price stratosphere, even 50 MHz models can be bought relatively cheaply today, brand new.

dc power supplies

Although often overlooked as test equipment, the simple dc power supply is definitely part of the test bench. In addition to powering units being repaired that either lack or have a defective dc supply, the bench power supply can be used for a variety of biasing and other troubleshooting functions. In chapter 5 you will learn the basic theory of dc power supplies, and in chapter 6 you will find construction projects that allow you to work the projects in this book.

Signal tracer

The signal tracer is a high-gain audio amplifier that can be used to examine the signal at various points along the chain of stages in a radio or audio amplifier. A demodulator probe permits the signal tracer to "hear" RF and i-f signals. The signal tracer is a backup replacement for the oscilloscope if it is absolutely impossible to obtain the latter. However, the signal tracer is such a poor backup to the 'scope as to be deemed insufficient. However, there are advantages to these instruments, and their cost is so low as to be a worthwhile addition to the bench.

3

Electrical conductors, insulators, and semiconductors

THE STUDY OF SOLID-STATE ELECTRONICS ALWAYS (AND QUITE PROPERLY) starts with a consideration of fundamentals. In this chapter, you will learn about electrical conductors, insulators, and semiconductors. Understanding the differences between these materials is necessary to understand the active electronics devices—transistors and integrated circuits—that are used in amplifier circuits.

Matter, atoms, and such

Everything in the physical world is made of matter, and there are many different kinds of matter, but all sorts of matter are all made up of smaller units called *atoms*. The atomic concept originated with the ancient Greeks, some of whom thought that all matter was made up of tiny, indivisible particles. The Greek atomic concept died for many centuries, but was rediscovered in the nineteenth century when science began to notice that certain things did not fit well with their knowledge of the day. For many decades thereafter some physicists ("atomists") believed that atoms were the indivisible particles of the Greeks. But in the first two decades of the twentieth century, scientists such as Ernest Rutherford (1911) and his student, Danish scientist Neils Bohr, (1913) discovered that the "indivisible" atom was actually composed of smaller particles that are now named *electrons, protons*, and *neutrons*. Today it is known that there are even more elemental particles, but for the purpose of electronics the three particles of the "Bohr model" of the atom are sufficient. These particles will be discussed in detail shortly.

Elements and compounds

Elements are materials whose chemical properties remain the same even when only one atom is present. To reverse the definition, the atom is the smallest portion of an element that remains chemically identifiable: all of the chemical attributes of the material exist in even a single atom. The atoms of each of the 100+ natural and man-made elements are all different from each other. The difference between the different elements is a result of differing numbers of electrons and protons comprising the atom.

A *molecule* is a compound made of at least two, often many more, atoms. For example, hydrogen (H) and oxygen (O) are elements (single-atom chemicals), but when an atom of oxygen combines with two atoms of hydrogen, the compound water (H_2O) is formed. The molecule is the smallest portion of a compound that remains chemically identifiable.

Electrons and protons

Of the three atomic particles that are important to electronics, two of them possess an electrical charge. The charges on the electrons and protons are equal to each other, but of opposite polarity: the electron has a negative electrical charge, while the proton has a positive electrical charge. The elemental electrical charge has a magnitude of 1.602×10^{-19} Coulombs, and is sometimes called the *unit charge*.

Electrons and protons have a considerable difference in mass and size. The electron is the smaller of the two, having a mass of 9.1×10^{-31} kilograms, while the proton's mass is 1.67×10^{-27} kilograms. The proton is therefore 1,835 times heavier than the electron.

The Bohr model of the atom

No one knows for sure what atoms look like, although modern tunneling scanning electron microscopes are gaining in magnification for us to begin to speculate about seeing atoms. There are several models for atoms, but the one created in 1913 by Bohr remains sufficient for our discussion of electron devices. Keep in mind, however, that physicists now use more sophisticated models.

Figure 3-1 shows the Bohr model of the atom. In this view, the atom consists of a *nucleus*, containing protons and neutrons, surrounded by a cloud of orbiting electrons. If

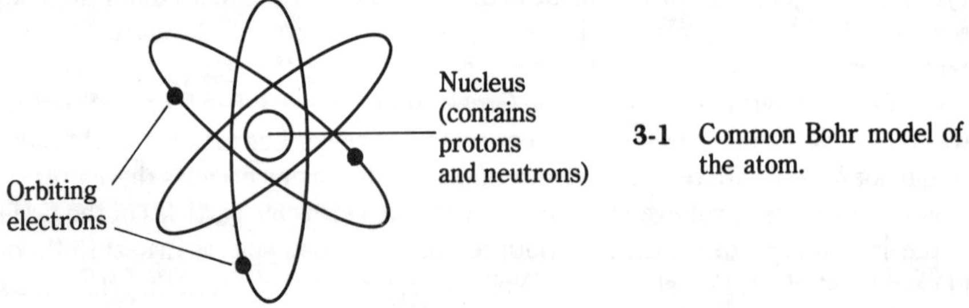

3-1   Common Bohr model of the atom.

you see a little of the solar system in this model, then you are right; the sun is analogous to the nucleus, and the planets are the electrons. A more detailed view is seen in Fig. 3-2. In this illustration we see the *helium* atom; it contains a nucleus with two protons (P_r^+) and two neutrons (N_u), orbited by two electrons (e^-).

The number of electrons orbiting the nucleus is identical to the number of protons in the nucleus. Because each proton's positive charge is balanced — i.e., neutralized — by an electron of negative charge, the whole atom is electrically neutral; i.e., it has no charge. Electroneutrality exists only when the number of electrons and protons is equal. When an

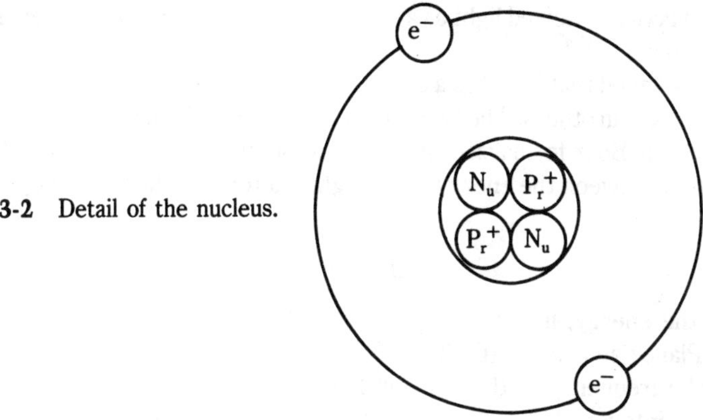

3-2 Detail of the nucleus.

orbiting electron is stripped away from the atom by an external force, the net electrical charge is +1, and the atom is said to be a *positive ion.* Conversely, when the atom gains a spare electron, it has a net electrical charge of −1, and is called a *negative ion.*

The *atomic number* of the atom is the number of protons in the nucleus, while the *atomic weight* is approximately the sum of the masses of the protons, electrons, and neutrons. The atomic weight and atomic number of all known elements are placed on a *Periodic Table of Elements,* which can be found in most high school physics or chemistry textbooks. Many dictionaries also include a listing of the elements, either in an appendix or alphabetized under *elements* in the main section. Alternatively, some dictionaries list the atomic weight and atomic number as an attribute of the element under its own name in the main body of the book.

Quantum mechanics

The quantum mechanics revolution, which brought us the atomic theory on which the entire field of electronics is based, began in the late fall of 1900 at a meeting of the German Physical Society in Berlin. Professor Max Planck presented a paper that revolutionized the world of physics. For decades scientists had been working on the problem of *black body radiation*; i.e., the radiation given off by a black body (such as a lump of iron) when it is heated. The theories that scientists believed at that time did not fit the experimental data. Planck demonstrated that the equations worked well if he allowed only certain discrete amounts of energy. These discrete energy levels are defined by:

$$U = n\text{h} \tag{3-1}$$

Where: U is the energy, in ergs
$\quad$ h is Planck's constant (6.63×10^{-27} erg-sec)
$\quad$ n is an integer (1, 2, 3 . . .)

In 1905, Albert Einstein demonstrated that Planck's theory applied to the photoelectric effect. In this phenomenon, the energy of electrons emitted when photosensitive metallic plates are exposed to light is not a function of light intensity (as expected), but rather of light frequency. This discovery opened a disquieting question for scientists who had (especially since Maxwell) viewed light as a wave: the photoelectric phenomenon made light look like a particle.

It is now believed that light has a dual nature: it behaves as a particle in some experiments and as a wave in others. The two properties are complementary, not contradictory. In the 1920s Neils Bohr formalized this proposition in his *Complementarity Principle.* The relationship between the energy of the light photon and the frequency (color) of the light is:

$$U = nh\nu \tag{3-2}$$

Where: U is the energy, in ergs
  h is Planck's constant $(6.63 \times 10^{-27}$ erg-sec)
  ν is the frequency of the light, in hertz
  n is an integer $(1, 2, 3 \ldots)$

It had been noted in the nineteenth century that certain materials would emit light when excited above the ground state. Applying a certain amount of energy to a system would cause the emission of specific colors of light as the system returned to equilibrium. Several sets of light spectra were noted, most famously the Rydberg series, the Lyman series, the Paschen series, and the Balmer series.

Bohr explained the several series by applying the Planck's theory of quantized energy to the electrons orbiting the nucleus in an atom. When an electron at rest receives energy from an external source, its energy level (hence orbital radius, r) increases only in certain allowable discrete levels defined by Planck's constant (U_2, U_3, etc. in Fig. 3-3). The energetic state is unstable, and the electron soon returns to ground state. Because of conservation of energy, however, that energy must go somewhere, so it is emitted as a photon of light with a frequency according to Eq. 3-2.

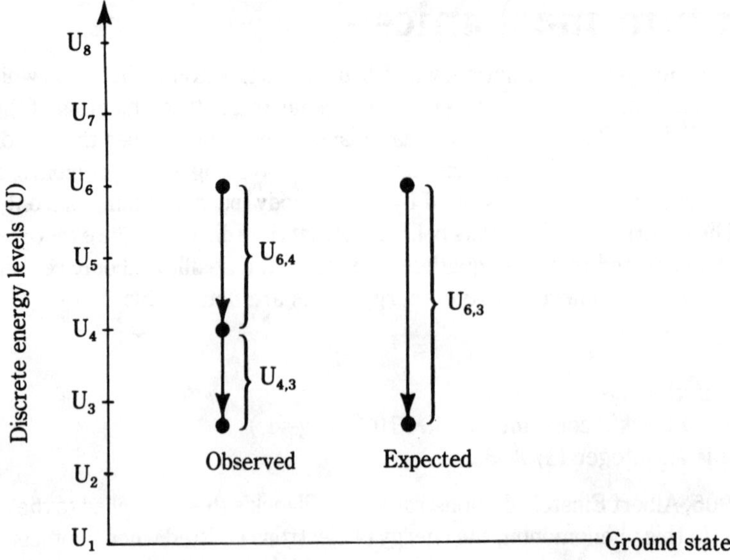

3-3   State transition diagram shows expected and observed energy transitions.

The intuitive expectation was that an electron would fall from the excited state back to its initial state in one movement. This motion would cause a single color of light to be emitted equal to $U_{6,3}$ (in the case of Fig. 3-3); i.e., the difference of energy levels U_6 and U_3. But in the experiments there were more than one color in the spectra, indicating a different scenario taking place. Bohr explained the existence of multiple colors by postulating that the electrons dropped back to the original state in more than one step. In Fig. 3-3 there are two steps ($U_{6,4}$ and $U_{4,3}$) that are, from a conservation point of view, functionally equivalent to $U_{6,3}$, but that allow for the existence of the multiple color emissions.

The phenomenon explained by Bohr is the basis of a number of devices today, including maser (microwave amplification by stimulated emission of radiation) and laser devices. It is also the operant phenomenon in such mundane devices as neon glow lamps, fluorescent lamps, and certain semiconductor devices that we will explore further in later chapters. In the laser, the simultaneous emission of energy from an extremely large number of electrons is coordinated by an external energy source (such as a xenon flash tube) so that the emissions occur in phase with each other, leading to the phenomenon of "coherent light" for which the laser is famous.

The Bohr model of the atom assigned electrons to orbits around the nucleus at specific quantized distances prescribed by their respective energies. Bohr's solar system model of the atom is still accepted today for naive (but useful) explanations, even though modern physicists know that the real situation is somewhat more complex. During the 1920s, quantum mechanics became more formalized into the theory that is essentially used today (with modifications).

In 1924 Prince Louis de Broglie proposed that not only light, but matter also, has a dualistic complementary wave-particle nature. The wave function of matter is usually symbolized by λ. According to de Broglie, the wavelength of a particle such as an electron is given by:

$$\lambda = \frac{h}{mv} \tag{3-3}$$

or,

$$\lambda = \frac{h}{\sqrt{2\,Um}} \tag{3-4}$$

Where: λ is the wavelength
       h is Planck's constant
       U is the energy of the particle
       m is the mass of the particle
       v is the velocity of the particle

Combining the Bohr and de Broglie theories produced a model in which the simple orbital electron actually became a standing wave in which an integer number of de Broglie waves fit into the space allocated by the Bohr theory for an electron of a given energy level (Fig. 3-4). According to de Broglie's theorem, we can deduce the following relationship in which the circumferences of the orbits are integer multiples of the de Broglie wavelength:

$$2\pi r = \frac{nh}{mv} \tag{3-5}$$

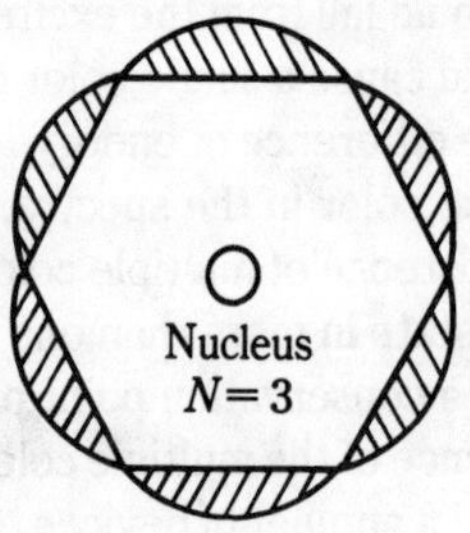

3-4 DeBroglie waves for $N = 3$ state.

Erwin Schrödinger disputed Bohr's solar system model of the atom. Instead of a billiard ball nucleus surrounded by billiard ball electrons, Schrödinger proposed an entirely new model based on a probabilistic interpretation of the wave function. Like light waves, elementary subatomic particles sometimes behave like particles and other times like waves. Again we have a complementary system, but in this case the waves are the "matter waves" postulated by de Broglie. According to Schrödinger's view, the atom consists of a matter wave nucleus surrounded by matter wave electrons. Schrödinger's wave equation describes the matter waves (Ψ) in terms of probability (Ψ^2).

It is important to realize that matter-wave equations do not describe a real chain of events the way water-wave equations describe real movement by real water particles. The equations describe only the probabilities of finding a real particle at a specific place at a given time.

To these factors we must now add another facet: the Uncertainty Principle. In 1927, Werner Heisenberg proposed the Uncertainty Principle, which holds that certain pairs of properties of atomic particles cannot both be measured with accuracy simultaneously. For example, it is impossible to measure precisely both the position and momentum of an electron. Momentum is the product of mass and velocity, so it follows that we cannot know both where the electron is located and how fast it is traveling. Stated mathematically, Heisenberg's Uncertainty Principle is:

$$\Delta P_x \Delta X > h \tag{3-6}$$

Do not confound the Uncertainty Principle with the mere inability to measure some parameters because of some kind of disturbance effect. Many physical measurements are inaccurate because the act of measurement or the nature of the instruments disturbs the system, and thereby changes the value of the measurement enough to introduce very large errors. For example, a low-impedance voltmeter disturbs a high-impedance circuit enough to introduce serious errors.

What the quantum mechanics scientist is telling us, however, is that the electron actually does not possess both a precise location and a precise momentum. Truly astounding!

The probability interpretation of Schrödinger leads to some disquieting problems. Consider the tunnel diode, for example (Fig. 3-5). The usual tunnel diode explanation goes something like this: In an ordinary pn junction, the transition region between n-type and p-type semiconductor material consists of a dipole layer created by relatively immobile unneutralized electrons and holes. This dipole layer has an associated electric field of up to 10 kV/cm in an unbiased junction, and even more in a reverse-biased junction. Although the band-gap energy is the same on both sides of the junction, a difference in the potential

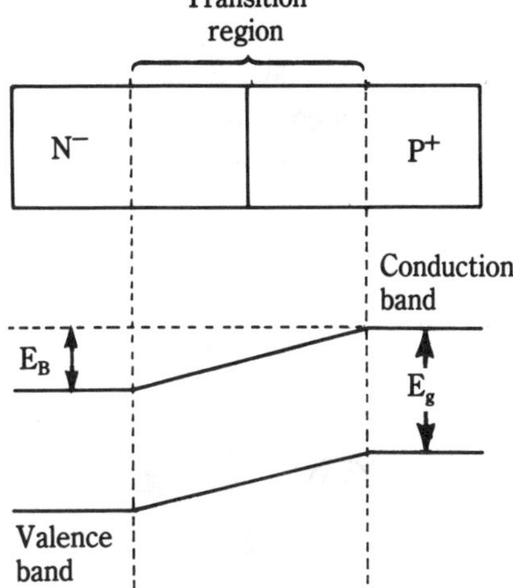

3-5 Tunneling action energy diagram.

energy on the two sides results in an electrical potential barrier (E_B) to energy. The potential barrier forms a blockade—an electrical brick wall—to electrons trying to pass over the junction. Unless an electron has sufficient energy, it cannot leap over the potential barrier.

During the first three decades of the twentieth century, Bohr and a relatively small band of scientists working in Copenhagen devised a system that seems to explain atomic phenomena. Physicists developed a world of weirdness where matter (including electrons)—the hard stuff of reality—has a complementary wave-particle nature. Instead of a mass of tiny ping-pong balls, electrons and protons are a ghostly dance of probability waves. That system is quantum mechanics (QM), and it displaced the comfortable cause-and-effect definitions of classical physics, and put in their place a new set of definitions that are based on probabilities and "tendencies to exist." The seeming paradox is that "unpredictability and uncertainty appear intrinsic to the universe at the deepest levels." Yet QM is accepted by scientists as the mathematical construct that best predicts the behavior of matter at the subatomic level. Even though problems with the theory persist, QM became for scientists the operant paradigm.

Part of the problem faced by the scientists is the utter inadequacy of human language to express the realities of the quantum world. It's not that something mystical happens—as asserted by certain popular writers on science—in quantum phenomena; the equations work, or else transistors wouldn't.

Shells

The electrons orbiting the nucleus in an atom cannot take up just any position. There are certain rules that they must obey. The principal rule is that the electrons can only exist in certain numbers at specified distances from the nucleus. Each discrete distance is a radius and forms a *shell*. Associated with each shell is an energy level that is a function of the distance of the shell from the nuclear center and the momentum of the electrons in the shell. Thus, the energy levels correspond to the shells.

As a matter of convenience, it is common practice to only consider the outermost shell. This shell is called the *valence shell,* and its electrons are called *valence electrons.* The reason that this is done is that the outermost shell determines chemical and electrical activity. It is also a common practice to show only the outermost electrons—i.e., the valence electrons—and a central core consisting of all other electrons plus the nucleus (Fig. 3-6).

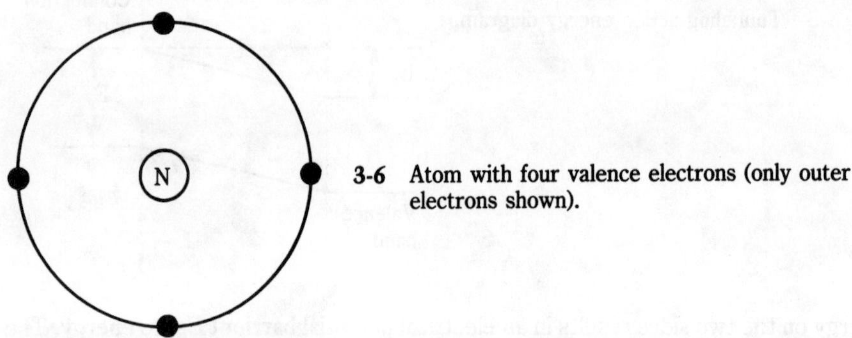

3-6   Atom with four valence electrons (only outer electrons shown).

Each shell has a discrete number of electrons that totally fill it. Although fewer than this number can occupy a shell, it is not possible for a larger number to fit in. Once the shell is filled, all additional electrons are added to the next higher shell. In the case of the helium atom described in Fig. 3-2, there is only one shell, yet it is completely filled. Each shell has a number of electrons that fill it, and make it electrically and chemically stable. For the first shell it was two. This difference can be seen in the two elements hydrogen (H) and helium (He). Hydrogen has but one orbital electron, so its one shell is not filled. As a result, hydrogen is very active chemically and will burn explosively in the presence of oxygen. Helium, on the other hand, has both of its electrons in the single shell, so it is stable. Helium is not chemically active and will not burn under any ordinary circumstance. That reason is why helium was preferred for dirigibles ("blimps") and other balloon airships over the earlier hydrogen—something that the operators of the *Hindenburg* learned to regret.

Other shells than the first are filled with greater numbers of electrons. For example, the second shell is filled with eight electrons, and is stable. This configuration of eight electrons is prevalent, and is called a *stable octet.* The third shell is filled with eighteen electrons. Higher-order shells than about the fourth are not normally considered in semiconductor electronics because of the nature of the elements used in solid-state electronics, although conductors have more shells than four.

Conducting electricity

Ordinary electrical conduction is the flow of electrons from one point to another. The electrons that become part of an electrical current are "free" electrons that have been stripped away from their associated atoms. They will wander until they find a positive ion that needs an electron. In due course, we will discuss electrical conduction further, especially as it applies to semiconductor devices.

One of the things that distinguishes conductors, semiconductors, and insulators from each other is the ability — or lack thereof — to conduct electricity. It is the valence-band electrons that determine this property for any given material.

The electrons can be divided into two groups. Those that are in the lower energy levels, and are thus more tightly bound to the nucleus, are said to be in the *valence band,* while those in the higher *conduction band* are less tightly bound to the nucleus (Fig. 3-7). Only small amounts of energy are needed to separate electrons in the conduction band from the associated atom.

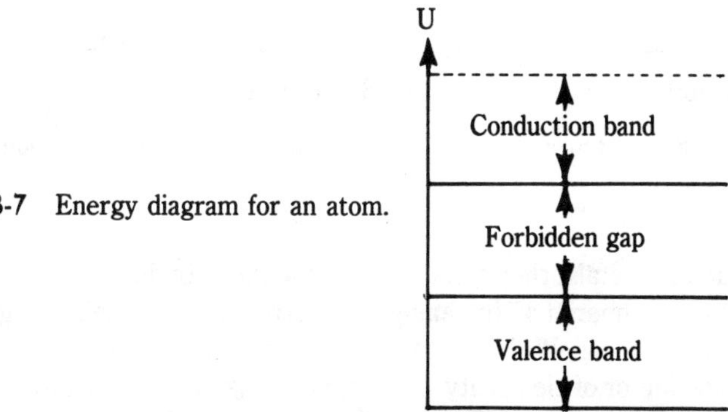

3-7   Energy diagram for an atom.

The conduction and valence bands are separated by an energy gap, called the *band gap* or *forbidden gap.* The width of the forbidden gap is expressed in electron volts (eV), and is different for different materials. For example, in the two semiconductors that are most popular in electronics, germanium (Ge) and silicon (Si), the forbidden gap energies are 0.785 eV and 1.2 eV, respectively.

The *band-gap energy* is the amount of energy that must be added to the electron in order to move it from the valence band to the conduction band. Thermal energy from the environment is relatively low level, especially at room temperature. In conductors, the thermal energy is quite sufficient to raise large numbers of electrons across the forbidden gap. As a result, there are many loosely bound electrons available to support an electrical current.

In semiconductors, there are some thermal electrons in the conduction band, while in insulators there are almost none. Figure 3-8 demonstrates these differences in terms of the width of the forbidden gap. Notice that the gap is widest in insulators (Fig. 3- 8A), indicating that a large energy must be input to the system in order to strip away electrons and cause conduction. When this situation occurs, the insulator is said to have "broken down," a situation that can occur under sufficiently high voltages. The semiconductor has the next widest forbidden band (Fig. 3-8B), which indicates that some free electrons can exist, but not many (without an external energy source). Finally, in conductors all of the available electrons are in the conduction band (Fig. 3-8C), and there is no forbidden gap.

Another factor that distinguishes the insulator, the semiconductor, and the conductor from each other is the number of electrons in the valence shell. An insulator material will have the shell filled so that the inherent stability makes it more difficult to separate an electron. Conductors have one, two, or three electrons in the valence shell. Because this

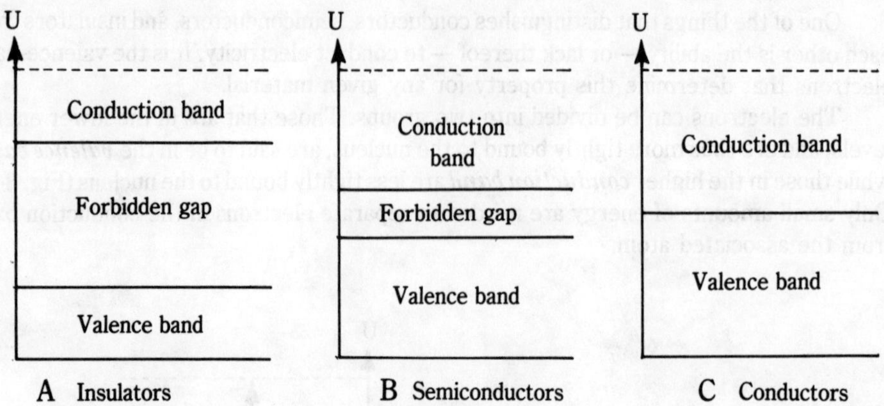

A   Insulators   B   Semiconductors   C   Conductors

3-8   Three different types of energy diagram: A. insulators, B. semiconductors, C. conductors.

property is found in metals, they tend to make the best conductors. For example, silver (Ag) has an atomic number of 47 (meaning 47 electrons), so its shell configurations are: 2, 8, 18, 8, 8, and 3 electrons. With only three electrons in the valence shell, the silver atom is a very good conductor of electricity. The copper (Cu) atom has an atomic number of 29, so it has only one electron in the valence shell. Gold (Au), with an atomic number of 79, like silver has three valence electrons.

Semiconductors have four valence electrons, so are said to be *tetravalent* materials. The two most popular elements used in solid-state electronic devices are silicon (Si) and germanium (Ge). The atomic number of Si is 14, so the silicon atom (Fig. 3-9) has a shell

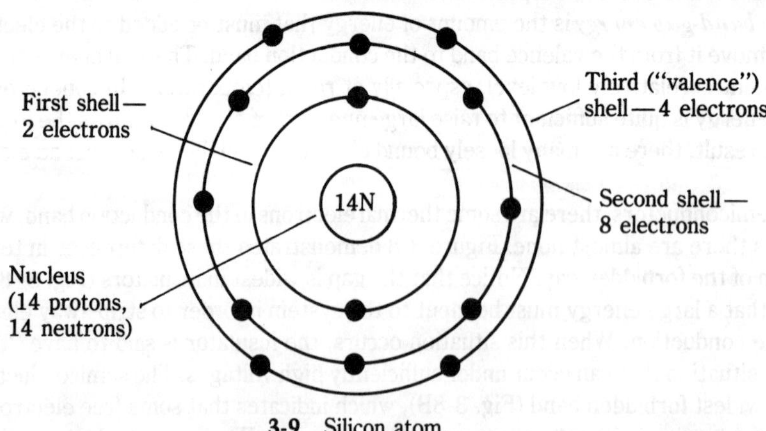

3-9   Silicon atom.

electron configuration of 2, 8, and 4. Similarly, Ge has an atomic number of 32, so its shell configuration (Fig. 3-10) is 2, 8, 18, and 4. In both cases, the valence shell contains four electrons. The simplified model shown earlier in Fig. 3-6 is of a semiconductor because it shows a nucleus and electron core surrounded by four valence electrons.

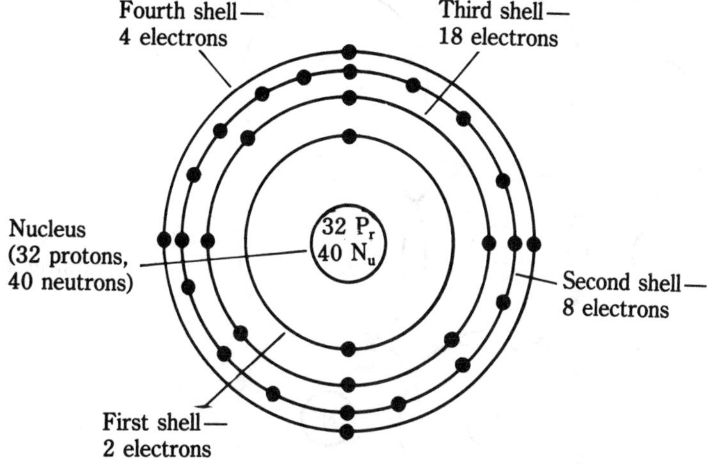

3-10 Germanium atom.

Bonds between atoms

When atoms interact with each other, they often form bonds. Indeed, it is in the bonding of atoms that compounds are made from elements. However, like atoms also bond with each other, as seen in semiconductor materials. These bonds are what hold the material together. There are several different kinds of bonds: *ionic, metallic,* and *covalent.*

The ionic bond is found in insulator materials. In this form of bonding, an electron removed from one atom attaches itself to another, forming a pair of atoms that are not electrically neutral. The atom that lost an atom is a positive ion, while the atom that gained an electron is a negative ion; the attractive force between them makes the bond very difficult to break.

Metallic bonds are found in conductors. The arrangement in these materials has been described as an extensive cloud of electrons surrounding positive metal ions, or, as others put it, an electron sea with islands of positively charged metal ions. The force of attraction between the electron cloud and the ions holds the material together.

Covalent bonds are those found in semiconductor materials, and are thus important for understanding solid-state devices. In a covalent bond, two adjacent atoms will share electrons (Fig. 3-11). The atoms of semiconductors tend to be tightly packed, and arrayed

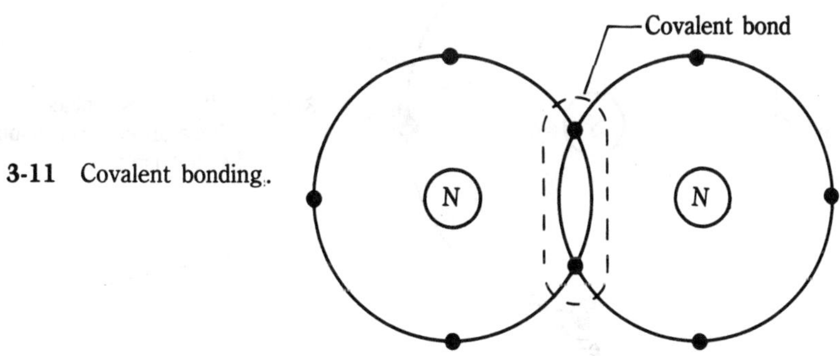

3-11 Covalent bonding.

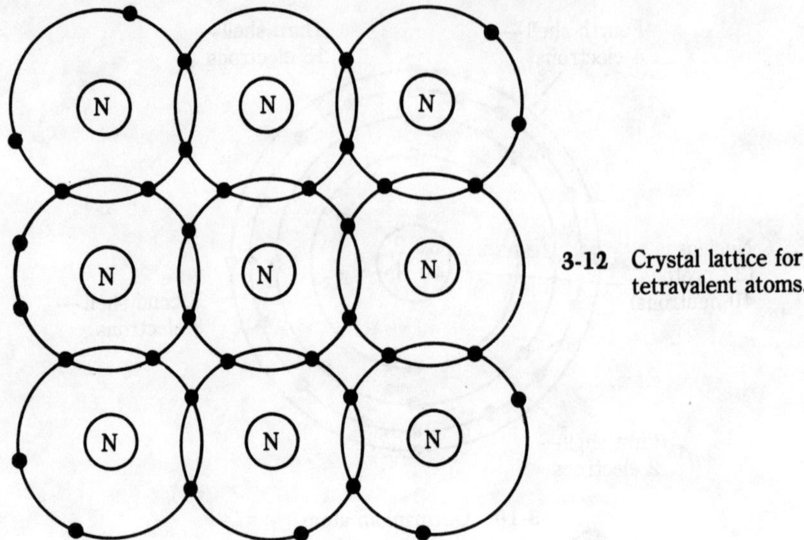

3-12 Crystal lattice for tetravalent atoms.

in a *crystal lattice* structure (Fig. 3-12), because each of the four valence electrons in each atom is shared with other atoms. Thus, the four electrons of tetravalent semiconductors are shared by four other atoms in the crystal array. By sharing its electrons, each atom is able to fill its valence shell with four extra electrons, forming a stable octet. Keep in mind that the actual structure is three dimensional, even though it is shown only two dimensionally in Fig. 3-12.

Holes and hole conduction

It was stated previously that ordinary electrical conduction is the flow of electrons. There is another form of conduction seen in semiconductor materials: *hole conduction*. A *hole* is, quite simply, a place where an electron should be, but isn't. If an electron is stripped away from the valence shell of an atom, it leaves a positive ion because the atom is no longer electroneutral. The spot where the electron is missing is called a hole (Fig. 3-13). The nu-

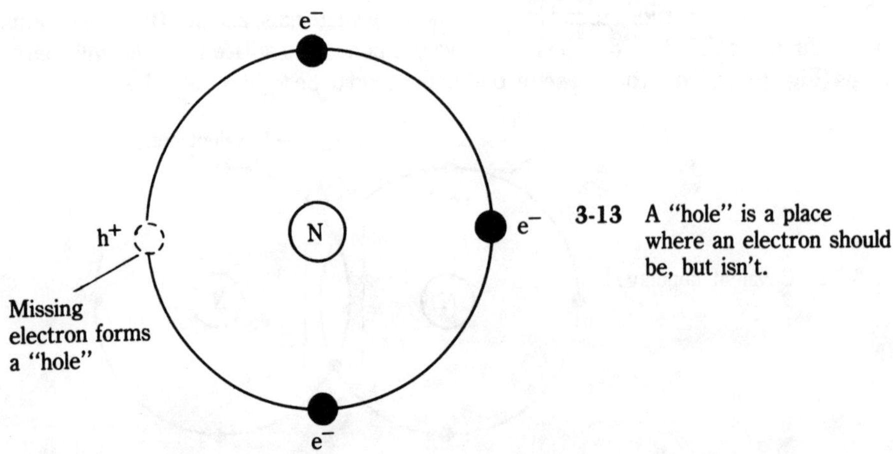

3-13 A "hole" is a place where an electron should be, but isn't.

cleus contains one more proton than the number of orbiting electrons, so it will have a net unit electrical charge of +1. Mathematically, we can treat the hole as if it were a real physical entity that has a charge equal to that of the electron, but of opposite polarity, and the same mass as an electron. The hole does not exist in the same sense that the electron does, but for our purposes we can pretend that it does: the math still works out.

In electronic textbooks, hole conduction is sometimes confused because scientists legitimately consider it as the flow of electron-sized, positively charged particles with a net electrical charge of +1. However, hole conduction is nothing more than electron conduction in reverse. Figure 3-14 shows this mechanism. When an electron vacates its space

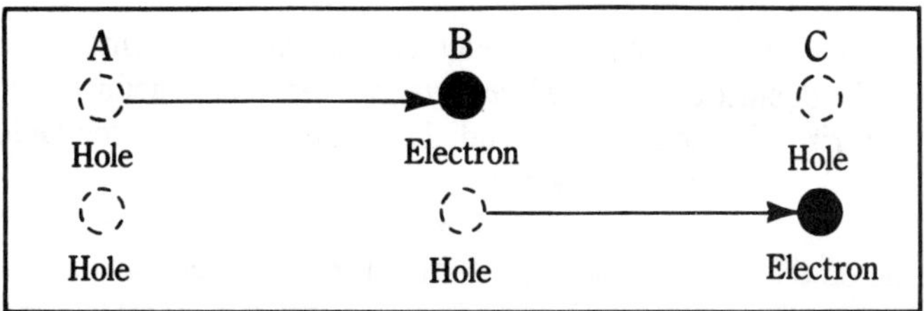

3-14   Hole conduction is merely electron conduction in reverse.

at point "A" and moves to fill a hole at point "B" (electron/hole annihilation), we can view it either as an electron flow of A → B, or a hole flow of B → A; similarly when the electron moves on, creating a new hole at "B" and filling one at "C." The overall net effect can be modeled as either electron flow ABC, or hole flow CBA.

Figure 3-15 shows how holes are created in semiconductor crystal lattices. When tetravalent semiconductor atoms form bonds and one of them is less one electron, there will be one less covalent bond than ordinarily would be the case. A hole is said to exist at the spot in the crystal lattice where the missing electron should be located. When an electron fills the hole, strengthening the covalent bond, it does so at the expense of creating another hole elsewhere (*hole creation*), which appears to be hole flow similar to that just discussed.

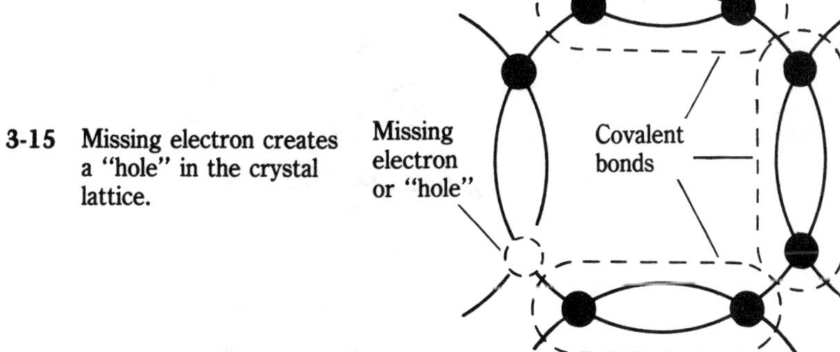

3-15   Missing electron creates a "hole" in the crystal lattice.

Making semiconductors conduct

Ordinarily, semiconductor materials in their pure state do not conduct electrical current very well. The covalent bonds of the crystal lattice mimic stable octet configurations and so limit current flow. The degree of current flow in semiconductors is seen by comparing the resistivity (ρ) of semiconductors with those of conductors and insulators. The resistivity is measured in ohm centimeters (Ω-cm), and the smaller the number, the better the conductivity of the material. For our three classes:

- Conductors 10^{-6} Ω-cm
- Semiconductors 10^{1} Ω-cm
- Insulators 10^{14} Ω-cm

Semiconductors can be made into conductors by adding small amounts of impurities to the material. The impurities are called *dopants*, while the pure semiconductor is called the *intrinsic* material. When dopants are added to the intrinsic semiconductor (the process is called *doping*), the dopant atoms form covalent bonds with the intrinsic atoms; the newly created combination material is called *extrinsic*. The amount of dopant required is very small, being on the order of one dopant atom for every 200 million intrinsic atoms.

There are two different generic categories of doping. *Acceptor doping* creates enough valence band holes in the crystal lattice of the intrinsic material to sustain conduction. *Donor doping* is exactly the opposite: it creates an excess of conduction band electrons.

Figure 3-16 shows the situation for acceptor-doped semiconductors. The impurity, or dopant, added to the intrinsic material is *trivalent* (boron, gallium, and aluminum); that is, it has three valence electrons. These electrons form covalent bonds with the atoms of the crystal lattice, as shown. However, because only three electrons are available in the valence shell of the dopant atom, one atom of the intrinsic material is short-changed, and a hole is created. Because there will be an excess of holes in the material, the trivalent-doped extrinsic material carries a net positive charge, so is called *p-type* semiconductor material.

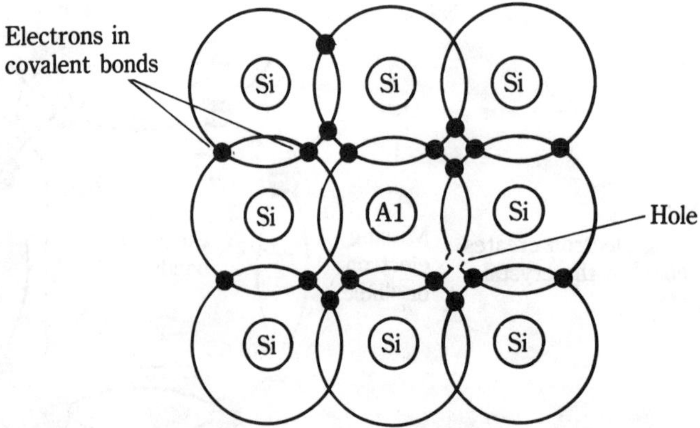

3-16 Crystal lattice showing position of hole in p-type material.

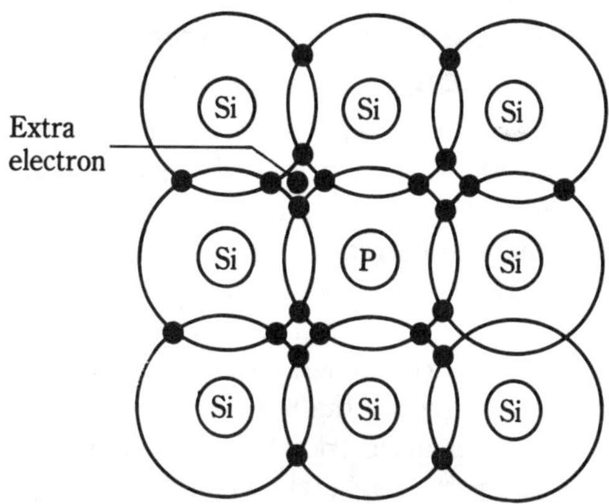

3-17   Extra electron in n-type material.

Figure 3-17 shows a donor-doped material. In this case, the dopant is *pentavalent*, so has five valence shell electrons. Typical donor dopants include antimony, phosphorous, and arsenic. In donor-doped material, the electrons of the dopant form covalent bonds with nearby semiconductor atoms. However, because the semiconductor atoms are tetravalent, only four of the available five dopant electrons find a mate. The result is an excess electron for each dopant atom which is available for electrical conduction. Donor-doped materials have a negative net electrical charge, so are called *n-type* semiconductors.

Majority and minority charge carriers

The electrons and holes in extrinsic semiconductor materials are collectively known as *charge carriers* because, not too surprisingly, they carry the electrical charge of 1.602×10^{-19} coulombs each. In no semiconductor are there only holes or only electrons, but the balance between them is greatly altered. For example, in n-type semiconductors, free electrons greatly outnumber the holes, so electrons are the *majority carriers* and holes are the *minority carriers*. In p-type material, exactly the opposite occurs: holes are the majority carriers and electrons are the minority carriers. The majority carriers are created through the operation of the dopant, as described previously. The minority carriers in each case result from thermal agitation of the material causing covalent bonds to break, a phenomenon that occurs at any temperature greater than Absolute Zero ($\approx -273.16°$ C).

At a certain temperature, which is material dependent ($200°$ C for Si, and $85°$ C for Ge), a sudden increase occurs in the number of carriers. This increase is caused by the breaking of otherwise stable covalent bonds by the thermal energy. For this reason, among others, semiconductor devices tend to be rather heat sensitive, and precautions should be taken in their use.

At one time, people working with diodes and transistors would heat-sink the leads being soldered with alligator clips to avoid heat damage to the device. That is no longer considered necessary for most cases, but you are still admonished to use a low-powered (20 to 50 watts) soldering pencil, rather than a 75-watt (or more) blowtorch when soldering solid-state parts.

Drift and diffusion currents in semiconductor materials

Charge carriers in a piece of semiconductor material move about in a random fashion, colliding with other particles and rebounding in odd directions. The randomness of room-temperature charge motion, in the absence of any kind of external ordering force, means that the sum current is essentially zero. An *electrical current* is defined as electrical charges in motion, so when the number of carriers traveling in any one direction is equal to the number of carriers in the opposite direction (which will reliably occur when extremely large numbers of carriers move in random directions), the overall effect is for mutual cancellation to reduce the net current to zero.

Without an electrical field applied (to make an external ordering force), the motion of a single particle might take a path such as the dotted line in Fig. 3-18. When, however, an electrical potential (V) is applied across the semiconductor, it creates an electrical field

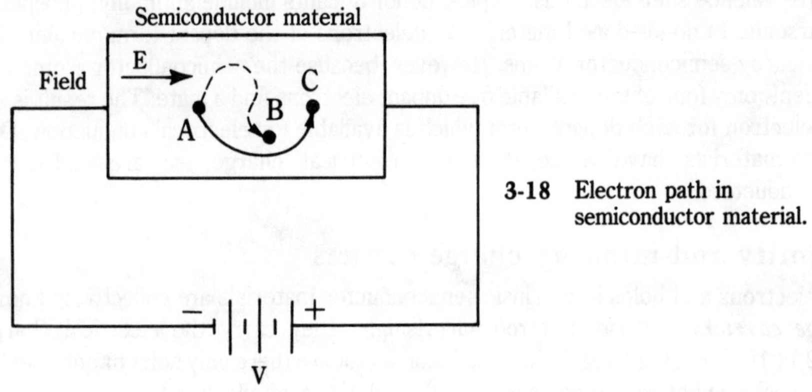

3-18   Electron path in semiconductor material.

(E-field) that causes the carriers to drift in one direction; i.e., toward the opposite polarity side of the field. Ordinarily, in a perfect vacuum, the electrons would travel in a straight line parallel to the E-field lines of force. In a solid material such as semiconductor, though, there are other atoms and electrons present, and the random collisions still occur, but now the field reduces the tendency to randomness, and biases the direction of the carriers toward the ends of the field (as determined by their own polarities). Any one electron may travel only a short distance, but the statistical net motion of large numbers is toward the ends of the field: holes migrate toward the negative end of the field, and electrons migrate toward the positive end. This kind of current is called a *drift current*.

There is also another type of current found in semiconductor materials. Figure 3-19 shows the situation for a *diffusion current*. When the concentration of either type of carrier is greater in one section of the material than in other sections, the mutual repulsion

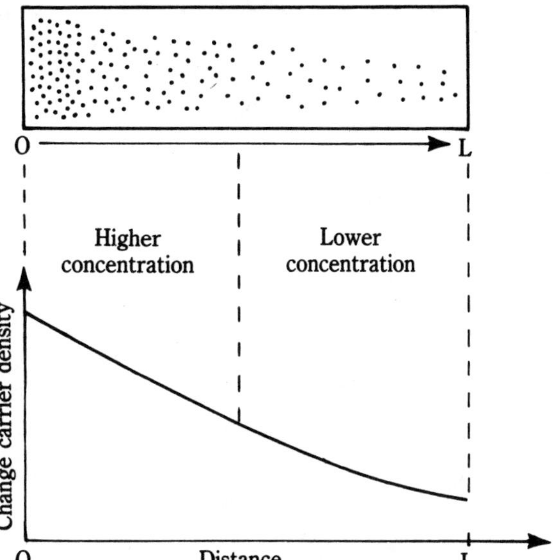

3-19   Charge carrier profile in semiconductor.

force of nearby charge carriers causes some of them to diffuse toward regions where the charge density is not as great. The graph aligned with the block of material in Fig. 3-19 shows the charge density vs. length for the path being considered. This kind of current does not require an external electrical field.

4

dc power sources

ELECTRONIC EXPERIMENTERS FREQUENTLY NEED A SOURCE OF LOW-VOLT-age direct current (dc) electrical power for their projects, especially if solid-state electronic devices are being used. You might build simple, but effective electronic circuits based on easy-to-apply devices such as the operational amplifier or 555 integrated circuit (IC) timer chip. Such circuits typically require either a monopolar dc source in the +4.5V to +15 V range, or a bipolar supply in the ±4.5V to ±15V volts range.

In this chapter you will find a construction project for a versatile workbench and laboratory dc power supply that is based, in part, on a commercially available kit. Unless otherwise noted, the experiments and projects in this book were tested on this bench power supply. It is *bipolar*, meaning that there are two separate outputs: one produces a positive output voltage, and the other produces a negative output voltage. Furthermore, the two output voltages can independently be either fixed or adjustable over the range ±1.25V to ±17V. Up to 500 milliamperes (0.5 amperes) of direct current can be drawn from the power supply. The power supply is voltage regulated, so it can be used in a variety of crucial electronic or instrumentation applications.

Some readers might want to wire the circuit on perforated circuit board such as Vectorboard® (available at electronic supplies outlets), but most readers are well advised to buy the kit. This project is based on the Jameco Electronics* Model JE-215 Dual Power Supply Kit. This kit contains a printed circuit board and all of the small parts that are mounted on the board. (Additional parts are needed for the cabinet, but more of that in a moment.) In addition, for the novice electronic builder there are detailed instructions on assembling the kit.

Experienced builders also might want to buy the Jameco JE-215 kit. First, the kit is a lot less hassle to assemble than point-to-point "rat's nest" wiring on perforated board.

---

*1355 Shoreway Rd., Belmont, CA, 94002. 415–592–8097.

Second, when I priced out the individual components in a catalog, even at discount prices, the total cost was from 0.9 to over 2 times the retail price of the kit, depending on the sources of the parts. In other words, there is no real savings under any circumstance where the parts must be bought, and it might well be twice as expensive to supply your own parts — and you don't have the advantage of a professional printed circuit board. The parts list is shown at the end of this chapter.

The power supply circuit

The circuitry for the dc power supply (Fig. 4-1) is simple and straightforward. A step-down transformer (T1) is used to reduce the 115V alternating current (ac) received from the wall outlet in your home to the low voltage level required for the dc power supply. A pair of rectifier diodes (CR1 and CR2) are used to convert the alternating current to an impure form of dc, called *pulsating dc* in electronic technical textbooks, while filter capacitors (C1 and C2) are used to make the rectifier output nearer to pure dc. The working of these components is beyond the scope of this book, but many basic electronics books cover these topics in depth.

A pair of integrated circuit voltage regulators (IC1 and IC2) are used to stabilize the output voltage levels under a variety of different conditions. The LM-317T (IC1) device is the positive voltage regulator, while the LM-337T (IC2) is the negative voltage regulator. A typical package for these devices is shown in Fig. 4-2. It is a plastic "power transistor-like" package with a metal mounting tab and three electrical connection pins (labeled 1, 2, 3). Note in Fig. 4-1, however, that the pins have different functions on the two different types of regulator.

The LM-317T and LM-337T devices are adjustable three-terminal integrated circuit voltage regulators. The adjustable feature is programmable by the ratio of two resistors: R2/R3 on IC1, and R4/R5 on IC2. In the Jameco kit, the variable resistors (R3 and R5) are used to set the required output voltage. Jameco supplies these parts as small "trimmer" potentiometers (variable resistors) that mount on the printed circuit board. You might prefer to replace these parts with separate output controls that are mounted on the front panel.

The output terminals for the dc power supply printed circuit board (Fig. 4-3) are a set of four #6 machine screws and nuts mounted on one end of the board. The two center terminals are connected together to form a common electrical connection that serves as the ground points for the two supplies. The other two screws are the negative (−) and positive (+) output voltage terminals.

Follow the printed Jameco instructions when assembling the printed circuit board, except for one point. The integrated circuit voltage regulators (IC1 and IC2) should be mounted slightly differently than shown in the instruction. Figure 4-4A shows the method recommended in the instructions: the two outer pins (1 and 3) are bent away from the center line. This method has the potential for causing internal damage to the voltage regulator due to stresses placed on the internal connections. I prefer the method shown in Fig. 4-4B. You will have to be careful to leave enough length to pass through the printed

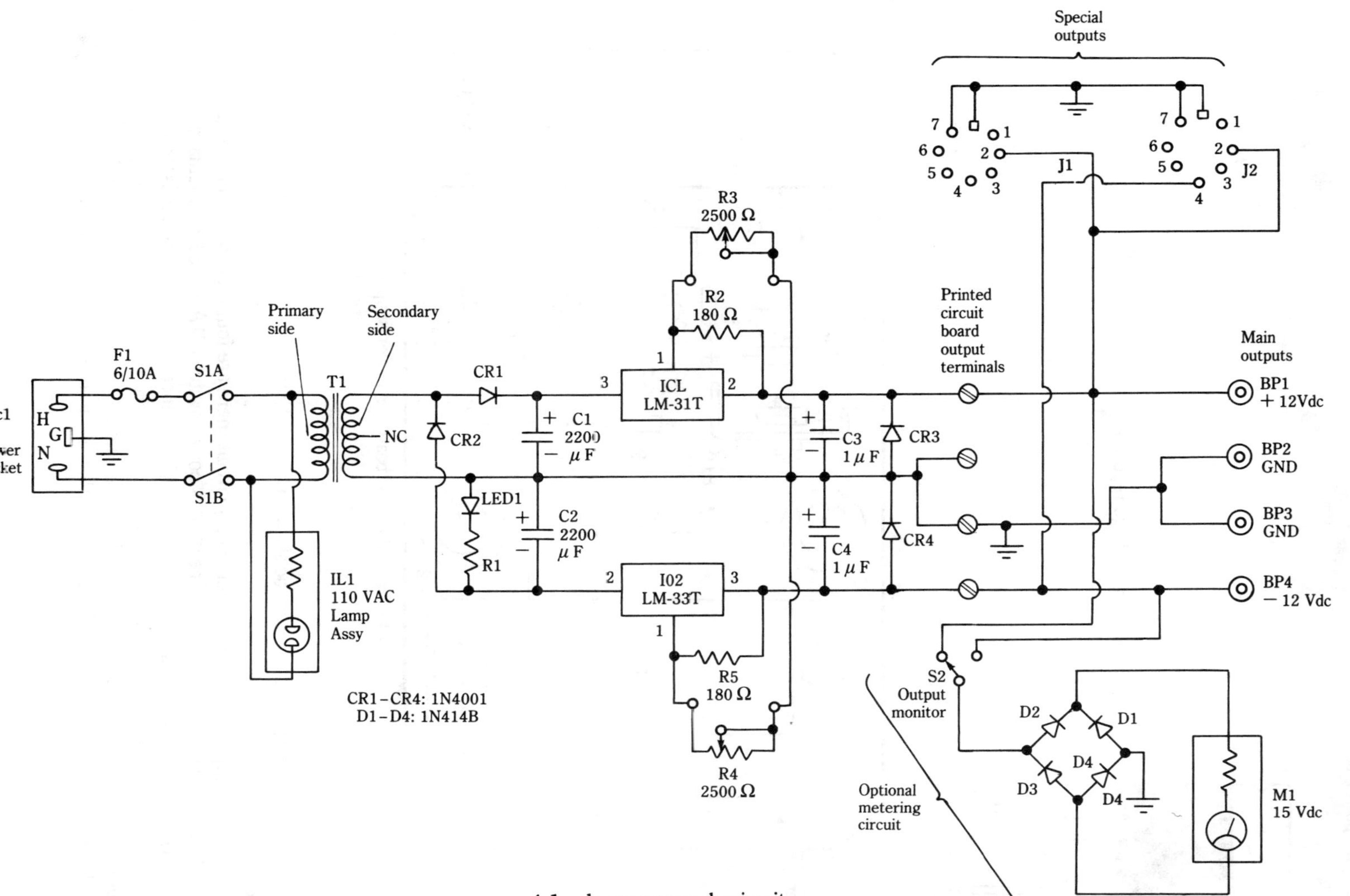

4-1 dc power supply circuit.

4-2 Plastic package voltage regulator IC.

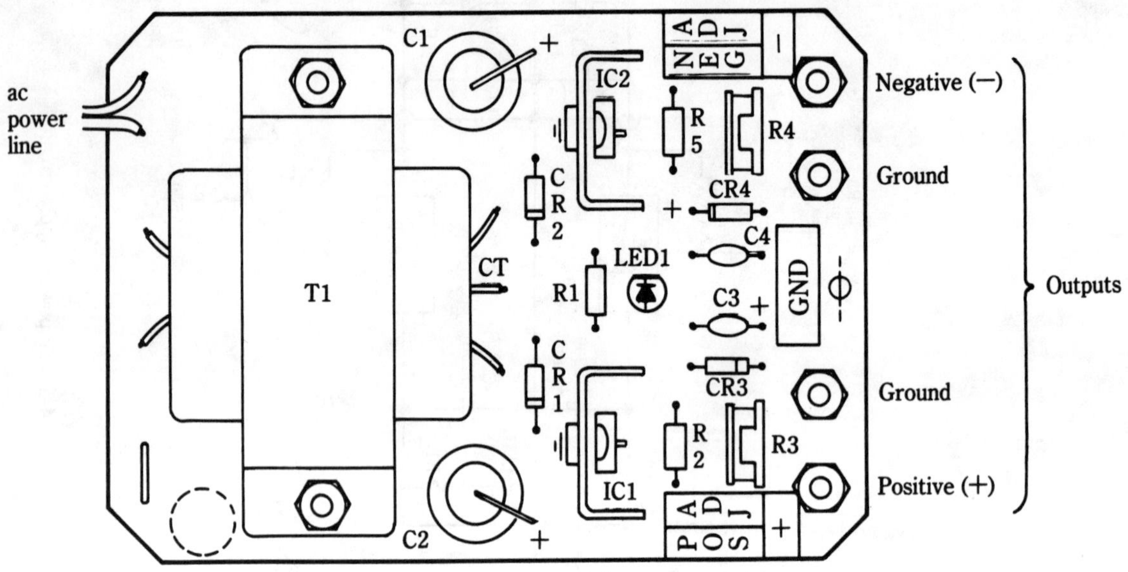

4-3 Printed circuit board for power supply.

circuit board. Nonetheless, the method shown in Fig. 4-4B results in a healthier future for the regulator.

At the ac input side of the power supply circuit there are four external components added (i.e., not included in the kit). The ON/OFF power switch (S1A and S1B) is actually a dual switch so that both sides of the ac power line are opened up when the circuit is turned off. You can get away with a single switch, but there are safety reasons for using two switch sections operated by a single handle. Buy either a double-pole single-throw (DPST) or double-pole double-throw (DPDT) switch that is rated for 125Vac, or higher.

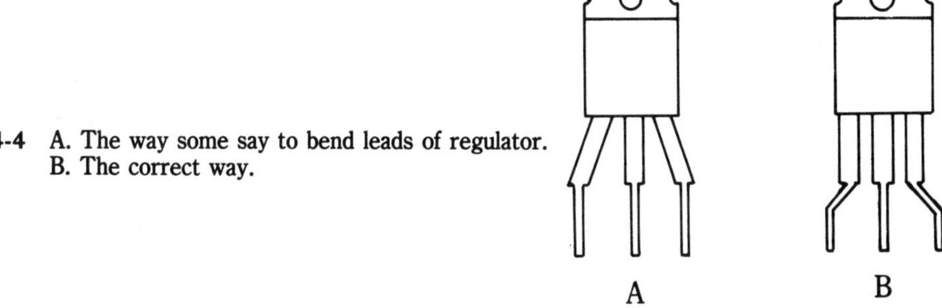

4-4 A. The way some say to bend leads of regulator.
B. The correct way.

The ac power line from the wall socket has three wires — (don't use a two-wire ac power cord — those are not always safe in electronic instruments!) — labeled *hot, neutral,* and *ground.* The hot and neutral are connected through the fuse and switch to the primary side of the power transformer (T1). The fuse is used to protect the circuit in case of a catastrophic failure. Perhaps more important, however, is that the fuse protects your home in case of such a failure. Not using a fuse can cause a fire if the power supply shorts and the power cord heats up enough from overcurrent to flame. Buy a good fuse holder, and use a 6/10 ampere slow-blow fuse.

The panel light (IL1) tells you when the power supply is turned on. It is a single assembly containing a neon glow lamp and a series current-limiting resistor. I used a 105Vac to 125Vac Linrose type B2150A1 that was purchased on a blister pack display at a local electronic parts distributor. These devices are cylindrical in shape and are attached to a metal front panel or box (as the case may be) using a "Tinnerman speed nut" (supplied).

The connection of the ac power line is always a matter of concern for any power supply project. Please understand one thing if nothing else: *Alternating current (ac) from the wall socket is terribly dangerous — it will kill you if you are careless!* The Jameco JE-215 kit comes with a two-wire ac power cord, but I don't recommend its use. Obtain a three-wire ac power cord and a matching chassis-mounted socket (SOC1). There are very potent safety reasons for using the three-wire cord.

Several other parts are added outboard to the printed circuit board. These include output binding posts (BP1 through BP4), an optional metering circuit, and a pair of special output jacks. The binding posts (Fig. 4-5) are sometimes called *five-way binding posts,* after the supposed number of ways you can make connection to them. They provide an insulated path for the dc power through the metal cabinet. In addition, binding posts will accept a banana plug, which is a convenient, if temporary, way to carry power from the power supply to your project or experiment on the bench.

Construction of the project

The dual-polarity dc power supply should be built inside of a metal cabinet. You can either use a decorative cabinet such as I used, or a plain "battleship gray" aluminum flange box,

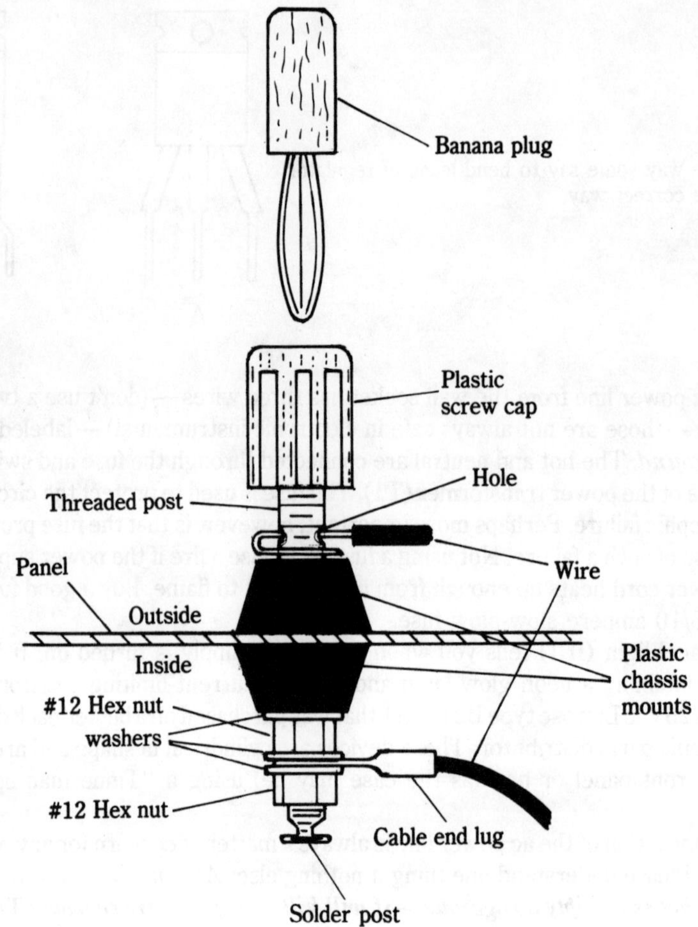

4-5   Connection to the five-way binding post.

as you please. Keep in mind, however, that the decorative cabinet looks a lot better. I used a Hammond 1458D4B-1184 (Fig. 4-6), which is 8 × 8 × 4 inches, and painted light blue with a white front panel and black rear panel. There are any number of cabinets on the market, and you will find a good selection at most local or mail-order electronic parts distributors.

Assembly of the printed circuit board follows a straightforward path. Use a pencil-type soldering iron and either 60/40 or 50/50 resin core solder (sometimes marked *radio-tv, electronic*, or something similar). Do not use acid core or coreless solders! Those types are intended for plumbing use, and are not suitable for electronics. Acid core solders will corrode and destroy the printed board and wiring.

Small components such as resistors and rectifier diodes are mounted by passing the leads through the hole in the printed circuit board. Press the body of the part gently against the printed circuit board so it seats firmly, and then turn the board over (holding

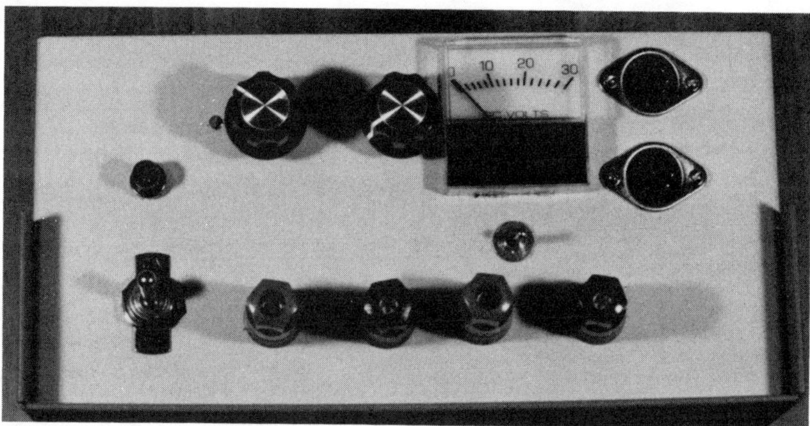

4-6   Front panel of power supply.

the part in place) and bend over the leads. Make sure to position those leads so that they don't short-circuit to nearby conductors of the printed circuit board. Clip off the excess, leaving ⅛ to ³⁄₁₆ inch of lead, and solder. Hold larger components, such as the potentiometers (R3 and R4), in place with your fingers while soldering on the other side of the printed circuit board.

At this point you have to make a decision on whether you want the power supply to be a fixed-output (i.e., only one output voltage available from each side) or a true variable-output type. If you are satisfied with having only one output voltage (e.g., ±12 volts), then mount the potentiometers supplied with the kit at J3 and J4 on the printed circuit board. If you want the supply to be variable, then you must discard the potentiometers supplied with the kit and use a pair of potentiometers mounted on the front panel instead. If you opt for the latter, don't install the original potentiometers. Instead install two wires at each potentiometer spot on the printed circuit board. There are three holes in the board for each potentiometer; use the two outermost holes in each case.

Mount the printed circuit board approximately in the center of the bottom plate of the Hammond cabinet. Drill three ⁵⁄₃₂ holes for #6 machine screws mounting hardware in the bottom plate. These three screws correspond to the three holes in the printed circuit board, two of which are also used to mount the power transformer (T1). It is probably a good idea to mark these holes on the bottom panel prior to assembling the printed circuit board—it's simply easier. Use either nylon or metal standoffs at each mounting point (Fig. 4-7). These stand-offs are cylindrical devices that are either threaded for screws or hollow all the way through to accept a machine screw (the latter are preferred). Use a stand-off that is designed for #6 machine screw hardware and is ¼ to ⅜ inch long.

If you cannot find stand-offs, or if the offered stand-offs are only sold in a large-quantity package (at high price), then a substitute strategy is possible. I sometimes use two or three hex nuts stacked on top of each other (Fig. 4-8) as a de facto stand-off. If you are really precise about drilling the mounting holes in the bottom plate, then it is possible to use #6 hex nuts for the stand-off. However, if you are less precise, then use the next larger

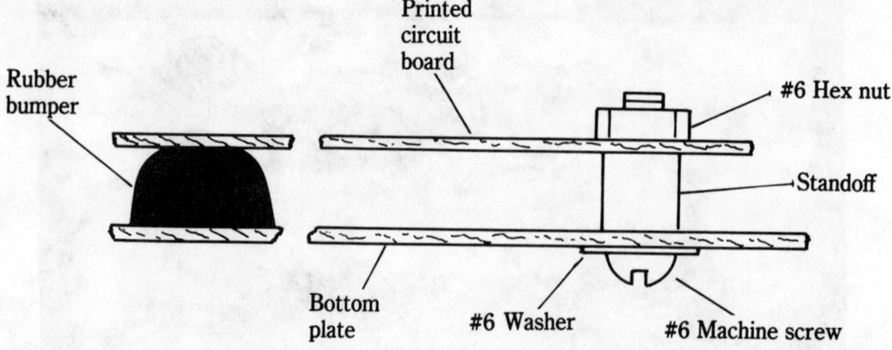

4-7   Rubber bumper used to stabilized PC board.

size hex nuts as the stand-offs so a little "slop" will make up for the tolerance problem when mounting the board. For example, in my power supply I used #6 machine screws and hex nuts to secure the printed circuit board and #8 hex nuts to perform the stand-off function.

Jameco supplies an adhesive-backed rubber bumper with the kit. Place it between the printed circuit board and the bottom plate (Fig. 4-7) at a point indicated in the Jameco instructions. Its function is to add support to the heavy power transformer in order to prevent warping of the printed circuit board over time.

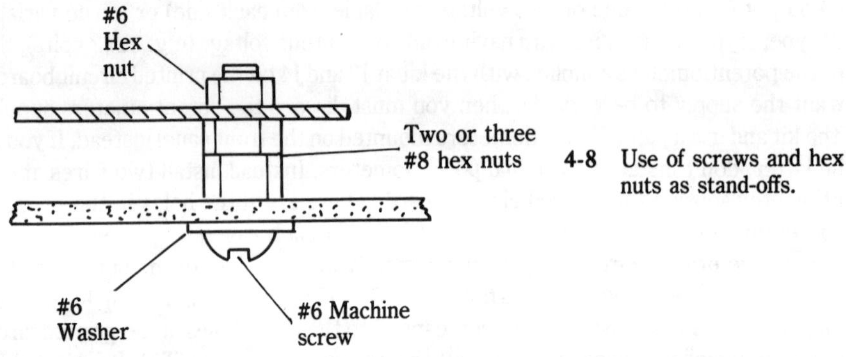

4-8   Use of screws and hex nuts as stand-offs.

The rear panel is shown in Fig. 4-9. Mount the ac electrical power connector (SOC1) and fuse holder on this panel. The connector (SOC1) is used to bring ac from the power line wall socket safely inside the cabinet. Some people like to use a simple grommet and strain relief knot to do this job, but there are safety problems with that approach if the power line is strained or the grommet (or panel) is damaged. Use an Underwriter's Laboratory (UL) approved socket. The kind that I selected is the heavy-duty square type similar to those used on the back of IBM-PC® computers and most of the clones. These sockets are easily available at electronic parts distributors. Don't forget to buy the matching ac power cord,

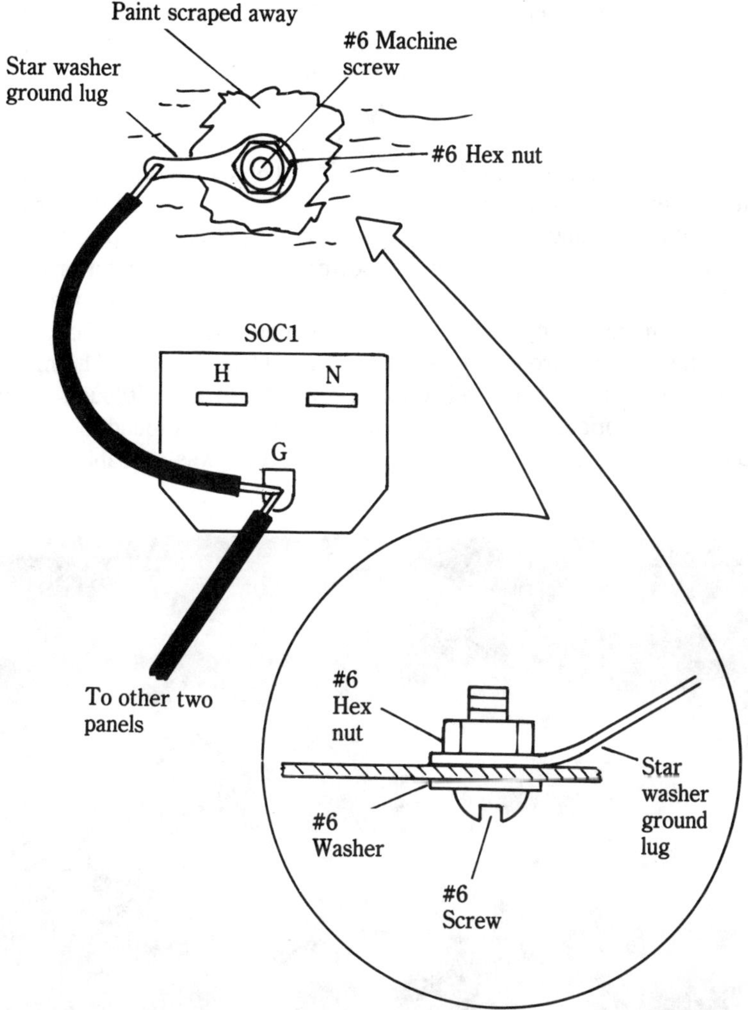

4-9 115 Vac socket wiring. *Danger!* ac will kill you if touched. Disconnect the power supply at the wall outlet when working on this circuit.

or you'll wind up robbing the family personal computer when you want to use the power supply.

Mount the fuse holder (F1) on the rear panel alongside the ac power connector (SOC1). Install the 6/10 ampere slow-blow fuse in the fuse holder after you finish soldering the connections, or else some damage might occur to the fuse. The pins on the back of SOC1 will be marked *H* (hot), *N* (neutral), and *G* (ground). Connect the fuse wire to the hot terminal.

Connect the ground terminal of SOC1 to a ground screw that passes through the rear panel (Fig. 4-9 inset). Drill a $^5/_{32}$-inch hole in the rear panel at a convenient point close to

SOC1. Scrape away the paint on the inside of the panel around the hole until only bright, shiny metal shows. Use a ⅜-inch or longer #6 machine screw and matching hex nut for the ground screw. Make the electrical connection with a ground lug beneath the hex nut. Do not use the smooth form of ground lug, but rather the "star washer" type that bites into the exposed metal.

Make two connections to the ground terminal of SOC1, one directly to the rear panel, the other routed to the bottom panel and front panel. That way, all three metal surfaces that are used are grounded as a safety measure against catastrophic ac failure.

The internal side of the front panel is shown in Fig. 4-10 (the exterior side was shown previously in Fig. 4-6). The front panel contains the controls, switches, meter, and all output terminals. In the prototype of this project, the meter was an afterthought, so I left too little room for it on the front panel—hence it is a little crowded. The meter that was selected is small, and is able to read dc voltages up to 30 volts. Although any available dc voltmeter of appropriate range (20 volts or higher) will do, I found the Modutec 1064-OMS-DVV-030 to fit nicely into the limited space that was available.

4-10   Back of front panel showing wiring.

If you build the adjustable version of the power supply, then a meter on the output line is advisable. Some people might want to use two meters, one each for both positive and negative outputs. Indeed, that is the preferred approach if money is no object. But those meters will cost from $10 up (with most appropriate types being in the $17 range), so you might want to consider a circuit such as shown as the optional meter circuit in Fig. 4-1, and also in Fig. 4-11. In this circuit, a bridge circuit consisting of four 1N4148 (or equivalent) diodes is used as an automatic polarity-sensitive steering switch for the meter. When switch S2 is connected to the negative power supply (V−), the meter current will flow

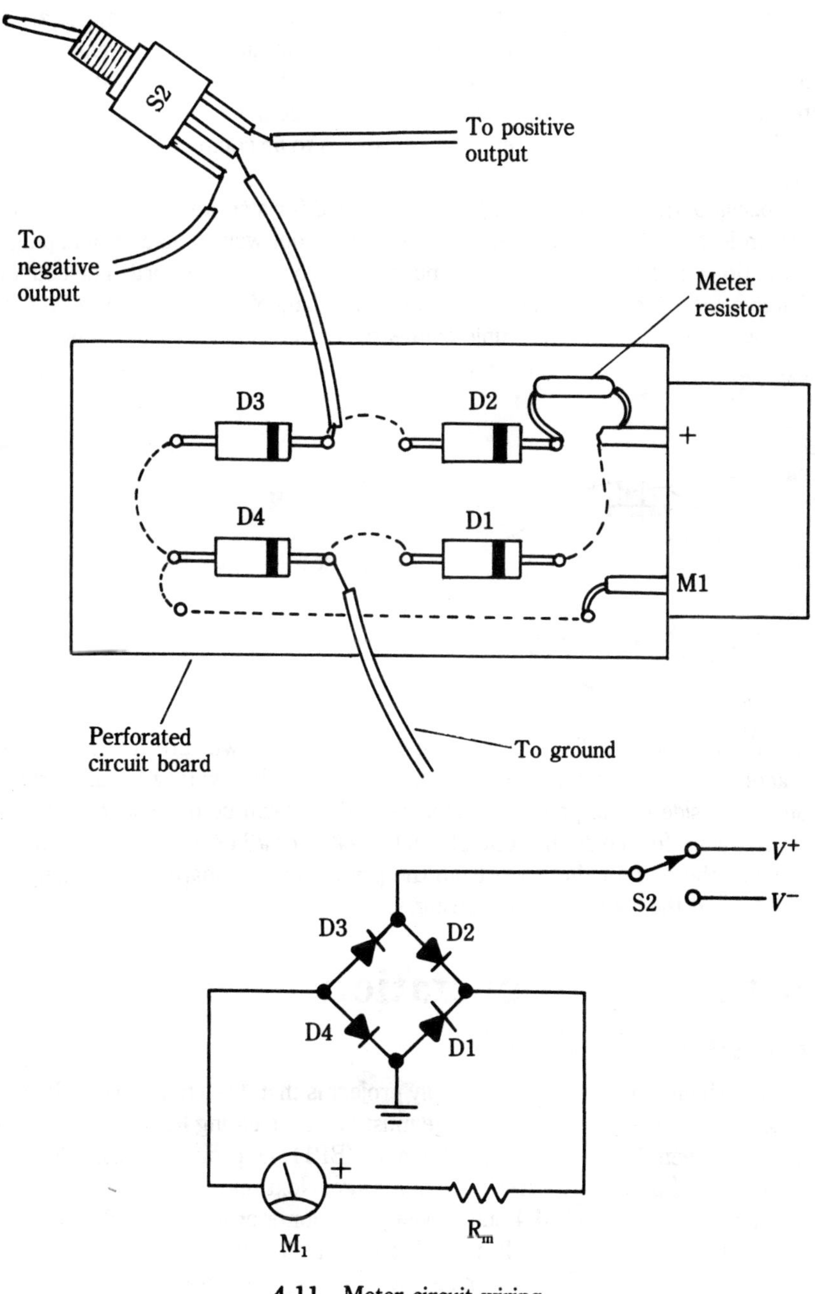

4-11   Meter circuit wiring.

through D3; the meter M1 and diode D1, to ground. Alternatively, when S2 is set to read the positive voltage (V+), the meter current flows through diode D2; the meter M1 and diode D4, to ground. The diodes cause a slight inaccuracy in the meter reading, but this is tolerable for most applications, especially at potentials above ±6 volts or so.

The meter board containing the diode bridge can be mounted anywhere, but it turned out to be convenient to build it so that the electrical connections from the meter are used to mount the board. The board is made of perforated circuit board. To make a pair of mounting points, place loops of #22 solid wire at one end of the board, then solder them to the meter terminals. In some cases (such as the Modutec meter selected here), a small resistor will be packed with the meter. This resistor is to be connected in series with the meter (see R_m in Fig. 4-11).

The special output jacks (J1 and J2) are intended for a series of homebrew instruments and projects that will be built in the future. I did not want to use banana plugs for those applications, and opted instead for a multiconductor dc power cord. The connector selected was a low-cost chassis-mounted 8-pin DIN "audio" connector (Fig. 4-12) because it is low in cost (seemingly simple connectors can cost an arm and a leg!), and is widely available at electronics parts outlets and stores that sell audio supplies.

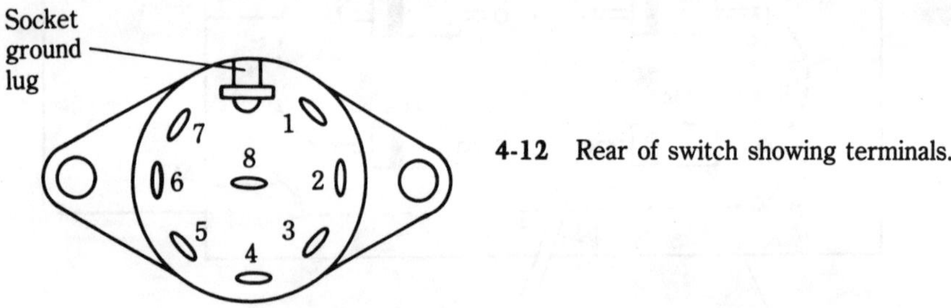

4-12   Rear of switch showing terminals.

The overall connection diagram is shown in Fig. 4-13. Shown here are the rear panel, bottom panel, circuit board, and front panel, with appropriate wiring for all parts. The wiring on the ac side of the power transformer (T1) should be done with #14 or #16 stranded wire, while for the dc and control functions #22 or #24 solid wire is useful. In the latter case, use the #22 for the wires from the printed circuit board to the main output binding posts, and the #24 for other wiring.

Adjustment and operation of the power supply

One of the nice things about this power supply project is that it is simple to use. If you opt for the fixed-voltage type, the output voltage must be set adjusting R3 and R4. Connect a dc voltmeter between the positive output terminal (BP1) and ground (BP2). Adjust trimmer potentiometer R3 for the desired output voltage. Next, move the voltmeter to read the potential between BP3 and BP4, and adjust potentiometer R4 for the desired output voltage. When these voltages are adjusted, disconnect the voltmeters and place the top cover on the power supply. The power supply is turned on and off using S1.

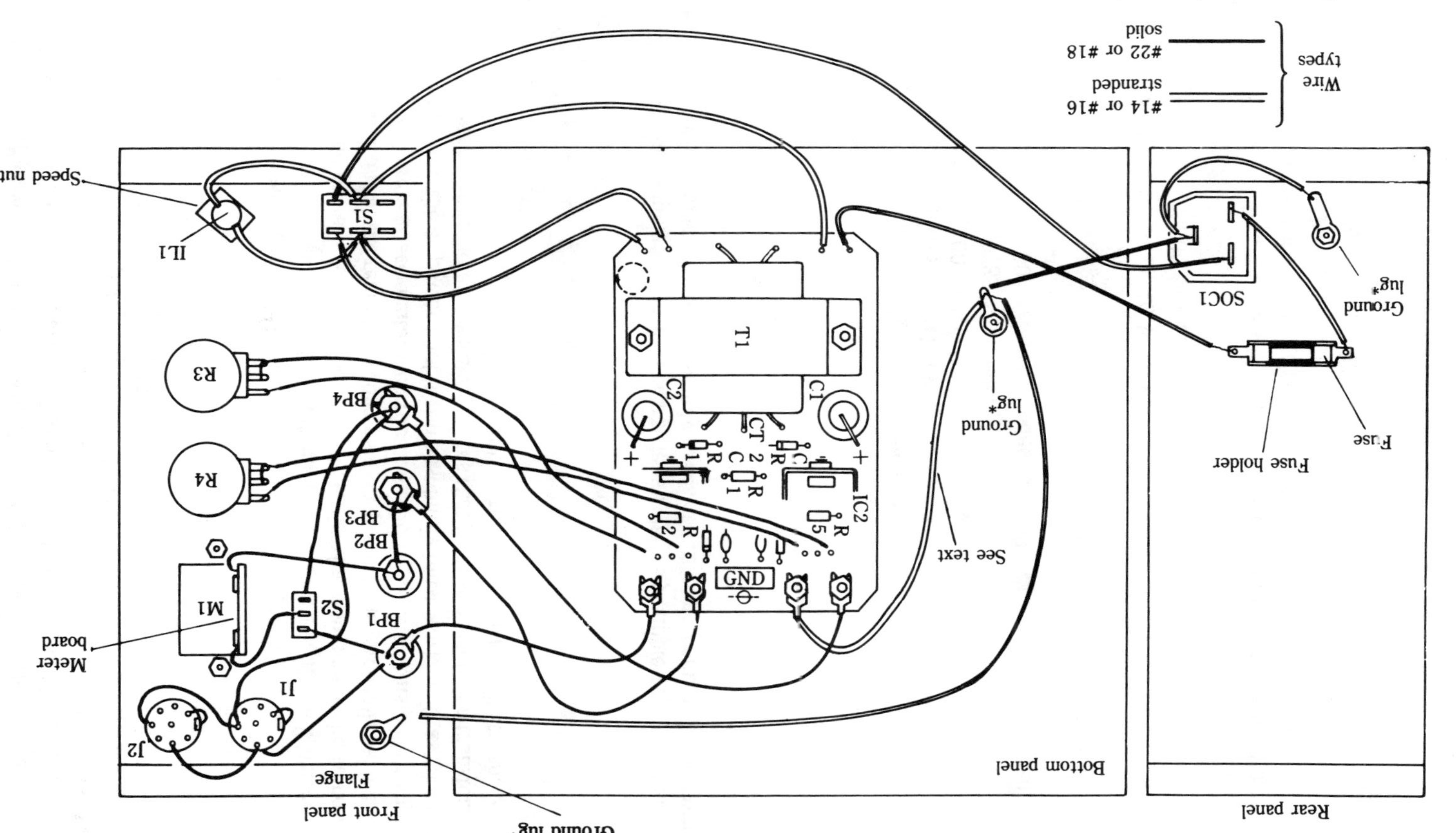

4-13 Wiring layout.

If you opt for the adjustable voltage form of this project, then no trimmers are used. The adjustment is made using knobs attached to R3 and R4 on the front panel. Set S2 for either V− or V+ and adjust the associated potentiometer on the front panel (R3 or R4 as appropriate) for the desired output voltage level. Reset S2 to the other polarity, then adjust the remaining potentiometer.

Safety note

This power supply project is designed to operate from a 115Vac power source, such as the wall outlets in your home or shop. The 115-volt line can be *fatal* if contacted! Whenever you need to open the power supply cabinet, it is essential that you unplug the ac power cord both at the wall and at the rear of the power supply project and physically set it aside. This procedure will prevent accidental electrocution. Don't work on this power supply while it is connected to the ac power line! If you absolutely must work on it (as when adjusting the trimmer potentiometers), safety dictates that you use an isolation transformer to power the project. Also, you must be absolutely sure to use plastic adjustment tools, never a metallic screwdriver.

Conclusion

The bipolar dc power supply shown in this article is relatively simple to build, and will serve as a low-cost solution to the power supply needs of a large percentage of electronic experimenters.

Parts list

Jameco JE-215 power supply kit:

IC1	LM-317T IC voltage regulator
IC2	LM-337T IC voltage regulator
C1, C2	2,200 μF/16 V electrolytic capacitor
C3, C4	1 μF/35 V tantalum capacitor
CR1-CR4	1N4001 rectifier diodes (1N4001 through 1N4007 can be used)
LED1	Red light-emitting diode
R1	1,000 Ω ¼ W resistor
R2, R5	180 Ω, ¼ W resistor
R3, R4	2,500 Ω trimmer potentiometer (if fixed supply is desired).

Other parts:

R3*, R4*	2,500 Ω, 2-watt linear taper chassis mounted potentiometer with ¼-inch shaft (if adjustable version is needed).
S1	GC Electronics 35-132 DPDT switch
IL1	Linrose B2150A1 neon lamp, 105-125 volts
S1	SPDT miniature toggle switch

BP1, BP4   Red 5-way binding posts
BP2, BP3   Black 5-way binding posts
SOC1   ac socket, chassis mounted

M1   0 to 20 (or 0 to 30) Vdc panel meter
J1, J2   8-pin, 270-degree DIN female chassis mount connector
Power cord
  to match AC socket
F1   Fuse holder
                  $^6/_{10}$ ampere slow-blow fuse to match F1

Augat 92008 p/n PKE5-90B-1.4 knob (two required)
Hammond 1458D4B-1184 8-x-8-x-4-inch cabinet

5
pn junction diodes

NOW THAT WE'VE DISCUSSED SEMICONDUCTOR MATERIAL, IT'S TIME TO PUT the knowledge to work in an actual electronic component: the pn diode. These devices are based on pn junctions, which are also the basis for a common class of transistor (which we will study shortly). In this chapter you will become familiar with the operation of the pn junction and learn about the pn junction diode.

The word *diode* means "two electrodes," and in this sense it is taken to mean two different types of semiconductor material (n and p), each forming one electrode of the device. You will discover just how these materials form electrodes in the next section.

What is a pn junction?

A piece of p- or n-type semiconductor is not terribly useful except as a kind of resistor. However, if we place a piece of p-type material in intimate contact with a piece of n-type material, as in Fig. 5-1, then some interesting things occur.

Keep in mind that Fig. 5-1 is a learning device, and does not imply that a pn junction can be manufactured by laying blocks of two different types of material on top of each other. Later in this chapter you will learn some of the ways that the diode is made, but for now please accept this graphical presentation as a learning exercise that is close to reality.

Recall from the previous chapter that both types of semiconductor material contain both majority carriers and minority carriers. The minority carrier in either type of material is always the opposite-charge carrier from the majority carrier. Thus, in p-type material, holes are the majority carriers and electrons are the minority carriers. Alternatively, in n-type material the majority carriers are electrons and holes are the minority carriers.

Diffusion current   At the interface between the p-type and n-type regions of the diode, there is an attraction between majority carriers in the two regions. Thus, some electrons from the n-type region will cross the boundary between regions in order to fill

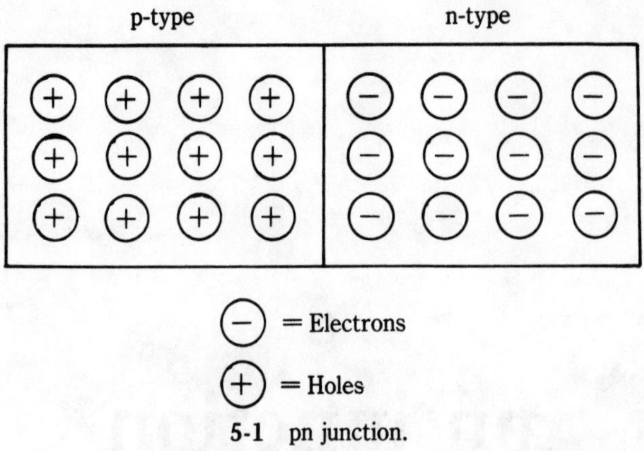

5-1 pn junction.

holes in the p-type material. Because these electrons flow from a higher concentration to a lower concentration, they are said to comprise a *diffusion current* across the junction.

Barrier potential   As the diffusion current continues, a negative charge of excess electrons is built up in the p-type material close to the boundary. This negative charge acts to repel further electron incursion because "like charges repel." It does so by creating an electrical field across the boundary (Fig. 5-2). The voltage across the junction that results from this field is called a *barrier potential*, or *junction potential.*

Depletion zone   A consequence of the movement of charge carriers, and the creation of the barrier potential, is that the area in the vicinity of the boundary becomes depleted of carriers. This action forms a *depletion zone* either side of the barrier. Because equal numbers of holes and electrons take place in the diffusion current, it is possible to predict the width of the depletion zone in either material by looking at the impurity doping profile. If both n-type and p-type materials are equally doped, then the depletion zone is uniformly distributed across the junction. However, if one side, say the p-type material, is more heavily doped than the other — in this example, n-type — then the depletion zone will intrude farther into the n-type region than in the p-type region in order to collect equal numbers of carriers.

Minority charge carrier action   The barrier potential prevents majority carrier movement across the junction. Minority carriers behave differently, however. Thermal electrons generated in the p-type material are attracted by the positive potential on the n-type side of the barrier, so can cross the zone. Similarly, holes generated by loss of thermal electrons on the n-type side "move" across the barrier to the p-type side. The movement of minority carriers across the pn junction is enhanced by the barrier potential. This action is exactly opposite the action of the barrier potential on majority carriers.

The biased pn junction

If an external voltage is applied to a pn junction, the junction is said to be *biased.* If the external bias voltage is applied such that its polarity aids the flow of majority carriers across the junction, then it is called *forward bias.* This type of bias occurs when the positive terminal of the voltage source is applied to the p-type side of the junction and the nega-

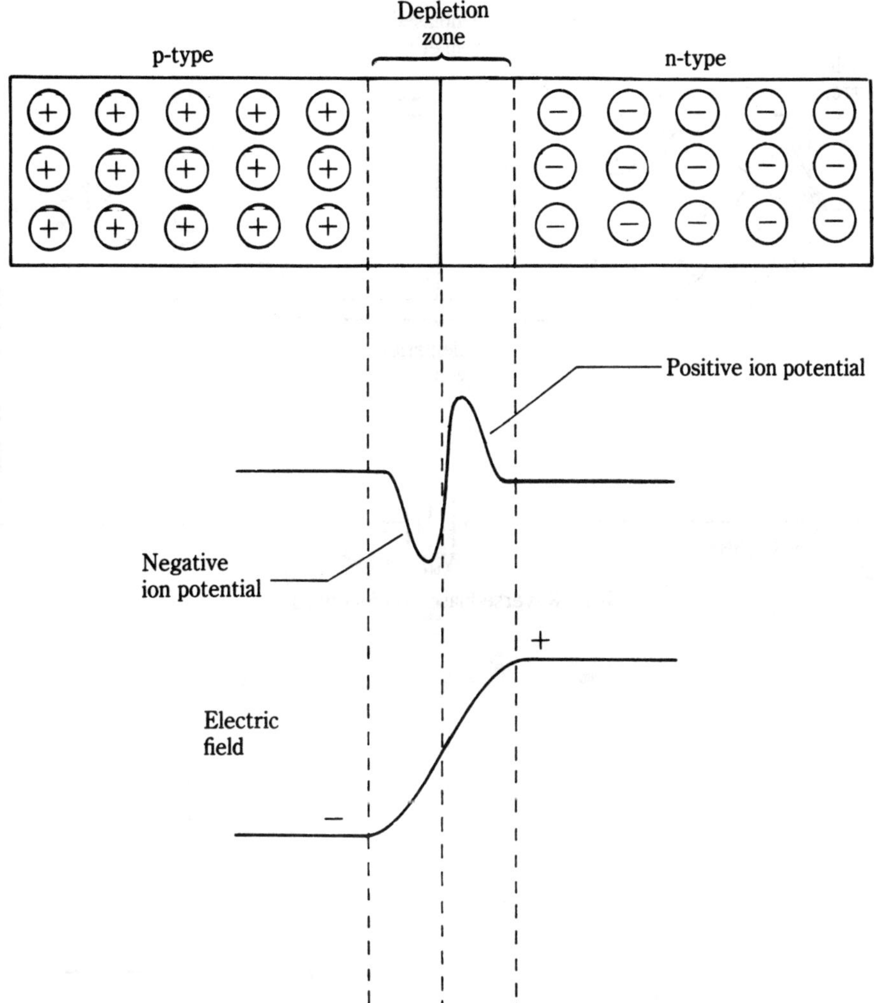

5-2   pn junction profile showing depletion zone and barrier potential.

tive terminal is applied to the n-type side of the junction. A reverse-biased pn junction exists when an external electrical potential is applied across the junction such that the negative terminal of the external potential is applied to the p-type side of the junction and the positive terminal is applied to the n-type material.

Reverse-biased pn junctions

Figure 5-3 shows a reverse-biased pn junction, while Fig. 5-4 shows the I-vs.-V characteristic. Holes in the p-type material are electrically attracted to the negative side of the battery, while electrons in the n-type material are attracted to the positive side. Thus, both types of majority carrier are attracted *away* from the junction.

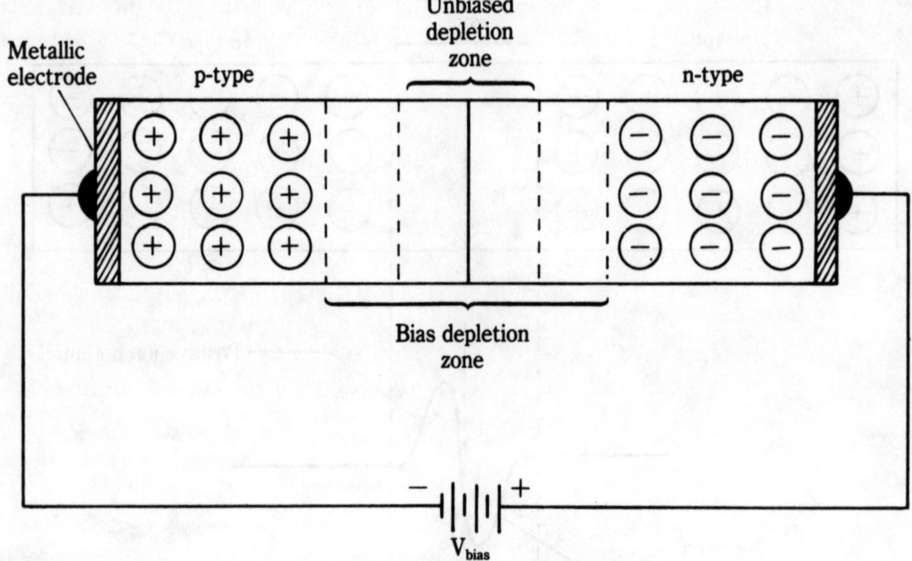

5-3　Reverse-biased pn junctions.

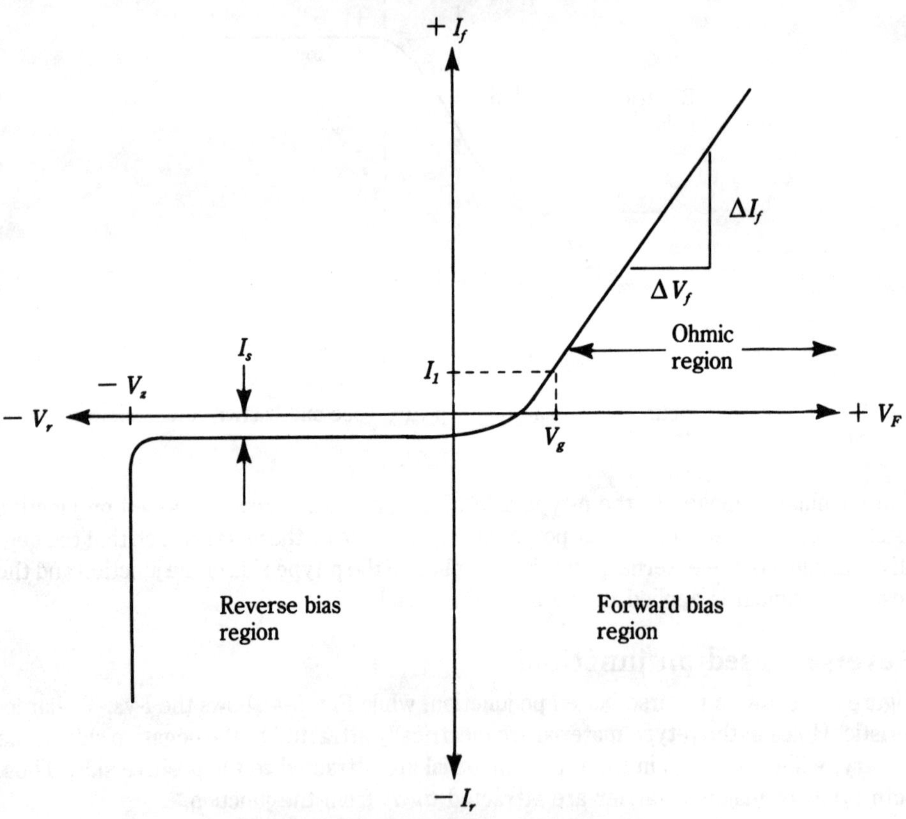

5-4　I-vs.-V curve for pn junction.

Reverse-biased depletion zone   One effect of reverse-biasing the pn junction is that the depletion zone widens over its unbiased width (compare the two depletion zones in Fig. 5-3).

Reverse saturation current (I_s)   Because the depletion zone is wider in a reverse-biased junction, the barrier voltage is higher, so it is even more difficult for majority carriers to migrate across the junction. Minority carriers can migrate, however, and are in fact aided by the reverse-bias potential. When the reverse-bias potential is raised, the reverse current is increased rapidly to a point at which all available minority carriers are being swept away. This current level is called the *reverse saturation current*, and is symbolized by I_s. The value of I_s will be on the order of 10^{-7} ampere (0.1 μA) for silicon and between 10^{-6} ampere (1 μA) and 10^{-5} amperes (10 μA) for Germanium.

Reverse resistance (R_s)   Electrical resistance is defined as the ratio of voltage to current (V/I). In the case of the reverse-biased pn junction, the current is I_s, and the voltage is the applied reverse bias voltage:

$$R_r = \frac{V_r}{I_s} \tag{5-1}$$

Example 5-1

Find the reverse resistance (R_r) of a pn junction if the leakage current is 10^{-7} amperes and the reverse-bias voltage is -4 Vdc.

Solution:

$$R_r = \frac{V_r}{I_s}$$
$$= \frac{4 \text{ volts}}{10^{-7} \text{ amperes}}$$
$$= 4 \times 10^7 \ \Omega$$
$$= 40{,}000{,}000 \ \Omega$$

Reverse breakdown or *avalanching*   Notice in Fig. 5-4 that the reverse current is constant from a small reverse-bias potential up to a certain point, marked V_z, where the reverse current abruptly shoots up to a very large value. This reverse-bias breakdown occurs because of a phenomenon called *avalanching*. When reverse current carriers are subjected to an electrical field, they acquire a certain kinetic energy level. If this electrical field is large enough — and applied potentials of only a few volts can make it too high under the right conditions — then avalanching can occur.

The avalanche process is relatively easy to visualize. In the depletion zone, energetic carriers can cause *collision ionization* by striking atoms and forcing them to give up an electron. This ionization adds carriers to the process on an exponentially growing scale, as each new carrier also becomes kinetic and strikes other atoms. The current very rapidly multiplies, causing a sudden and very drastic reduction in junction resistance. The value of R_r will drop from a level of several megohms to only a few ohms.

Effect of temperature   The previous discussion on pn junctions left out a very important consideration: temperature. The reverse saturation current of a pn junction is a strong function of temperature. Figure 5-5 shows the reverse current of a typical junction

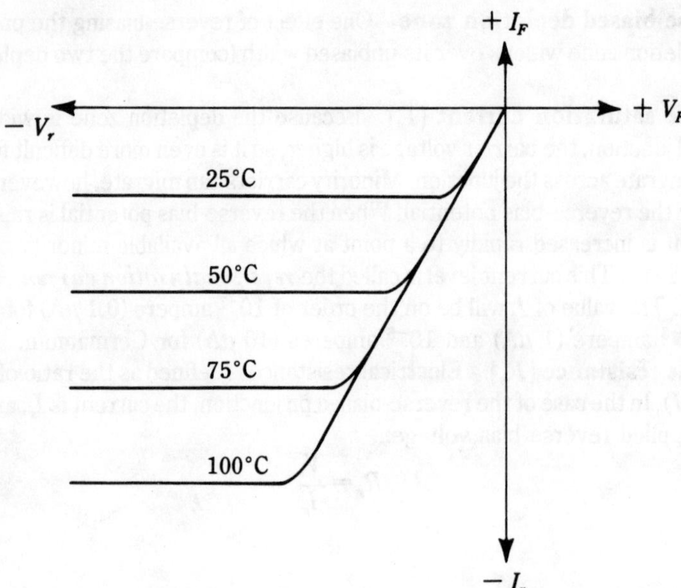

5-5   Leakage currents varying as a function of temperature.

at various temperatures. Notice that increasing the temperature tends to increase the reverse saturation current, thereby reducing the junction resistance. The value of I_s will approximately double for every 10°C rise in temperature. This phenomenon occurs because the minority carriers are largely thermally generated, so a larger number becomes available as a high temperature is applied.

The temperature dependence of pn junctions is so predictable, under the right circumstances, that it can be used to make a temperature sensor in modern electronic circuits. A number of commercial IC temperature sensors are on the market that use the reverse saturation current of pn junctions to sense the applied temperature.

Reverse-bias junction capacitance   When a pn junction is reverse-biased, a depletion zone forms between the p- and n-type materials. Because most of the charge carriers are swept out of the zone it is effectively an insulator, as illustrated previously by the values of reverse resistance, R_r, that can be achieved. Either side of the depletion zone are zones in the p- and n-type materials that are heavily populated with majority carriers, effectively forming highly conductive regions. This situation is very similar to a parallel plate capacitor; i.e., two conductors separated by an insulator. Thus, a pn junction will exhibit a capacitance in the reverse-bias condition. Figure 5-6 shows an equivalent circuit for a reverse-biased diode, showing the capacitance effect (C_{pn}, or C_p in some texts). The junction capacitance value is on the order of 10 picofarads (pF) to around 450 pF, depending on the device design (30 pF is a commonly seen value).

A feature of the pn junction capacitance is that it varies with the applied reverse-bias potential. Special-purpose voltage-variable capacitance diodes, called *varactors*, are manufactured specially for use in voltage-dependent automatic frequency control (AFC) in radios, as radio or transmitter tuners, or for certain special-purpose applications (usually at radio frequencies).

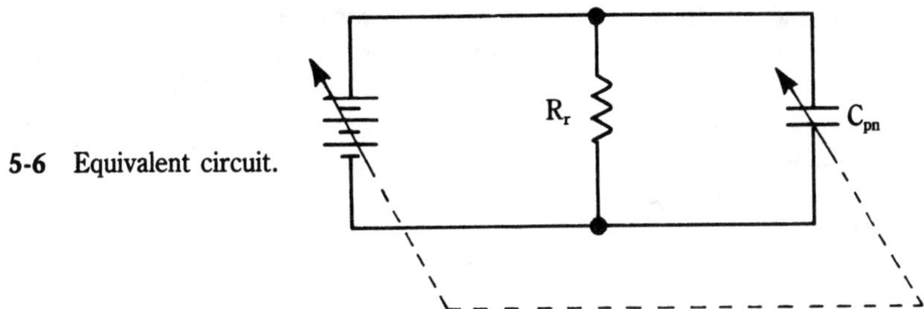

5-6 Equivalent circuit.

Forward-biased pn junctions

In the forward-bias condition, the positive terminal of the dc power supply is applied to the p-type end of the pn junction, and the negative terminal is applied to the n-type material. Because like polarity charges repel each other, all of the majority carriers are swept to the junction region, as shown in Fig. 5-7. This action reduces the barrier potential and facilitates the flow of charge carriers across the junction. In other words, when the diode is forward-biased, a forward current will begin flowing even though the bias voltage will have to

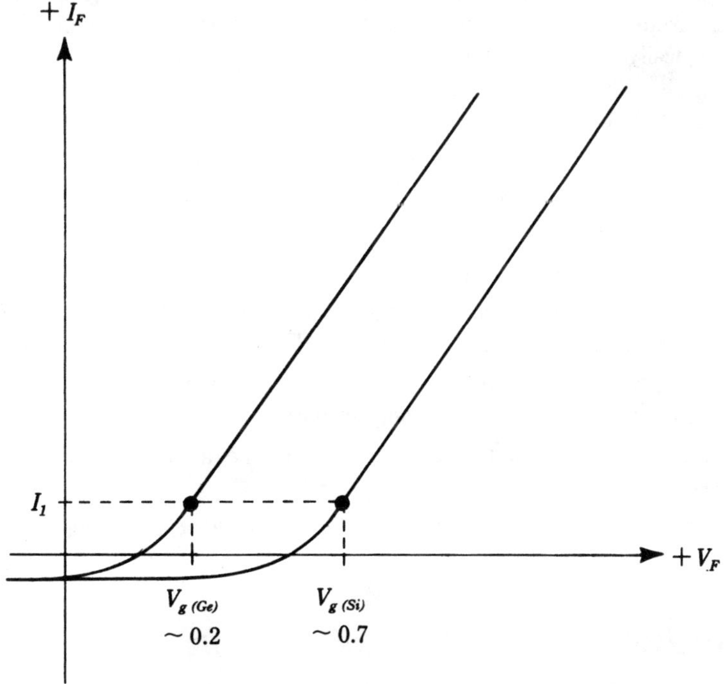

5-7 I-vs.-V showing knee potential for silicon and germanium pn junctions.

oppose the barrier potential at low values. This factor causes a nonlinear I-vs.-V characteristic at low values of forward-bias voltage V_f (see Fig. 5-7). Once the potential is overcome, however, the device behaves ohmically (i.e., the current increases linearly — or nearly so — with applied voltage).

The point at which the device goes ohmic is a *knee potential*, or *forward character-istic potential*, and is sometimes denoted either V_g or V_γ. The value of the knee potential is 0.2 to 0.3 volt for germanium devices and 0.6 to 0.7 volt for silicon.

Forward resistance   The forward resistance of the junction is defined as the quotient of the knee potential and the current that flows at that potential or, in terms of Fig. 5-7:

$$R_f = \frac{V_\gamma}{I_1} \tag{5-2}$$

Dynamic or ac resistance   The static, or dc, resistance defined by R_r is not used widely in practical electronics. Of more importance is the *dynamic resistance*, also called the *ac resistance*, and symbolized by r_d. The dynamic resistance is defined as the slope of the I-vs.-V characteristic in the ohmic region, or

$$r_d = \frac{\Delta V_f}{\Delta I_f} \tag{5-3}$$

Temperature effects   Temperature affects the forward-biased operating conditions of the diode. If the forward current is held constant (I_f = constant, ΔI_f = 0), then the knee voltage drops with increases in temperature. The temperature coefficient is around 2 mV/°C (typically 1.8 mV/°C for Si and 2.02 mV/°C for Ge).

Experiment 5-1

This experiment examines forward- and reverse-bias conduction in a pn junction diode. These diodes are presented in more detail in the following section, but for the present you need only know what to obtain in the way of a component.

1. Obtain a 1N4148 small signal diode, or its equivalent.

2. Obtain an ohmmeter. If you use an analog ohmmeter, set it to the "RX100" scale. If you use a digital ohmmeter, set it to "HIGH OHMS" or to the diode symbol.

3. Connect the leads of the ohmmeter across the leads of the 1N4148 diode, and write down the resistance reading (in ohms).

4. Reverse the leads of the ohmmeter and take a new measurement of the resistance.

5. Compare the two resistance readings. There should be a very high ratio between them, or the diode is not working properly.

pn diodes

The simplest product that is made with pn junctions is the pn diode. These devices are available in tiny glass-packaged versions that pass only a few milliamperes of current, to large 1- to 3-ampere epoxy body devices, to stud-mounted devices capable of passing forward currents up to hundreds of amperes.

Diodes are used in an extremely varied range of applications. Some examples include demodulation of radio and TV signals, frequency mixing, power supply rectification, digital logic, and switching. In this section, you will take a look at popular diode packaging styles, different methods by which diodes are fashioned, some of the ratings considered important on practical diodes, and some applications circuits.

The diode device

When the pn junction is fitted with ohmic contacts (ordinary metal) to permit electrical access from the "outside world," the device becomes a diode.

Both the circuit symbol for such a diode and its relationship to the junction is shown in Fig. 5-8. The n-type end of the semiconductor becomes the *cathode*, or negative electrode, of the diode, while the p-type end becomes the *anode*, or positive electrode.

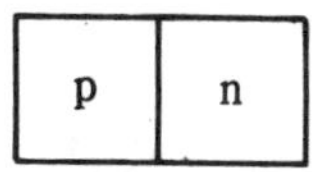

5-8 pn junction and diode symbol.

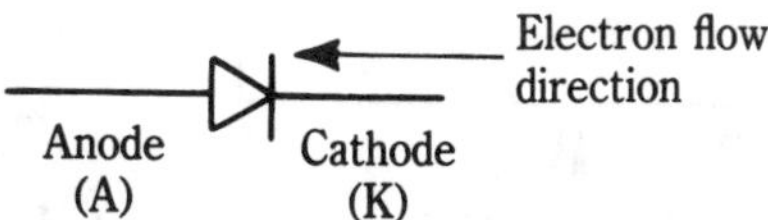

There is frequently some confusion about the diode symbol because some texts state that the arrow points in the direction of the current flow, even though electrons flow in the opposite direction. What is referred to in that context is so-called *conventional current flow*, which is used in some technical school texts. In most electronic circuits, however, it is more convenient to refer to *electron current flow*, which is exactly the opposite of conventional current flow. Both currents are the same, but a semantic difference causes the dual-flow situation.

As long as you remain consistent when solving problems, the difference is a moot point. For the purposes of this book, assume that electron flow is intended.

Diode current flow   Electrons (i.e., electrical current) pass through a pn diode in only one direction: from the n-type to the p-type, or, in terms of the diode symbol shown in Fig. 5-8, from cathode to anode. Figure 5-9 shows two different situations. In both cases, the diode is connected between a load resistance (R_L) and a voltage source (V).

In Fig. 5-9A, the polarity of the dc voltage applied to the circuit forward-biases the diode, so current (I_L) will flow. The value of this current will be:

$$I_L = \frac{V_R}{R_L} \tag{5-4}$$

Voltage V_L is the voltage drop across the load resistor and will be less than the applied voltage by the amount of the voltage drop across the diode (V_D).

In some cases, the diode voltage drop is so small compared with V that it is ignored. For example, the diode voltage drop might be only a few tenths of a volt in a 28-volt circuit.

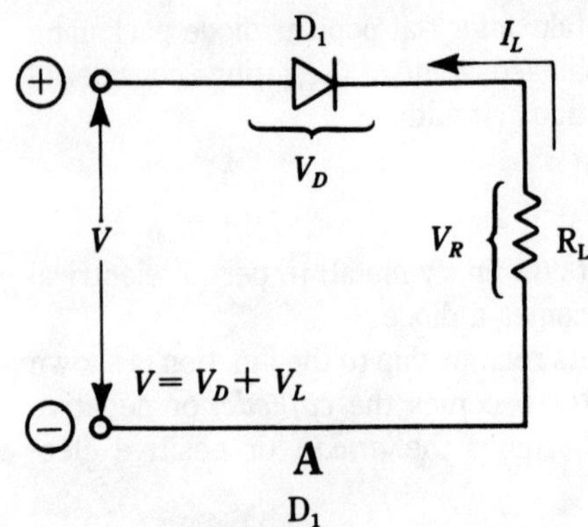

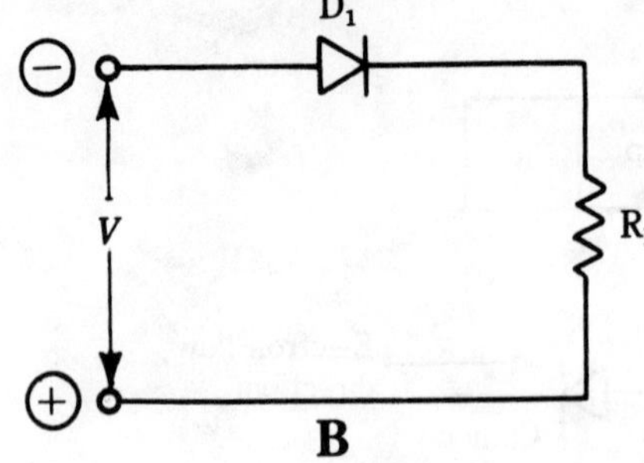

5-9 pn junction diode in A. forward-biased, and B. reverse-biased conditions.

However, if V_D is of sufficient magnitude to have a practical effect on the outcome of the calculation, then it should be included:

$$I_L = \frac{V - V_D}{R_L} \tag{5-5}$$

When the polarity of the applied voltage is reversed (Fig. 5-9B), then the anode of the diode becomes negative with respect to the cathode. Under these conditions, the diode is reverse-biased, so no current can flow. In actuality, there will be a tiny leakage current, which is the reverse current discussed earlier, but this current is usually negligible compared to the forward current.

The ability of the diode to pass current in only one direction makes it able to perform a number of jobs in electronic circuits. In dc power supplies, for example, it rectifies the bidirectional ac into unidirectional pulsating direct current. In AM radios, the diode acts as a demodulator to remove the speech or music signal from the radio frequency (RF) carrier signal. In some instrumentation circuits, diodes are used to clip or shape waveforms or to block signals of unwanted polarity.

Experiment 5-2

1. Connect the circuit of Fig. 5-10. The diode is a rectifier type and can be any of the series 1N4001 to 1N4007. The meters can be either analog or digital types. The power source is a 9Vdc transistor radio battery.

2. Note the current reading. It will be in milliamperes. Convert to amperes:

$$I_{amperes} = \frac{I_{ma}}{1,000}$$

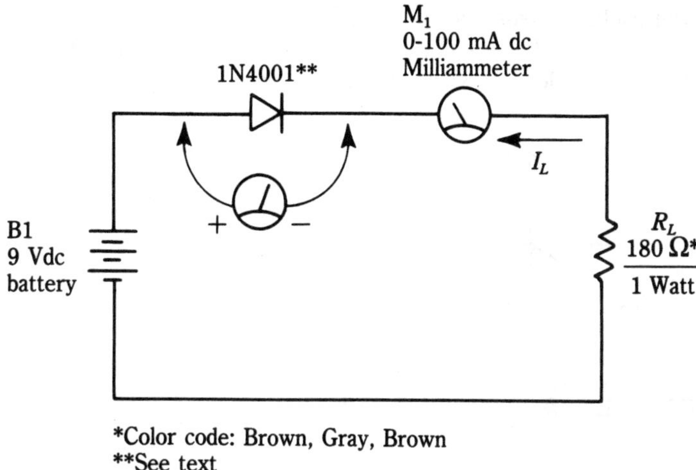

5-10   Test circuit for experiment 5-2.

3. Measure the voltage across the battery, B1. It should be close to 9Vdc.

4. Measure the voltage drop across the 180-ohm (brown-gray-brown) resistor.

5. Measure the voltage drop across the diode, D1. Observe the polarity.

6. Compare the readings obtained in this experiment with the theoretical expression of Eq. 5-5.

Diode ratings

Diode parameters of interest are the peak forward voltage, peak forward current, peak reverse voltage (PRV), and *leakage current.* All of these parameters, except PRV, are fairly ordinary, so they need not be amplified here.

The *peak reverse voltage*, also called the *peak inverse voltage* (PIV), is the maximum reverse-bias potential that can be applied to the diode for any period of time without causing permanent damage to the diode. It is generally true that destruction of the diode will occur if the PIV is exceeded for any length of time.

Diodes are used in many different applications in electronic equipment. In general, however, a diode designed to be optimum in one application or type of operation will not do well in other applications. Signal diodes, for example, are meant to be used primarily in low-level modulation, demodulation, clamping, or switching uses. If they are used as power-supply rectifiers, they will burn out immediately. Rectifier diodes, on the other hand, are designed to handle much higher applied voltages and forward current levels. They are almost useless for most signal applications because they are too slow (low frequency) and have too much leakage current.

Diode packages

Diodes come in a wide variety of packages, depending on the job that they are intended to do. Signal diodes (e.g., 1N60, 1N914, 1N4148) are most often manufactured in small glass packages (Fig. 5-11A and B) mounted with axial leads. A color band, often black, at one end of the package identifies the cathode end. Some signal diodes, however, are mounted in radial lead packages similar to those of Fig. 5-11C and D. You will see these packages again in three-lead versions, since they are also very popular for transistors.

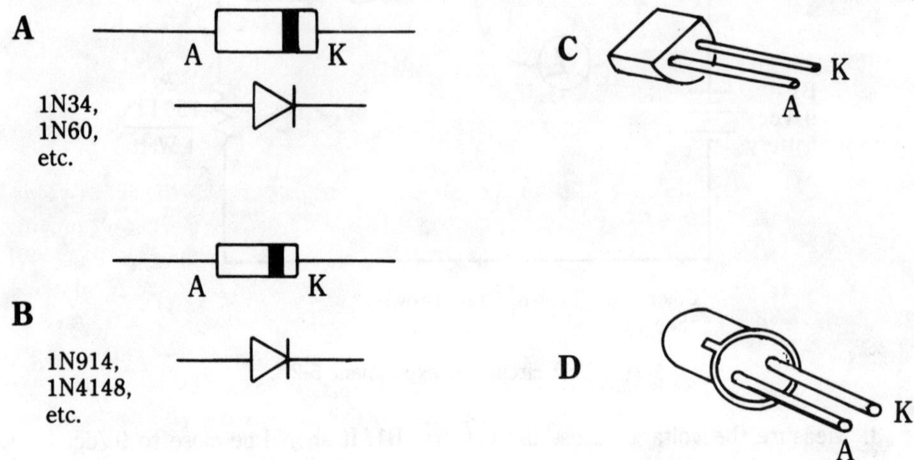

5-11 Popular diode packages (small signal).

Rectifier diodes are usually designed to work at power line frequencies (50/60 Hz), and must handle considerably higher current levels. For example, while the small signal diode typically handles 5, 10, or 50 milliamperes, depending on the type, rectifier diodes handle currents of 1 ampere to hundreds of amperes (again, depending on the type).

Small rectifier diodes (1 to 3 amperes) are found in glass or epoxy packages, such as Figs. 5-12A through D. These packages come in several different styles. In most cases the 3-ampere diodes are physically larger than the 1-ampere devices, even when in similarly shaped packages.

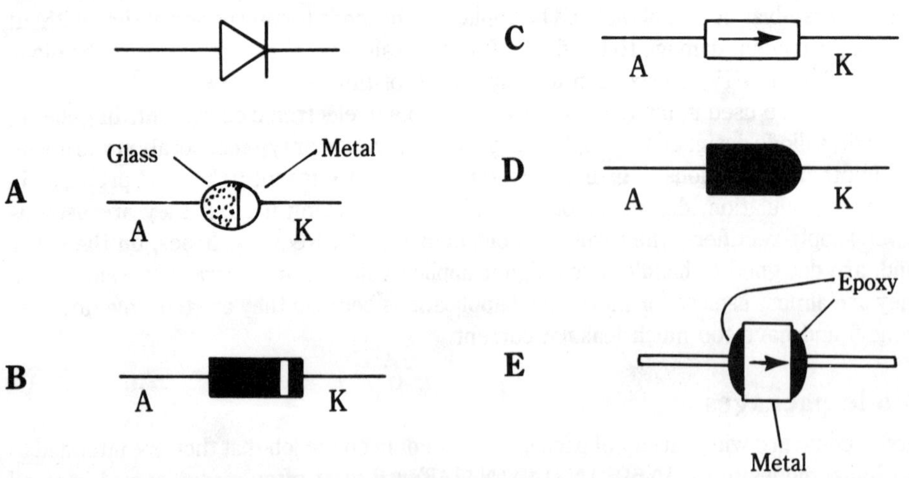

5-12 Popular rectifier diode packages.

Note in Fig. 5-12 that several different polarity marking methods are used. In the glass bead device (Fig. 5-12A), the cathode is marked by the metal end. In Fig. 5-12B, the case is a small cylinder of black epoxy; the cathode end is marked with a paint band (similar to the method used on small signal diodes). In still other cases, the cathode is marked with an arrow (Fig. 5-12C). Although not as popular as it once was, the package in Fig. 5-12D is sometimes seen. In these rectifier diodes, the cathode is marked by the bullet-shaped rounded end.

Larger axial lead diodes such as Fig. 5-12E are sometimes seen. These diodes are usually rated at 3 to 6 amperes forward current, so generally have considerably heavier lead wires than the 1- and 3-ampere diodes.

Once the current rating gets to somewhere around 6 to 8 amperes, the ability of the diode package to dissipate heat becomes very important. As a result of this requirement, diodes are made in stud-mounted packages such as Fig. 5-13A (medium current) and 5-13B (high current). The stud is mounted to an external heatsink, or a heat-conductive surface such as the chassis of the instrument, using a threaded hex nut fastener.

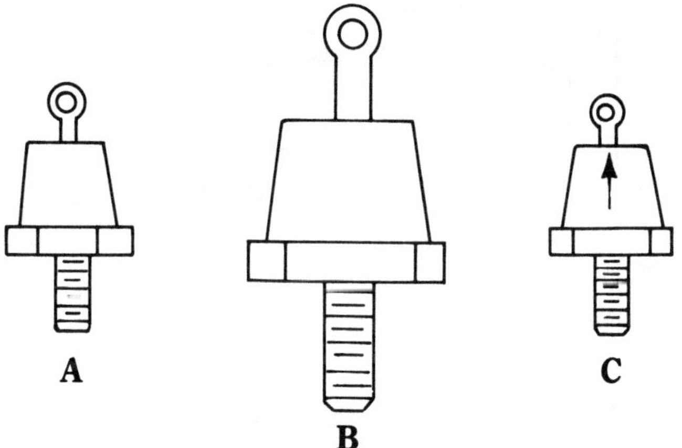

5-13 High-power rectifier diode packages.

Stud-mounted diodes are normally built with the cathode to the stud end of the package, as shown by the diode symbols in Fig. 5-13. In some cases, however, a reverse-polarity diode is seen, in which the anode is connected to the stud end of the package. These diodes are either marked with an arrow pointing toward the "wrong" end (Fig. 5-13C), or have an "R" suffix on the type number. In the latter situation, a "1Nxxxx" is the normal polarity version (e.g., Fig. 5-13A), while "1NxxxxR" is the same diode but with reversed polarity (e.g., Fig. 5-13C).

Types of diode construction

The methods used to form diodes from the basic materials have undergone a lot of change over the years, even though the oldest methods are still in use in some places. The basic forms of diode construction include point contact, grown junction, fused/alloy junction, diffused junction, and epitaxial. All of these are pn junction diodes, but their method of construction varies. Different methods of construction are used depending on the diode characteristics and the materials used.

Point contact   The point contact diode is shown in Fig. 5-14. This type of diode construction is reminiscent of the earliest crystal diodes that were made from certain natural crystalline elements for early "crystal set" radio receivers. The natural mineral galena (chemically, PbS) is a natural pn diode, but only certain spots on the diode surface work. As a result, a cat's whisker probe was used to find those spots. Early in World War II, microwave research showed the need for diodes that were reliable and predictable (unlike galena), so certain germanium and silicon diodes were developed using similar techniques (1N21, 1N23, 1N34, 1N60).

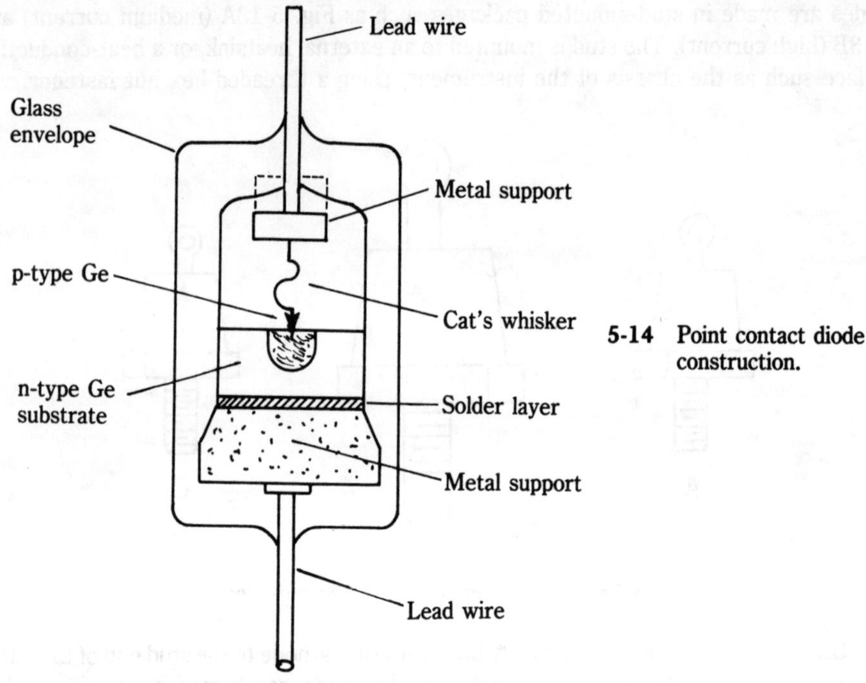

5-14   Point contact diode construction.

In the point-contact germanium diode (shown in Fig. 5-14), a substrate of germanium is doped with antimony to make n-type material. A cat's whisker of tungsten, molybdenum, or platinum is coated with indium and brought into contact with the n-type Ge. A brief electrical current pulse is passed through the cat's whisker with sufficient current to cause the indium to enter the semi-melted region around the point where the cat's whisker makes contact. The result is a small hemispherical section of p-type Ge within the n-type substrate. At the interface region between the p-type and n-type material, a pn junction is formed.

Diodes of this type are usually mounted in an axial lead glass package, as shown. They are still widely used, especially as video detectors in televisions and AM detectors in high-frequency radio receivers.

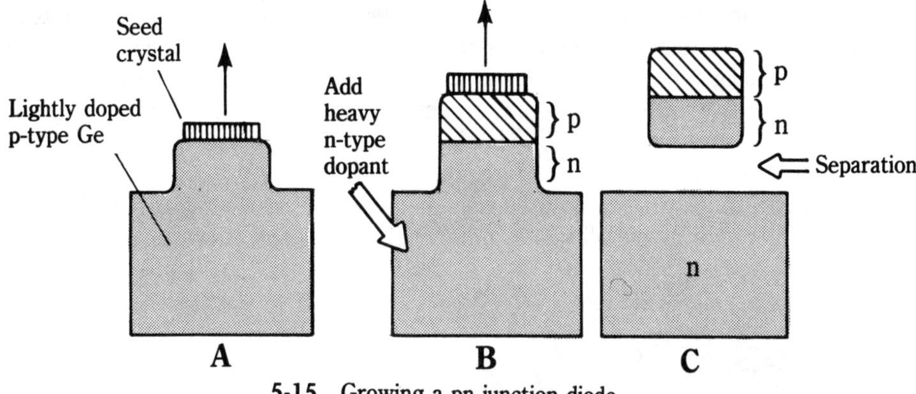

5-15 Growing a pn junction diode.

Grown junction   The grown junction is shown in Fig. 5-15. It is built using a crystal-growing technique that is not only standard in the semiconductor industry, but in others as well. (For example, certain synthetic gemstones are grown in this manner.) In all of these cases, a seed crystal is brought into contact with a molten supply of the material (see Fig. 5-15A). As the crystal is slowly withdrawn, an extrusion of the material crystalizes around the seed.

In making pn diodes, the molten material is lightly doped p-type germanium. When the crystal forms around the seed, then, it is the p-type section of the diode.

At the appropriate time in the process, a heavy n-type doping material is added to the molten mixture, causing it to become predominantly n-type (Fig. 5-15B). The rest of the crystal grown onto the seed will be n-type, with the interface between n- and p-type forming a pn junction. Once the crystal is grown, it is cut off (Fig. 5-15C), plated on the ends for electrical contacts, and sealed into an appropriate package.

Fused junction   A fused junction, or alloy junction, device is shown in Fig. 5-16. A wafer of n-type germanium is sliced to the correct size and shape to form a substrate. A

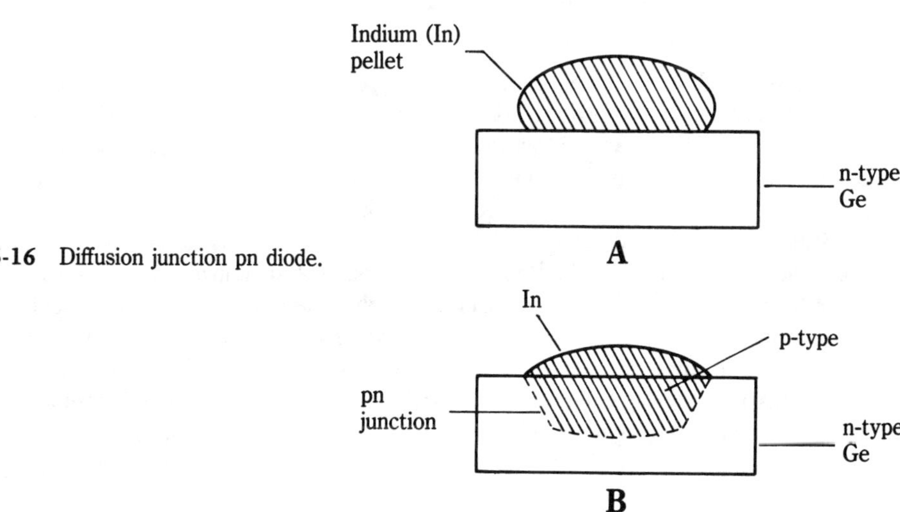

5-16   Diffusion junction pn diode.

p-type impurity, most often indium, is placed on the wafer, as shown in Fig. 5-16A. Indium melts at 155° C, while germanium melts at 987° C. If the wafer and indium pellet are heated in a reducing atmosphere to a temperature of 500° C to 600° C, the indium melts and alloys itself into a small region of the germanium (Fig. 5-16B), forming a p-type semi-conductor. Again, the interface between the n-type substrate and the p-type alloyed region forms a pn junction.

Diffused junction   The diffused junction diode is shown in Fig. 5-17A. It can be built using either p- or n-type substrate, although in the example shown an n-type silicon

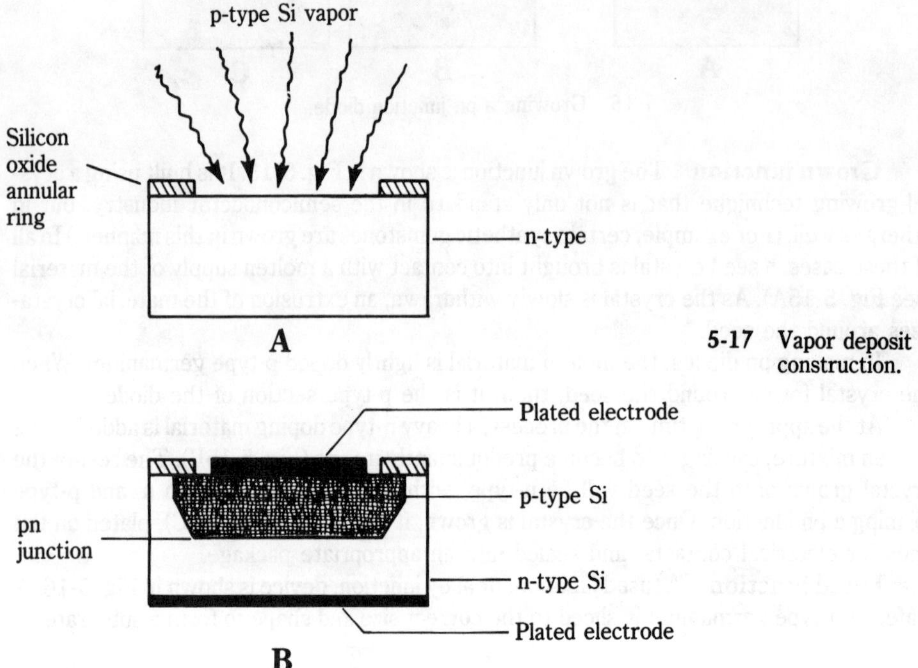

5-17   Vapor deposit construction.

substrate is selected. A silicon oxide annular ring protects a portion of the exposed surface, leaving only the center region available for processing. The process consists of exposing the surface of the n-type Si to a vaporized p-type impurity, which diffuses into the n-type substrate to form a pn junction. Plated metallic electrodes form ohmic contacts (Fig. 5-17B), and the device is then ready for packaging.

Epitaxial   The epitaxial diode is shown in Fig. 5-18. The substrate is made of heavily doped silicon (designated N^+ Si). It is heated to about 1,200° C in a nonoxidizing atmosphere of hydrogen and silicon tetrachloride. The chlorine ions in the gas disassociate, leaving Si atoms free to add themselves to the crystal lattice structure of the substrate. When phosphine dopant is added to the gas mixture, a regular n-type Si is produced, forming the epitaxial n-type Si layer shown in Fig. 5-18. When the expitaxial layer is completed, the wafer is exposed to oxygen, which causes it to form a protective silicon oxide layer. This layer is then etched away in the center region to permit the wafer to be exposed to a

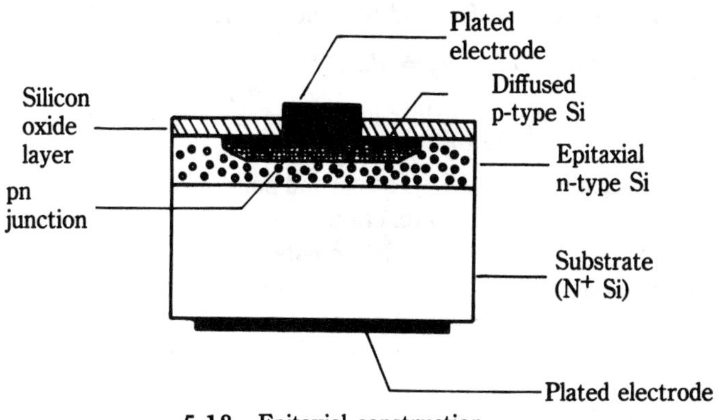

5-18   Epitaxial construction.

p-type dopant such as boron. This process forms a p-type region and the resulting pn junction. Plated-on metallic electrodes are added to make electrical contacts, and the device is ready for packaging.

The diode load line — A graphical solution

Consider Fig. 5-19, in which a pn diode (D1) is connected to a voltage source (V_B), which forward-biases D1, and a load resistance (R_L). Voltage V_R is the voltage drop across the

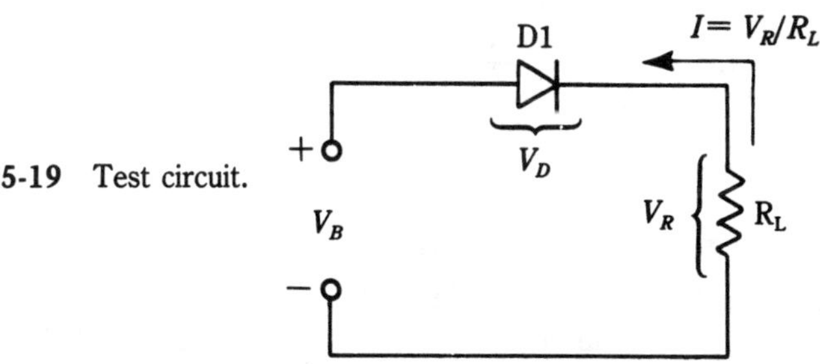

5-19   Test circuit.

load resistor, while V_D is the voltage drop across the forward-biased diode. The load current (I) is found from Ohm's law:

$$I = \frac{V_R}{R_L} \tag{5-6}$$

And the voltage drop V_R is:

$$V_R = IR_L \tag{5-7}$$

Kirchoff's voltage law (KVL) states that the algebraic sum of all voltage drops (e.g., V_D and V_R) and voltage rises (e.g., V_B) is 0, so:

$$V_B = V_D + V_R \tag{5-8}$$

Therefore, we can combine Eqs. 5-7 and 5-8 and write:

$$V_B = V_D + IR_L \qquad (5\text{-}9)$$

We can use Eq. 5-9 to create a dc load line that is superimposed on the diode's characteristic I-vs.-V curve (see Fig. 5-20). Point I_L on the vertical (current) axis is found by setting $V_D = 0$, and calculating V_B/R_L. The second point, along the horizontal (voltage) axis, is the supply voltage, V_B, and is functionally the voltage that exists when $I = 0$.

The straight line plotted between the two points has a slope of $1/R_L$, and is called the *dc load line*. The point at which the load line intersects the characteristic curve of the diode is called the *quiescent point*, or *Q point*.

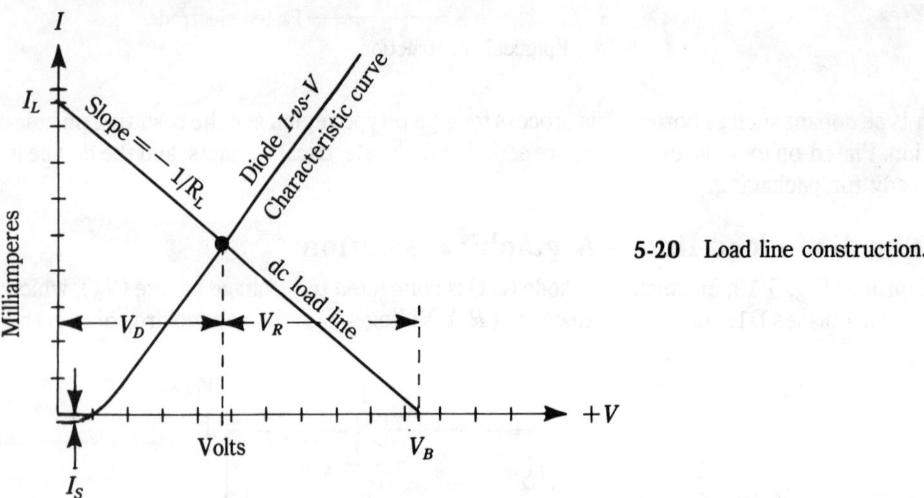

5-20 Load line construction.

Piece-wise linear approximation

The diode I-vs.-V characteristic curve is decidedly nonlinear, especially in the lower forward-bias regions. A piece-wise linear approximation (Fig. 5-21) of the characteristic curve is sometimes used to graphically solve problems. The piece-wise approximation line is drawn through the approximately linear ohmic portion of the I-vs.-V curve, intersecting the forward voltage axis at about the junction potential (≈ 0.3 volt for Ge diodes, and ≈ 0.7 volt for Si diodes). The dynamic resistance, R_d, of the diode (which is listed in manufacturer specification sheets) is the reciprocal of the slope ($\Delta I/\Delta V$) of the piece-wise linear approximation line.

The diode as switch

A *switch* is a device that permits the flow of current when it is closed and prevents the flow of current when it is open. Ordinary mechanical switches accomplish this action through the use of movable mechanical electrical contacts. Diodes, on the other hand, are naturally polarity-sensitive switches because of their unidirectional current-flow characteristic. In this section we will take a look at two basic cases — series and shunt — although the sub-

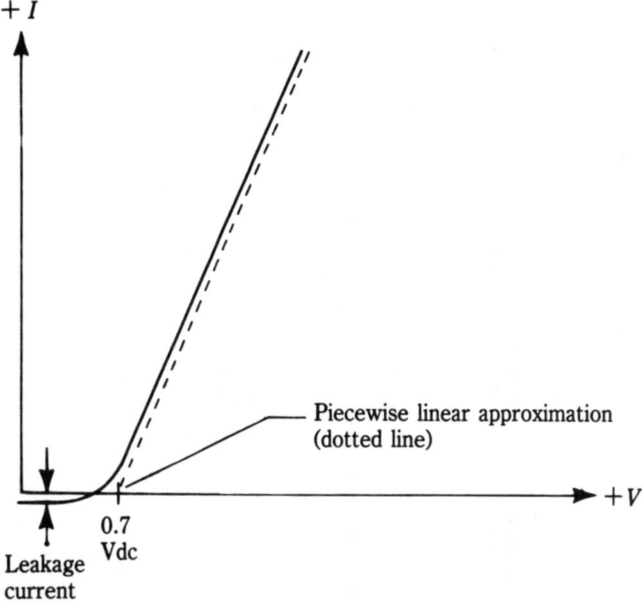

5-21 Piece-wise linear approximation of I-vs-V curve.

ject is considerably broader than can be accommodated within the scope of a book on amplifier circuits.

Series case The series case is shown in Fig. 5-22. You may recognize this circuit as the halfwave rectifier used in power-supply circuits. Rectifiers use the unidirectional property of diodes to convert bidirectional ac to unidirectional pulsating dc, which can then be filtered to produce near-pure dc.

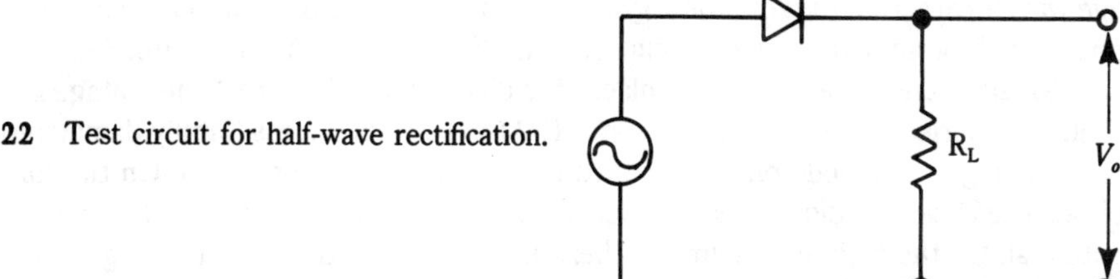

5-22 Test circuit for half-wave rectification.

The operation of the series switch, or halfwave rectifier, is shown graphically in Fig. 5-23. The alternating current input voltage (V_{in}) applied to the diode alternately forward-biases and then reverse-biases the diode's pn junction. For the sake of simplicity, only one complete cycle is shown. On the positive excursion of the applied voltage, the diode junction is forward-biased. Current flows in the diodes, through the load resistor, as soon as V_{in} exceeds the JUNCTION voltage (0.3 volt for Ge; 0.7 volt for Si). The output volt-

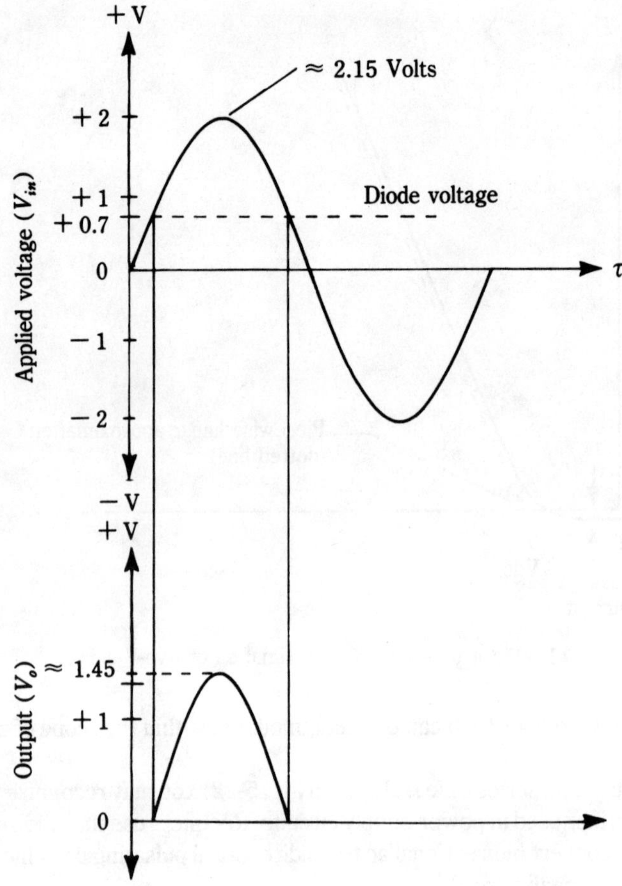

5-23   Waveform for half-wave rectification.

age waveform across the resistor is graphed as V_o in Fig. 5-23. An oscilloscope waveform photograph of an actual diode circuit such as Fig. 5-22 is shown in Fig. 5-24.

Shunt case   The shunt case places the diode in parallel across the voltage line, but requires a series current limiting resistor (R1) to prevent excess (fatally damaging) current flowing in the diode on the forward-biased half-cycle (Fig. 5-25). On the forward-biased half cycle, the diode conducts heavily, clipping the applied voltage to the JUNCTION potential. On the negative excursion there is no effect on the output voltage unless the diode breakdown voltage is exceeded.

Experiment 5-3

This experiment investigates the shunt switch properties of a pn junction diode. Two procedures are provided. The preferred procedure uses an ac signal generator and an oscilloscope, while the alternate procedure uses a dc source and a voltmeter.

Procedure I   Connect the diode shunt switch circuit of Fig. 5-26. Use a 1N4148 diode, or equivalent. The applied ac voltage is derived from an audio signal generator.

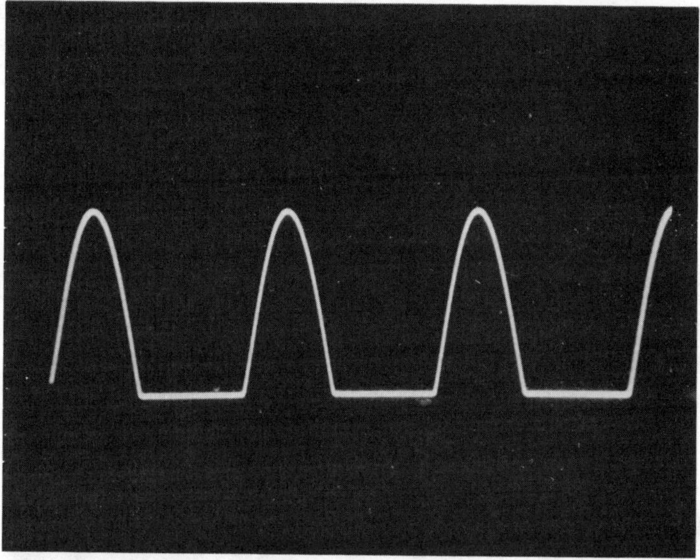

5-24 Oscilloscope trace showing half-wave rectification.

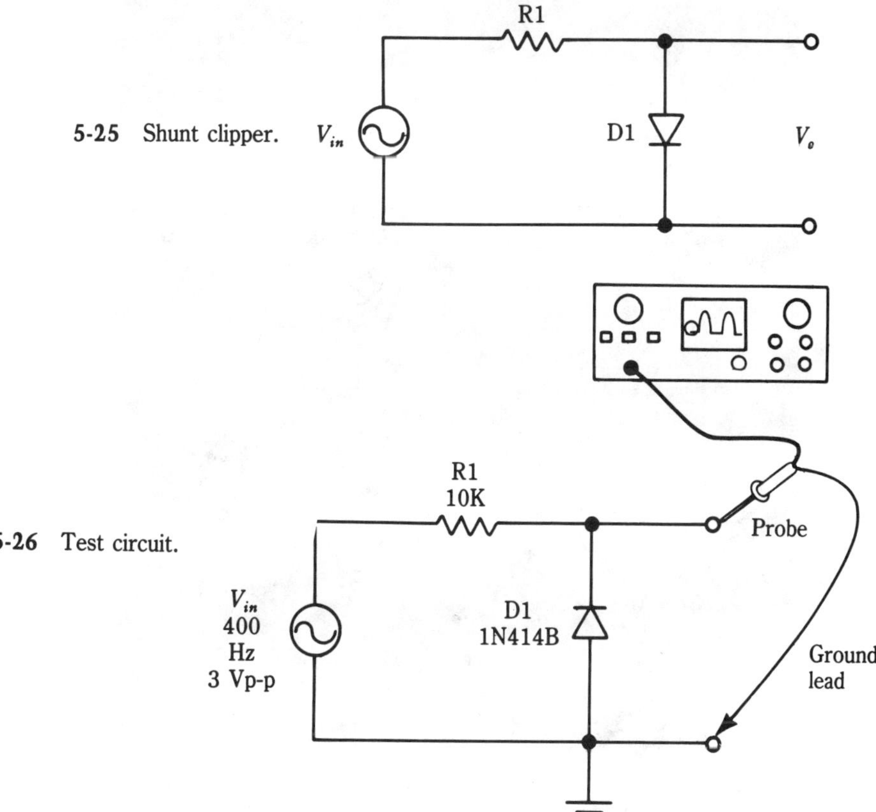

5-25 Shunt clipper.

5-26 Test circuit.

1. Set the oscilloscope input switch to GND (ground) and adjust the POSITION control to place the trace on the 0 vertical line in the middle of the screen. This position is the 0-volt line. If you have a dual-trace, (or *dual-beam*), oscilloscope, you might want to highlight the 0-volt line. Use the CH.2 trace (with input switch set to GND) to overlay the CH.1 trace. This action will show the 0 point easily, as shown in Fig. 5-27A). Set the input selector switch to DC.

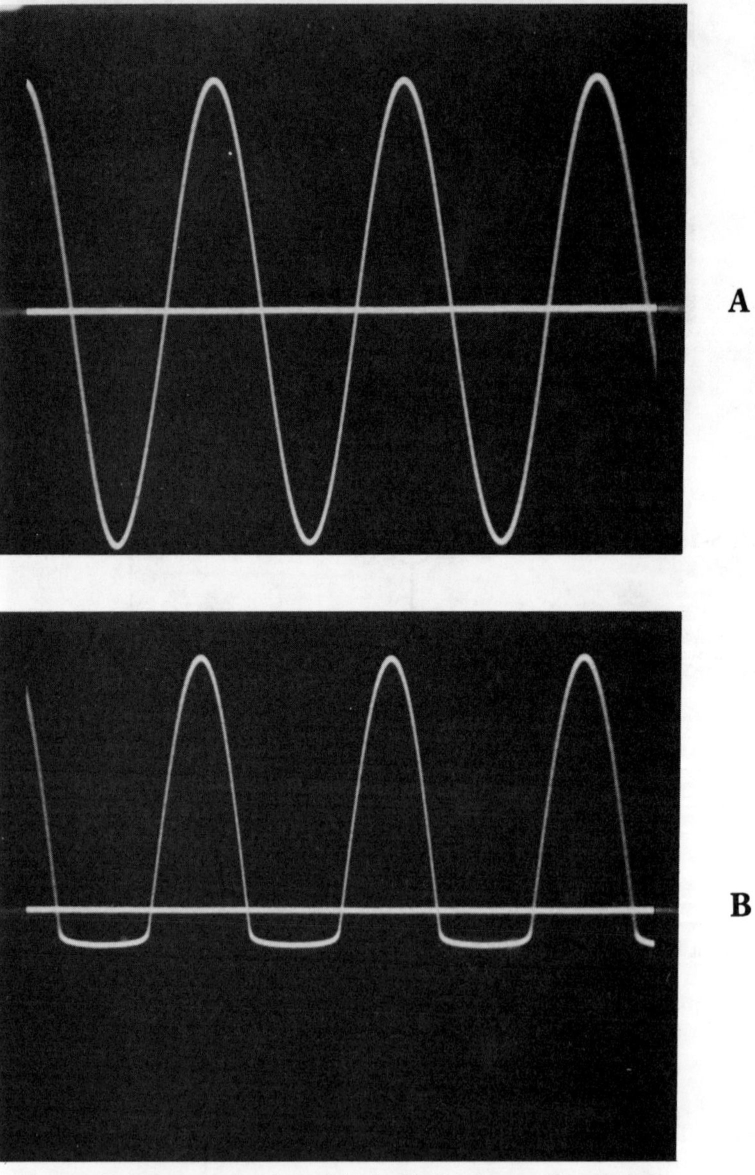

5-27   Oscilloscope waveforms: A. Applied sinewave; B. switch in position "A"; C. switch in position "B".

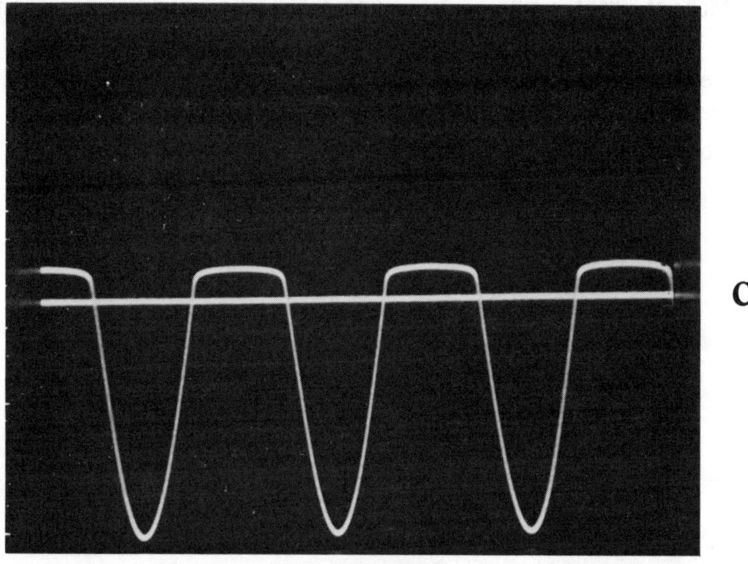

5-27 Continued

2. Set the frequency to a convenient frequency between 100 Hz and 1,000 Hz, and the output potential to 3 volts peak-to-peak (3 $V_{p\text{-}p}$).

3. Adjust the oscilloscope to produce at least two cycles of the audio sinewave (Fig. 27B).

4. The trace of Fig. 5-27B should appear. Can you explain why there is a trace below 0 volts?

5. Now, reverse the diode in the circuit. The trace should be like Fig. 5-27C.

Procedure II

1. Connect the circuit in Fig. 5-28.

2. Using a voltmeter, measure the voltage (and note its polarity) at points A and B.

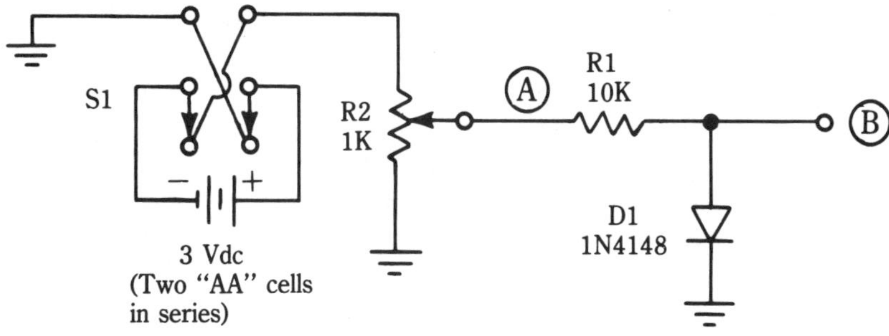

5-28 Test circuit II.

3. Make at least ten different measurements of the A and B voltages, and tabulate them.
4. Next, turn the DPDT switch (S1) to the opposite position, and repeat the ten measurements.
5. Note and explain any difference.

6

Introduction to amplifiers

AMPLIFIERS ARE DESIGNED TO AMPLIFY SIGNALS. HOWEVER, AMPLIFIERS fit into a number of different generic categories that define what they are and what they do. Certain concepts are common to all amplifiers, and still others are common to large classes of amplifiers using all forms of active devices — transistors, vacuum tubes, etc. — as the amplifying elements.

Classification of amplifiers

The classification of amplifiers as to the action element are well known. It is easy, for example, to see why some (now obsolete) amplifiers are *vacuum tube amplifiers* and others are *transistor amplifiers*. Transistor amplifiers are even broken into *bipolar* (npn and pnp) and *field-effect transistor* (JFET and MOSFET) types. However, there are also several other methods of classifying amplifier circuits that are not so dependent on the technology of the amplifying device. Several popular classification schemes are:

- Classification by common element
- Classification by active device conduction angle
- Classification by transfer function
- Classification by feedback vs. non-feedback
- Classification by frequency response

In this chapter we will consider some of the basics of amplifier circuits. Keep in mind that any of them could be constructed using transistors of either basic type, linear integrated circuits (ICs), or even vacuum tubes.*

---

*Vacuum tubes are obsolete except for very high power radio transmitter amplifiers. However, there is a sub-culture of audio enthusiasts who prefer tube amplifiers.

Classification by common element

All of the circuits that you will study in this book have distinct input and output circuits that share at least one common point. Active devices used in amplifiers—e.g., transistors—are designed so that at least one element is common to both input and output circuits. In the following chapter, you will discover that transistors have three elements: *collector, base,* and *emitter.* These elements form three different amplifiers: *common collector, common base,* and *common emitter.* These categories will be discussed in detail in chapter 7.

Classification by conduction element

When someone speaks of amplifier classes, he or she almost always means classification by conduction angle; i.e., the portion of the input signal cycle over which the output current flows in the amplifying device. We recognize three traditional classes, which are labeled *Class-A, Class-B,* and *Class-C,* and a combination class called *Class-AB.*

Class-A amplifiers   In this class, the output current (in the *collector* element of bipolar transistors, and *drain* in field-effect transistors) flows over the entire 360 degrees of the input cycle. This class is the least efficient class, but it is capable of low-distortion, or *linear,* operation using just one transistor.

Class-B amplifiers   In this class, output current flows over 180 degrees of the input cycle. This class is more efficient than Class-A amplifiers, but requires two or more transistors in order to make the amplifier linear.

Class-C amplifiers   This class is the most efficient of the classes and has output current flowing over less than 180 degrees of the input cycle. A typical conduction angle for a Class-C amplifier is 120 degrees, although the range 90 to 150 degrees can be found in practical examples. The Class-C amplifier is limited to radio frequency power amplifier applications, where fidelity is not a requirement.

Class-AB amplifiers   The Class-AB amplifier circuits are an attempt, usually successful, to obtain some of the benefits of both Class-A and Class-B circuits. When two transistors are operated in the Class-AB "push-pull" configuration, the circuit proves more linear than Class-B and more efficient than Class-A. Most modern hi-fi audio amplifiers are operated in one version or another of Class-AB.

The efficiency of an amplifier is dependent in part on the class of operation. Figure 6-1 shows a chart of amplifier efficiency versus class of operation and conduction angle. *Efficiency* is defined as the ratio of the dc input power consumed by the output circuit (collector path for bipolar devices) to the output power delivered to the load. Expressed as a percentage:

$$\eta = \frac{P_{out}}{P_{dc}} \tag{6-1}$$

The Class-A amplifier does not operate with great efficiency. In fact, most Class-A amplifiers operate around 25 percent efficiency. These amplifiers, therefore, consume large amounts of dc power from the power supply in order to generate any appreciable

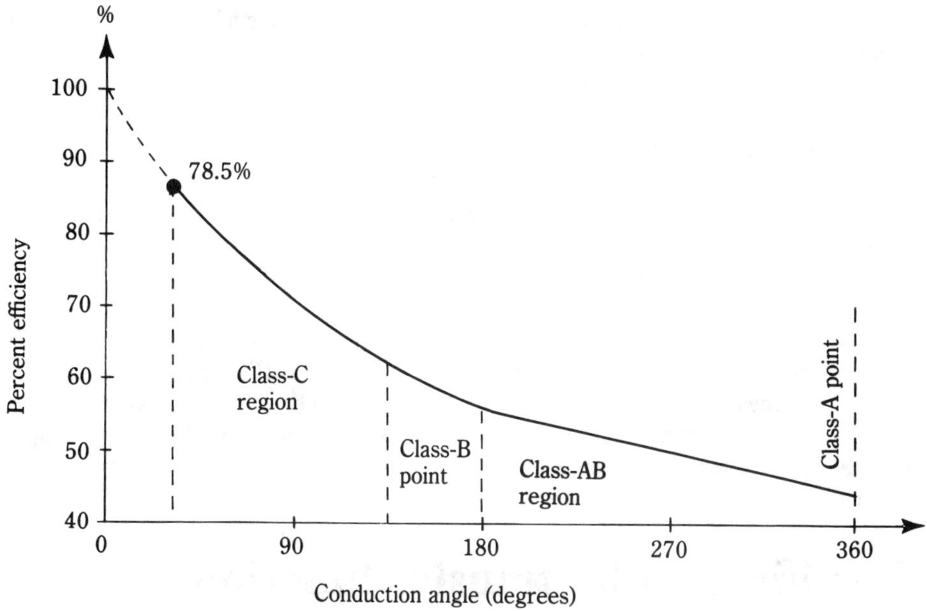

6-1 Regions of operation for amplifiers as a function of conduction angle.

amount of signal output power to the load. As a result, Class-A amplifiers are used almost exclusively as voltage amplifiers, which deliver very little actual signal power, but do build up the voltage level considerably.

One exception to the rule is where power is plentiful and component count is costly. In automobile and home broadcast radio receivers, for example, designers worry little about the available dc power, but can be terribly concerned over component count. If they don't use an integrated circuit, then they might select a Class-A amplifier circuit despite the low efficiency. Almost all car radios from 1957 until the late 1970s used single-transistor Class-A power amplifiers for the audio power stage; later designs used integrated circuits and so could take advantage of higher efficiency Class-B and Class-AB circuits.

Note: In the next chapter you will find a discussion on bipolar transistors. In this discussion, the collector terminal is often used as the output element for the transistor circuit. In this discussion, therefore, we will use the term *collector* to mean output circuit.

The *lost power*—the difference between dc power consumed and signal power delivered to the load—is given off as heat by the amplifier active device. Collector temperature is thus considerably higher in Class-A circuits than in the other circuits. Also, at zero input signal when there is no input signal, the collector power dissipation is maximum. In the Class-B and Class-C designs, on the other hand, the collector power dissipation drops to near zero when no input signal is present.

The Class-B amplifier develops collector power only when the input signal is nonzero. The reason is that the collector current flows over only 180 degrees, or one-half, of the input signal cycle. But it is impossible to make a linear amplifier using just one transistor in Class-B. It takes at least two transistors, driven 180 degrees out of phase with each other so that they will operate on alternate halves of the input signal cycle, to make a Class-B amplifier.

The Class-C amplifier cannot be made linear, regardless of the number of transistors used. As a result, the Class-C amplifier cannot be used in applications in which it is important to preserve the input waveform. There are no Class-C hi-fi audio amplifiers, for example. Similarly, there are no radio frequency (RF) amplifiers in Class-C if single-sideband (SSB) or amplitude-modulated (AM) signals are to be boosted. The Class-C amplifier is used for continuous wave (CW, or "morse code" telegraphy), frequency-modulated (FM), or high-level AM signals. The latter are AM signals in which the Class-C amplifier itself is modulated, rather than a preceding stage.

Because Class-C amplifiers severely distort the output signal, they also generate a lot of harmonics of the input signal (i.e., 2F, 3F, 4F, where F is the input signal). In radio transmitters, the harmonics are a distinct disadvantage, so they are eliminated by inductor-capacitor "tank" circuits that remove the harmonics. In addition, the RF pulse signal produced by the short conduction angle is reconverted to a sine wave by the flywheel effect of the LC tank circuit.

Classification by transfer function

The *transfer function* of any electronic network can be expressed as the ratio of the output signal over the input signal. For a voltage amplifier, for example, the transfer function is the ratio of the output signal voltage (V_o) over the input signal voltage (V_{in}). This ratio is often expressed as the *gain* of the circuit, and is symbolized by A_v.

There are four different general subclasses of transfer function: *voltage, current, transconductance*, and *transresistance*. Most readers are already familiar with the first two classes.

The transfer function of voltage amplifiers is:

$$A_v = \frac{V_o}{V_{in}} \qquad (6\text{-}2)$$

Because the units in both the denominator and numerator are volts, and therefore cancel each other ("divide out"), the voltage gain units are dimensionless.

The transfer function of the current amplifier is:

$$A_I = \frac{I_o}{I_{in}} \qquad (6\text{-}3)$$

The transconductance amplifier and transresistance amplifier are nearly as well known to most readers, even though the former is used extensively in integrated circuits. The transresistance amplifier has a transfer function that relates an output voltage to an input current. The units of electrical resistance apply to the transfer equation of this form of amplifier because in Ohm's law V/I equals resistance (R). The unit of the transfer equation is the *ohm* (Ω). The transfer equation of a transresistance amplifier is:

$$R_m = \frac{V_o}{I_{in}} \qquad (6\text{-}4)$$

The transconductance amplifier is only a little more familiar to most readers. Field-effect transistors use transconductance as the gain expression. In addition, there are transconductance linear ICs called *operational transconductance amplifiers* (OTAs), such as the CA-3080, on the market.

Transconductance is the reciprocal of resistance (I/V). The older unit of transconductance was "ohm" reversed, or *mho*. Today, however, by international agreement the mho is renamed the *siemen* after an early pioneer in electrical science.

Any transconductance amplifier relates an output current to an input voltage:

$$G_m = \frac{I_o}{V_{in}} \tag{6-5}$$

You will see more about transconductance amplifiers when you study field-effect transistors, or operational transconductance amplifiers.

Classification by feedback vs. nonfeedback

Feedback is a sample of the output signal that is rerouted back to the input port of the amplifier. Feedback can make a mediocre amplifier act like it is a much higher quality circuit. It is possible to cancel some distortion in the amplifier, and make it a lot more stable, by using a little negative feedback.

The basic block diagram for a feedback amplifier is shown in Fig. 6-2. The main amplifier will be a transistor or integrated circuit, and will have an "open-loop" (i.e., without feedback) gain of A_{vol}. The output signal voltage, V_o, is sampled, and the sample is passed back to the amplifier input via a feedback network. The transfer function, or *gain*, of the feedback network is designated by the Greek letter beta (β). In most cases, the gain of the feedback network will be negative, as the network is entirely made of passive components (resistors, capacitors, inductors), and no amplification is provided (indeed, some loss is provided). Of course, it is possible to make a feedback amplifier in which there is a main forward amplifier and a feedback amplifier stage.

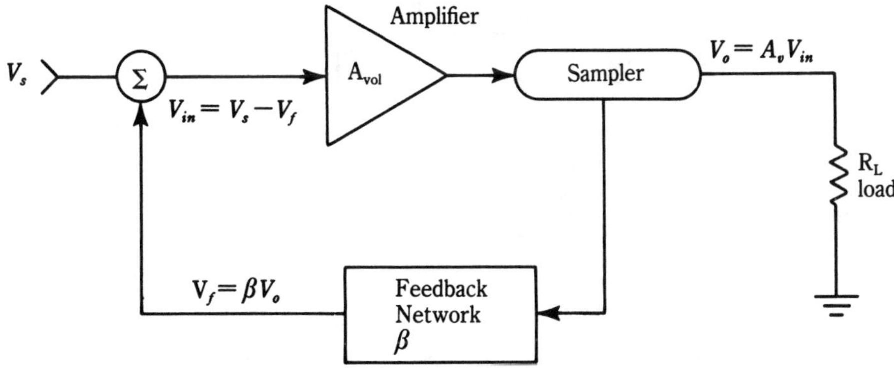

6-2   Feedback amplifier block diagram.

The feedback signal is summed with the source signal V_s to form an input signal $V_{in} = V_f - V_s$. The output signal is the product of this input signal and the voltage gain of the stage. In this section, we are dealing with a voltage amplifier system, but could just as easily call the circuit any of the other types of amplifiers by substituting the correct parameters.

The gain of an amplifier with feedback is given by the expression:

$$A_v = \frac{A_{vol}}{1 + \beta A_{vol}} \tag{6-6}$$

The use of feedback will improve the distortion situation of the amplifier. The total harmonic distortion (THD) will be reduced in the feedback amplifier, as in high-fidelity amplifiers used in audio. Very few, if any, audio amplifiers intended for hi-fi applications do not have rather extensive feedback. The degree of feedback is often expressed in decibels:

$$N_{db} = 20 \, \log \left(\frac{A_v}{A_{vol}} \right)$$

$$= 20 \, \log \left(\frac{1}{1 + A_{vol}\beta} \right) \tag{6-7}$$

Figure 6-3 shows an example of a feedback network involving only resistors. This type of network is commonly found with operational amplifiers, so you will see it later. The circuit shows how factor β is derived for this case. The feedback network is a voltage divider using two resistors, R1 and R2. The value of β is given by the transfer function of the feedback network.

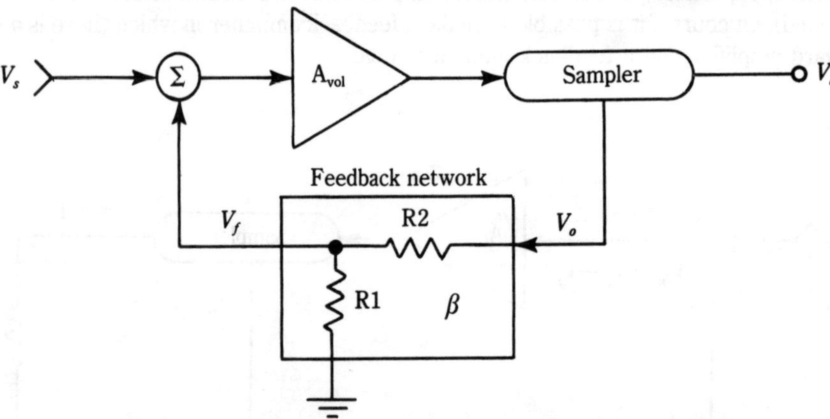

6-3   Feedback network in detail.

Recall that any transfer function is merely the output voltage divided by the input voltage. In this case, the output voltage from the network is feedback voltage V_f, and the input voltage to the network is the amplifier's output voltage, V_o. If this seems confusing, consider that the circuit is a feedback network. From our knowledge of voltage divider theory:

$$V_f = \frac{V_o R_1}{R_1 + R_2} \qquad (6\text{-}8)$$

The transfer function for the feedback network is obtained by dividing each side of the equation by V_o:

$$\frac{V_f}{V_o} = \frac{R_1}{R_1 + R_2} \qquad (6\text{-}9)$$

Because β is defined as V_f/V_o, we can state that:

$$\beta = \frac{R_1}{R_1 + R_2} \qquad (6\text{-}10)$$

From this equation, you can see that it is possible to set the gain of the amplifier by merely manipulating the resistors in the feedback network. This fact will not be lost on you once we get into chapter 11.

Classification by frequency response

Amplifiers are also classified as to approximate frequency response. This method is almost anecdotal, and there is considerable overlap of the various classes. One class is a *dc amplifier*. This type of amplifier will pass all ac frequencies within its range and dc levels. The fact that it is a dc amplifier does not mean that it won't pass ac signals, but that it will pass dc signals.

An *audio amplifier* is one that will generally pass ac signals in the audio frequency range of roughly 30 Hz to 20,000 Hz. A communications audio amplifier generally has a more limited audio response, such as 300 Hz to 3,000 Hz. High-fidelity amplifiers, on the other hand, must have the full 20 KHz response, and many of them have a response up to 50 KHz or 100 KHz.

A *wideband amplifier* is exactly what its name implies: it presents a very wide bandwidth. You might find a wideband amplifier presenting a bandwidth of dozens or hundreds of kilohertz, or even megahertz. An amplifier that has a response up to 100 KHz or 1,000 KHz is surely a wideband amplifier. We would also call an amplifier with a 100 MHz bandwidth *wideband*, but it is commonly the practice to refer to amplifiers with bandwidth greater than 1 MHz *video amplifiers*. This distinction is fading, however, as cable TV amplifiers are very wideband (2,000 MHz), but are called wideband amplifiers, rather than video amplifiers.

Radio frequency (RF) *amplifiers* are usually tuned to some specific radio frequency. These amplifiers are used in radio broadcasting and communications (receivers and transmitters) to select only the frequency of interest. These amplifiers might be wideband but will be centered about some specific frequency of interest. Most, however, are essentially narrow band circuits.

7

npn/pnp
bipolar junction
transistors

ALTHOUGH THE TRANSISTOR WAS NOT ACTUALLY INVENTED UNTIL THE late 1940s, it was predicted by physicists as theoretically possible for about ten years before that time. Natural semiconductor diodes were known as early as pre-World War I days when a lead-based mineral crystal called *galena* was used as the detector in radio crystal sets.

The semiconductor pn junction diode (see chapter 5) is a "blood relative" and precursor to the transistor, but it took metallurgy that grew out of World War II research efforts to stimulate the development of reliable, manufactured semiconductor diodes. Previously, it was not possible to get germanium or silicon that was pure enough to make these devices; intensive wartime research solved that problem. Once that hurdle was out of the way, scientists at Bell Laboratories were able to make the world's first working point contact transistor. Today, transistors are made in the same variety of ways as are diodes.

Simple bipolar transistors

The word *transistor* is derived from *trans*fer re*sistor*. Transistors are amplifying devices made from semiconductor materials. The simple bipolar transistor is often likened to a pair of diodes connected back to back. The story is much more complex, however, as can be seen if you attempt to obtain anything resembling "transistor action" by connecting a pair of signal diodes.

The basic transistor consists of three sections of semiconductor material, as shown in Fig. 7-1. There are two types shown here, called npn and pnp. (Figure 7-2 shows their respective circuit symbols.) Both consist of a central region of one type of semiconductor material sandwiched between two other regions of the opposite type of material. An npn device, for example, has two n-type regions sandwiching a region of p-type material. In the

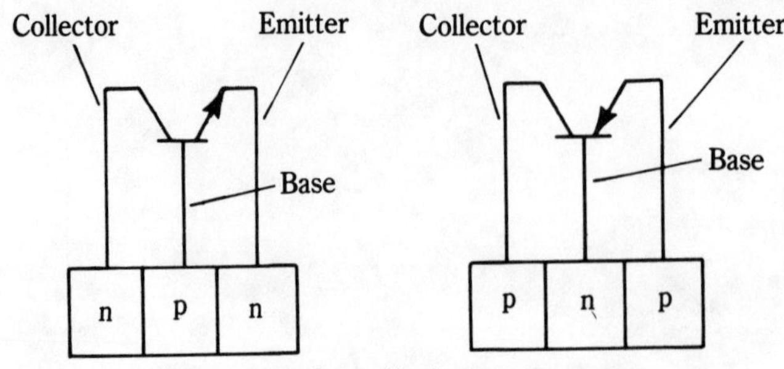

7-1 npn and pnp bipolar transistors.

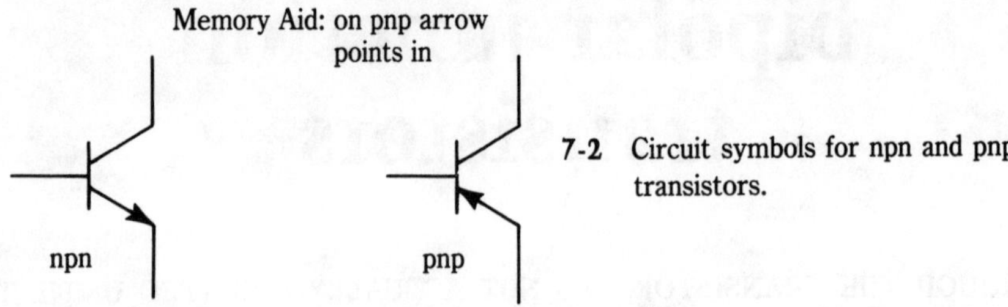

7-2 Circuit symbols for npn and pnp
transistors.

PNP transistor just the opposite situation obtains: there is an n-type central region between two p-type sections.

The central region is called the *base,* and it controls the activities of the entire transistor. The two end sections are called the *emitter* and the *collector,* respectively. You might be tempted to think of these regions as being interchangeable because of the apparent symmetry in the drawing. This was true in a very limited sense in the early days of transistors, but today's transistors use a physical geometry that is like our illustration electrically, but is quite different physically. The figure is only a graphical model of the actual situation.

The base region is typically much thinner than either the emitter or collector regions. In addition, the base region semiconductor material is considerably less heavily doped than either the emitter or base regions. As a result, charge carriers (electrons and holes depending on pnp or npn) crossing the base region have considerably less probability of recombining with the alternate type of charge carrier.

Our block diagram of the bipolar transistor is expanded in Fig. 7-3. Here we see an npn transistor, but it also serves as a model for the pnp type as well if the V_{EE} and V_{CC} power supply polarities are reversed.

There are two pn junctions in the bipolar transistor: one formed by the base and emitter (B-E), and the other formed by the collector and base (C-B). In normal operation, the B-E junction is forward-biased and the C-B junction is reverse-biased. At each junction there is a barrier potential.

Electrons in the emitter region are repelled by the negative terminal of the V_{EE} potential and attracted into the relatively more positive base region, forming emitter current I_E. Because of the thinness of the base region and its light-doping profile, few elec-

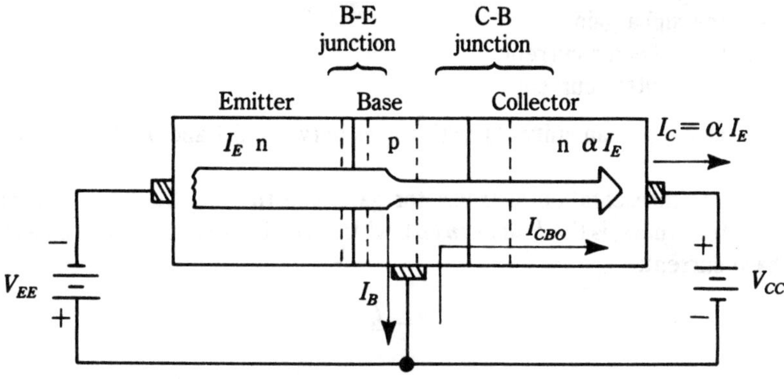

7-3 Charge carrier flow in a bipolar transistor.

trons recombine with holes. Most of them overcome the C-E junction barrier potential and continue into the collector region.

Once in the collector region, the negatively charged electrons come under the influence of the positive terminal of the V_{CC} potential, and are "collected" as current I_C. As current is drawn from the emitter region, additional electrons are attracted into the device from the V_{EE} power supply.

This discussion is similar to the pnp, except for the fact that in pnp devices the charge carriers are holes and the battery potentials are reversed. Charge carrier movement is the same as in Fig. 7-3, although electron flow is reversed.

There is a definite ratio between the emitter current (I_E), base current (I_B), and collector current (I_C). In normal operation:

$$I_E = I_B + I_C \tag{7-1}$$

and,

$$I_C \gg I_B \tag{7-2}$$

Normally, 95 to 99 percent of the emitter current flows in the collector circuit (i.e., $0.95 I_E \leq I_C \leq 0.99 I_E$), and only 1 to 5 percent of the emitter current flows in the base circuit.

The collector and emitter currents are different by the amount of the base current, as indicated by Eq. 7-1, and their ratio (I_C/I_E) is called the *alpha* (α) factor. The alpha factor is one way to express transistor gain.

Transistor gain

There are actually several popular ways to denote transistor current gain, but only two are of interest to us here: alpha (α) and beta (β). *Alpha gain* (α) can be defined as the ratio of collector current to emitter current:

$$\alpha = \frac{I_c}{I_e} \tag{7-3}$$

Where: α is the alpha gain
      I_c is the collector current
      I_e is the emitter current

Alpha has a value less than unity (1), with values between 0.7 and 0.99 being the typical range.

The other representation of transistor gain, and the one that seems more often favored over the others, is the *beta gain* (β), which is defined as the ratio of collector current to base current:

$$\beta = \frac{I_c}{I_b} \tag{7-4}$$

Where: β is the beta gain
      I_c is the collector current
      I_b is the base current

Alpha (α) and beta (β) are related to each other, and you can use the following equations to compute one when the other is known.

$$\alpha = \frac{\beta}{1 + \beta} \tag{7-5}$$

and,

$$\beta = \frac{\alpha}{1 - \alpha} \tag{7-6}$$

Example

What is the alpha of a transistor that has a beta of 120?

$$\alpha = \frac{I_c}{I_e}$$

$$\alpha = \frac{120}{1 + 120}$$

$$= \frac{120}{121}$$

$$= 0.99$$

The values given are for static dc situations.

In ac terms you will see ac alpha gain (h_{fb}) defined as:

$$h_{fb} = \frac{\Delta I_c}{\Delta I_e} \tag{7-7}$$

and ac beta gain (h_{fe}) is defined as:

$$h_{fe} = \frac{\Delta I_c}{\Delta I_b} \tag{7-8}$$

In both equations, the Greek letter delta (Δ) indicates a small change in the parameter with which it is associated. Thus, the term ΔI_c denotes a small change in collector current I_c.

Leakage currents

Another current flowing in the transistor of Fig. 7-3 is the collector-to-base leakage current. Recall from the earlier discussion of pn junctions that the *leakage current* is the movement of minority charge carriers across the pn junction under the influence of the barrier potential. This leakage is designated I_{CBO} in Fig. 7-3.

A note about notation is in order at this point. Several specified currents in the transistor are measured with one of the three terminals (C, B, and E) open-circuited. The basic notation is I_{CBE} with the letter for the open-circuit terminal replaced with an O. Thus, the I_{CBO} is the collector-to-base current as measured with the emitter open circuits.

Classification by common element

Classifying amplifier circuits by common element revolves around noting which element (collector, base, or emitter) is common to both input and output circuits. Although technically incorrect, this is sometimes referred to as the *grounded* element, as in grounded emitted amplifier. I tend to use *common* and *grounded* interchangeably, so bear with me if you are a purist. Figure 7-4 shows the different entries into this class.

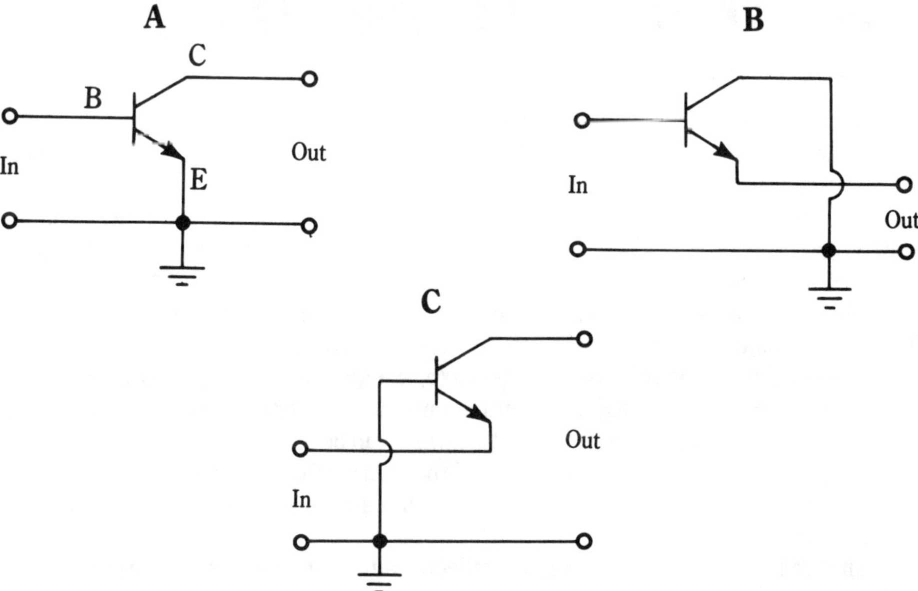

7-4   Amplifier configurations: A. Common emitter; B. Common collector; C. Common base.

Common emitter circuits

The circuit shown in Fig. 7-4A is the common emitter circuit. It gets its name from the fact that the emitter terminal of the transistor is common to both input and output circuits.

The input signal is applied to the transistor between the base and emitter terminals, while the output signal is taken across the collector and emitter terminals; i.e., the emitter is common to both input and output circuits.

The common emitter circuit offers high current amplification — the beta rating of the transistor. However, the common emitter circuit offers a substantial amount of voltage gain as well. The transistor is also a voltage amplifier, especially when a series resistor is placed between the collector terminal and the collector dc power supply. The values of gain for current and voltage are vastly different, however. The current gain is H_{fe}, but the voltage gain depends on other factors as well. Later, you will see that voltage gain depends on the R_L/R_E ratio in some circuits, and the product of that ratio and the beta in other cases.

The input impedance of the common emitter amplifier is medium ranged, or in the 1,000 Ω range. The output impedance, though, is typically high (up to 50 kΩ). Typical values will be determined by the specific type of circuit, but some approximations can be made. For most common emitter amplifiers, Z_{in} is equal to the product of the emitter resistor R_E and the H_{fe} of the transistor. The output impedance is essentially the value of the collector load resistor and will range from 5 kΩ to about 50 kΩ.

The output signal in the common emitter circuit is 180 degrees out of phase with the input signal. This phase shift means that the common emitter amplifier is an *inverter* circuit. The output signal will be negative-going for a positive-going input signal, and vice versa. The common emitter transistor amplifier is probably the most often used circuit configuration.

Common collector circuits

The common collector circuit is shown in Fig. 7-4B. In this circuit, the collector terminal of the transistor is common to both input and output circuits. This circuit is also sometimes called the *emitter follower* circuit.

The common collector circuit offers little or no voltage gain. Most of the time the voltage gain is actually less than unity (1), but the current gain is considerably higher ($\approx H_{fe} + 1$).

There is no phase inversion between input and output in the emitter follower circuit. The output voltage is in phase with the input signal voltage.

The input impedance of this circuit tends to be high, sometimes greater than 100 kΩ at frequencies less than 100 KHz. The output impedance is very low, however, because it is limited to the value of the emitter resistor, which can be as low as 100Ω. This situation leads us to one of the primary applications of the emitter follower: *impedance transformation*. The circuit is often used to connect a high-impedance source to an amplifier with low impedance.

The emitter follower, *née* common collector amplifier, is also frequently used as a *buffer amplifier*, which is an intermediate stage used to isolate two circuits from each other. One primary example of this application is in the output circuit of oscillator circuits. Many oscillators will "pull," or change frequency, if the load impedance changes. Yet some of the very circuits used with oscillators naturally provide a changing impedance situation. The oscillator proves a lot more stable under these conditions if an emitter follower buffer amplifier is used between its output and its load.

Common base circuits

Common base amplifiers use the base terminal of the transistor as the common element between input and output circuits (Fig. 7-4C). The output is taken between the collector and base.

The voltage gain of the common base circuit is high, on the order of 100 or more; however, the current gain is low, usually less than unity. The input impedance is also low, usually less than 1,000 Ω, because it is limited to the emitter resistance. On the other hand, the output impedance is quite high. Again, there is no phase inversion between input and output circuits.

The principal use of the common base circuit is in high-frequency (HF) and very high frequency (VHF) RF amplifiers in receivers. The circuit requires no neutralization at these frequencies, so it is superior to common emitter circuits. Neutralization prevents oscillation because of interelement capacitances that provide a feedback signal which is in-phase with the input signal.

Table 7-1 summarizes the properties of the common emitter, common collector, and common base amplifier circuits.

Table 7-1. Parameters of Amplifier Circuits.

Parameter	Common emitter	Common collector	Common base
Voltage gain	High	Low (≈ 1 or less)	High (>100)
Current gain	High	High ($\beta + 1$)	Low (<1)
Z_{in}	Medium (≈ 1)	High ($>100K$)	Low
Z_{out}	Medium to High ($\approx 50K$)	Low ($<100\Omega$)	High ($>500K$)
Phase inversion?	Yes	No	No

dc load lines, the "Q" point, and transistor biasing

The dc load line is a graphical method for expressing transistor operating conditions. The basis for the load line is the I_c-vs.-V_{ce} family of curves (Fig. 7-5). The curve is traced for a circuit similar to Fig. 7-6 at several different base currents, as the collector voltage is swept from zero to maximum.

Figure 7-6 shows a common emitter npn transistor circuit. The supply voltage is 30 Vdc; the collector current I_c can vary from 0 to a maximum of 12 mA (0.012 A). A 2,500 Ω collector load resistor (R_L) is connected between the collector of the transistor and the V_{CC} power supply. The collector-emitter voltage (V_{CE}) can rise to V_{CC} when $I_c = 0$, or drop to 0 when I_c is maximum. The collector current is set by the base current, I_B, hence also the base voltage V_B. At any given point, the value of V_{CE} is V_{CC} less the voltage drop across the load resistor (V_R). Because V_R is the product I_cR_L,

$$V_{CE} = V_{CC} - I_cR_L \tag{7-9}$$

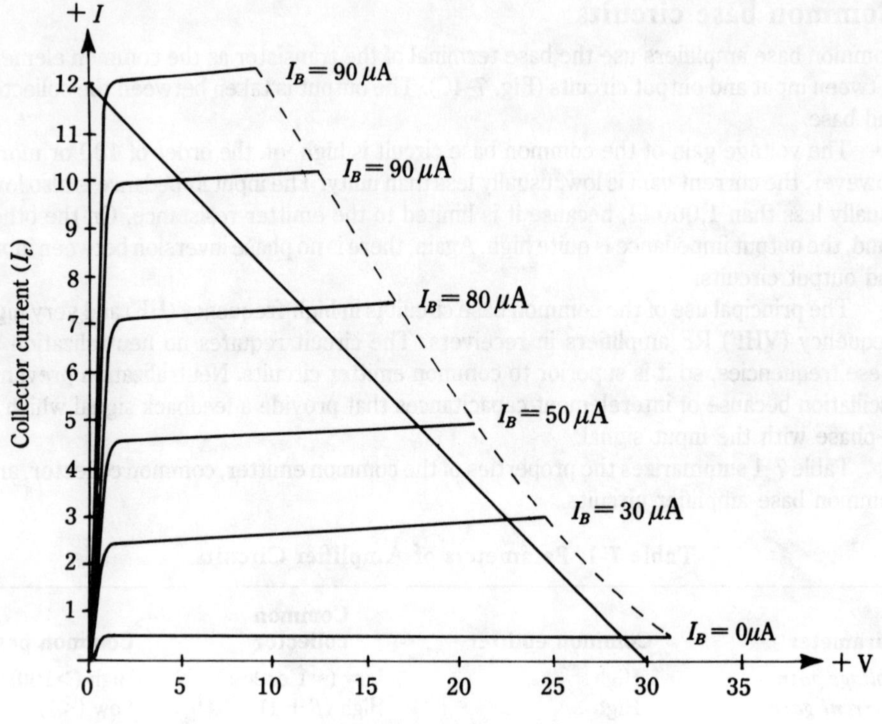

7-5 Family of operating curve and dc load line.

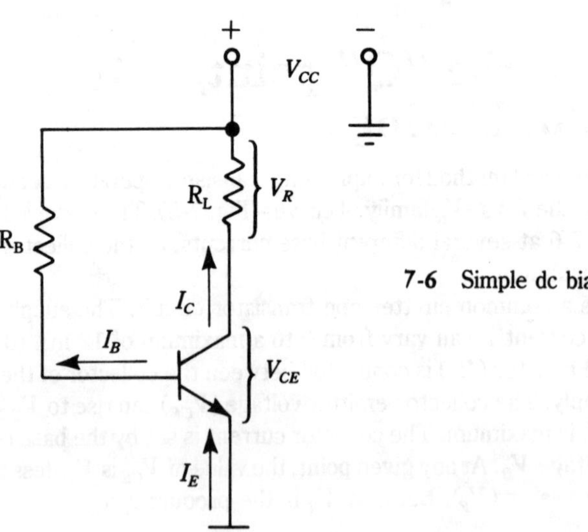

7-6 Simple dc bias network.

Example

For Fig. 7-6, calculate V_{CE} when $I_C = 7$ mA.

Solution:

$$V_{CE} = V_{CC} - I_C R_L$$
$$= (30 \text{ volts}) - ((0.007 \text{ A}) (2{,}500 \text{ }\Omega))$$
$$= (30 \text{ volts}) - (17.5 \text{ volts})$$
$$= 12.5 \text{ volts}$$

Transistor biasing

Biasing sets the operating characteristics of any particular transistor circuit, and is usually set by the current conditions at the base terminal of the device. Several different bias networks are commonly seen in transistor circuits, and they are summarized here.

Fixed-current bias

A fixed-current bias circuit (Fig. 7-7) sets a base current in the transistor at a fixed resistor between the V_{cc} power supply and the base terminal of the transistor. That current, I_B, is defined by:

$$I_B = \frac{V_{CC} - V_{BE}}{R_B} \tag{7-10}$$

In order to see how values for components are reached in a circuit such as Fig. 7-7, let's work an example.

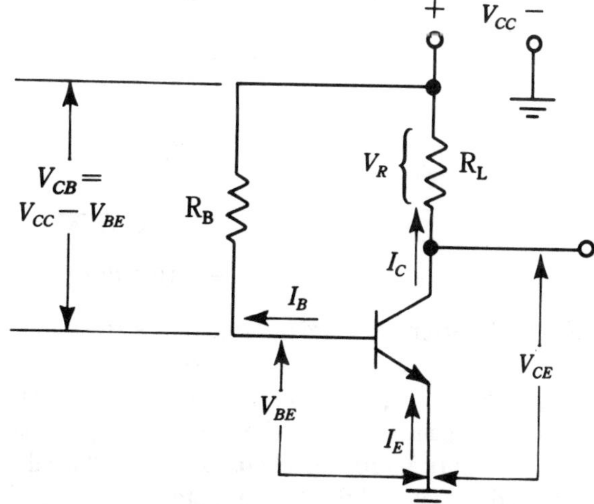

7-7 Voltage and current relationships in biased transistor circuit.

Example

A silicon npn transistor has a current beta gain, H_{fe}, of 90, and is to be operated from a 12 Vdc power supply, with a no-signal quiescent Q-point V_{CE} potential of 6 volts and a collector current of 2 mA (i.e., 0.002 A).

1. Write down what we know:
   $V_{cc} = 12$ Vdc
   $V_{CE} = 6$ Vdc
   $I_c = 0.002$ A
   $V_{BE} = 0.6$ volts (Si transistor is being used)

2. Calculate the value of load resistor R_L by Ohm's law:

$$R_L = \frac{V_{cc} - V_{ce}}{I_c}$$

$$= \frac{(12) - (6)}{0.002 \text{ A}}$$

$$= \frac{6}{0.002}$$

$$= 3{,}000 \, \Omega$$

3. Calculate base current I_B. By definition $H_{fe} = I_c/I_B$, so:

$$I_B = \frac{I_c}{H_{fe}}$$

$$= \frac{0.002 \text{ A}}{90}$$

$$= 2.2 \times 10^{-5} \text{ A}$$

$$= 22 \, \mu A$$

4. Calculate the value of base resistor R_B by rearranging Eq. 7-10 to solve for R_B:

$$R_B = \frac{V_{cc} - V_{be}}{I_B}$$

$$= \frac{(12) - (0.6)}{2.2 \times 10^{-5} \text{ A}}$$

$$= \frac{11 \text{ V}}{2.2 \times 10^{-5} \text{ A}}$$

$$= 518{,}182 \, \Omega$$

As a practical matter, standard resistor values are used, so R_L would be selected as either 2,700 Ω or 3,300 Ω, and R_B as 510 kΩ.

The circuit of Fig. 7-7 presents several problems, including a dependence on the transistor beta gain (H_{fe}) and the value of V_{be}. Variations in actual—vs. published "typical"—H_{fe} are approximately 0.55 to 0.7 volt in silicon transistors (0.2 to 0.3 volt in germanium devices), and this barrier potential is a function of temperature. A variant of the fixed bias circuit, shown in Fig. 7-8, helps solve some of these problems.

The principal difference between Figs. 7-7 and 7-8 is the use of an emitter resistor, R_E. The voltage drop across this resistor must be added to V_{BE} when calculating R_B, and is:

$$V_E = I_E R_E$$
$$= (I_c - I_B) \, R_E \tag{7-11}$$

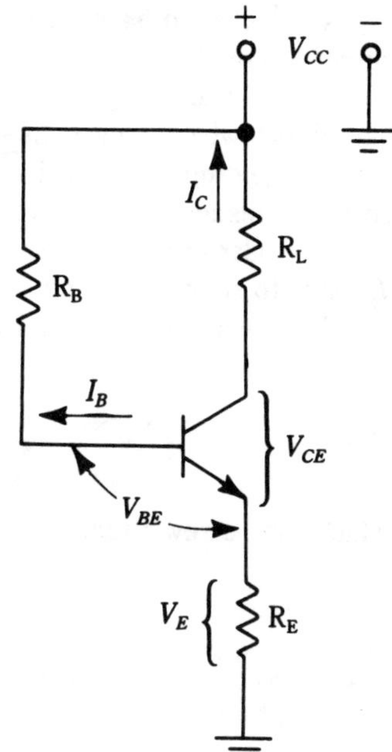

7-8   Same circuit as in Fig. 7-7 with emitter bias resistor.

or:

$$V_E = \left(I_c - \frac{I_c}{H_{fe}}\right)R_E \qquad\qquad (7\text{-}12)$$

A rule-of-thumb for R_E is that it should be approximately one-tenth of R_L, and in no case be more than one-fifth R_L. Most of the time $10\ \Omega \le R_E \le 1,000\ \Omega$.

Returning to our example, let's set $R_L = 3,000\ \Omega$; $R_E = 220\ \Omega$; $I_c = 0.002$ amperes. In that case,

$$V_E = (I_c - I_B)\,R_E$$
$$= (0.002\ \text{A} - 0.000022\ \text{A})\,(220\ \Omega)$$
$$R_B = \frac{V_{cc} - (V_{BE} + V_E)}{I_B}$$
$$= \frac{(12\ \text{V}) - (0.6 + 0.44\ \text{V})}{2.2 \times 10^{-5}}$$
$$= \frac{(12\ \text{V}) - (1.04)}{2.2 \times 10^{-5}\ \text{A}} = 498{,}182\ \Omega$$

For the circuit of Fig. 7-7, the following relationships obtain:

$$Z_o = R_L$$
$$Z_{in} = R_E H_{fe}$$
$$A_I = H_{fe}$$
$$A_v = \frac{R_L H_{fe}}{R_E}$$

Collector-to-base bias   In this type of bias network, the resistor supplying bias current to the base (R_B) is connected to the collector of the transistor, rather than V_{cc} (Fig. 7-9). An advantage of this circuit is that the quiescent (no-signal) conditions are stabilized somewhat because I_B is set by V_{CE}, rather than V_{cc}. Thus, when I_c tries to increase, the voltage drop across R_L increases, and because $V_{ce} = V_{cc} - V_{RL}$, the value of V_{ce} decreases. This action, in turn, reduces I_B, so, by $I_c = H_{fe}I_B$, the collector current decreases.

A similar action takes place when I_c tries to decrease. The result in both cases is that I_c tends to stabilize around the quiescent value.

In terms of the previous example, current I_B is set by V_{ce}, rather than V_{cc}, so:

$$V_{ce} = V_{cc} - R_L\,(I_B + I_c) \tag{7-13}$$

and,

$$V_{ce} = I_B R_B + V_{BE} \tag{7-14}$$

which can be rewritten,

$$R_E = \frac{V_{ce} - V_{be}}{I_B}$$

$$R_B = \frac{(6\ \text{V}) - (0.6\ \text{V})}{2.2 \times 10^{-5}\ \text{A}}$$

$$= \frac{5.4}{2.2 \times 10^{-5}\ \text{A}}$$

$$= 245{,}455\ \Omega$$

As was true in the previous bias circuit, it is sometimes prudent to use an emitter resistor to gain further stability by inserting an emitter resistor, as in Fig. 7-10. For the circuit of Fig. 7-10:

$$Z_o = R_L$$
$$Z_{in} = R_E H_{fe}$$
$$A_I = H_{fe}$$
$$A_v = \frac{R_L H_{fe}}{R_E}$$

Emitter bias or "self-bias"   Figure 7-11 is recognized as the most stable configuration for transistor amplifier stages. This circuit uses a resistor voltage divider (R_1/R_2) to set a fixed bias voltage (V_B) on the transistor. As a general rule, the best stability usually occurs when $R_1 \parallel R_2 \approx R_E$.

Because there is a substantial voltage drop across R_E, the V_{cc} voltage required for Fig. 7-11 is a bit higher than for the previous circuits. For purposes of this discussion let $V_{cc} = 18$ Vdc, $V_c = 9$ Vdc, and $V_E = 6$ volts. As in the previous examples, let $H_{fe} = 90$ and $I_c = 0.002$ A. The base current is:

$$I_B = \frac{I_c}{H_{fe}} \tag{7-16}$$

$$= \frac{0.002\ \text{A}}{90}$$

$$= 2.2 \times 10^{-5}\ \text{A}$$

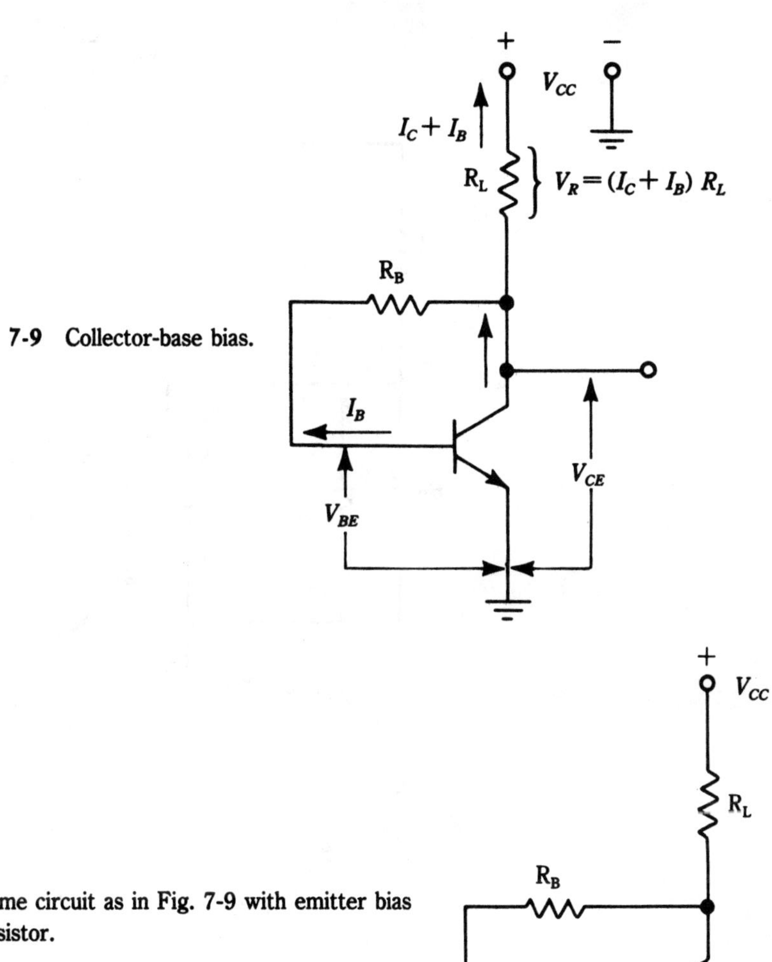

7-9   Collector-base bias.

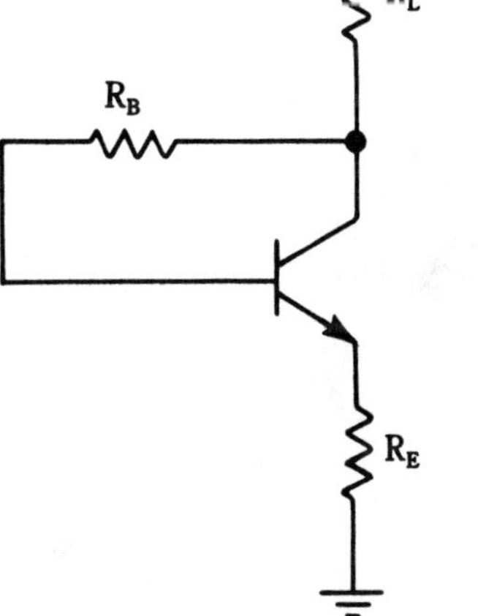

7-10   Same circuit as in Fig. 7-9 with emitter bias resistor.

Load resistor R_L is:

$$R_L = \frac{V_{cc} - V_c}{I_c} \qquad (7\text{-}17)$$

$$I_E = I_c - I_B$$
$$= 0.00198 \text{ A}$$

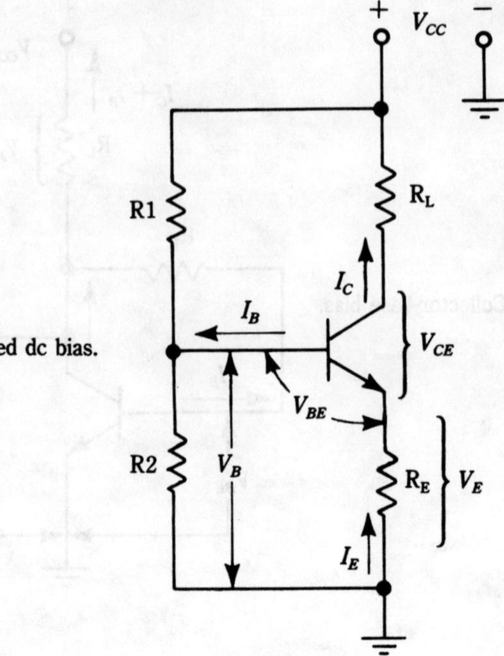

$$R_E = \frac{V_E}{I_E}$$

$$= \frac{6\ \text{V}}{0.00198\ \text{A}}$$

$$= 3{,}030\ \Omega$$

$$V_B = V_E + V_{BE} \tag{7-18}$$
$$= (6\ \text{V}) + (0.6\ \text{V})$$
$$= 6.6\ \text{Vdc}$$

$$V_B = \frac{V_{cc} R_2}{R_1 + R_2} \tag{7-19}$$

$$\frac{V_B}{V_{cc}} = \frac{R_2}{R_1 + R_2} \tag{7-20}$$

$$\frac{6.6\ \text{V}}{18\ \text{V}} = \frac{R_2}{R_1 + R_2}$$
$$= 0.37$$

or, when the arithmetic is done:

$$R_1 = 1.7\ R_2 \tag{7-21}$$

Further, $R_1 \parallel R_2 \approx R_E$, so:

$$\frac{R_1 R_2}{R_1 + R_2} = 3{,}030 \ \Omega \qquad (7\text{-}22)$$

Solving Eq. 7-22 for R_2 and then plugging in R_1 shows that $R_1 = 4{,}812 \ \Omega$ and $R_2 = 8{,}181 \ \Omega$.

Frequency characteristics

Transistors, like most other electron devices, operate only over a certain specified frequency range. Here are three basic cutoff frequencies that may interest us: *alpha, beta,* and the *gain-bandwidth product* (F_t).

The alpha cutoff frequency F_{ab} is the frequency at which the ac current gain h_{fb} drops to a level 3 dB below its low-frequency (usually 1,000 Hz) gain. This is the frequency at which $h_{fb} = 0.707 h_{fbo}$, where h_{fbo} is the ac current gain at 1,000 Hz.

The beta cutoff frequency is similarly defined as the frequency where the ac beta, h_{fe}, drops 3 dB relative to its 1,000 Hz value. In general, this frequency is lower than the alpha cutoff, but is considered somewhat more representative of a transistor's performance.

The frequency specification that seems to be quoted most often is the gain-bandwidth product, which is given the symbol F_t. This parameter is usually accepted only for transistors operated in the common emitter configuration. It is defined as:

$$F_t = \text{Gain} \times \text{Bandwidth} \qquad (7\text{-}23)$$
$$F_t = h_{fe} F_o \qquad (7\text{-}24)$$

Where: F_t is the gain-bandwidth product
$\quad\quad\ h_{fe}$ is the ac beta
$\quad\quad\ F_o$ is the frequency at which gain is measured

The value of F_t quoted in specification sheets is the frequency at which h_{fe} drops to unity.

If the beta cutoff frequency F_{ae} is known, then the gain-bandwidth product can be approximated from:

$$F_t = F_{ae} h_{feo} \qquad (7\text{-}25)$$

Recognize, however, that this is an approximation that might not hold up in every case. Also, you can often get away with assuming that F_o is approximately equal to, but usually slightly less than, the alpha cutoff frequency.

Transistor packages

The classic small transistor case is the TO-5 package shown in Fig. 7-12. This package was used as early as 1954 and consists of a metal can in which the transistor die is mounted. Plastic transistors soon replaced the metal, although metal is still used in high-reliability and military applications. A similarly sized plastic package is sometimes advertised as "similar to TO-5."

The TO-92 metal package is even smaller than the TO-5, but contains the same sort of small signal transistors. Like its TO-5 cousin, the TO-92 inspired plastic versions (Fig. 7-13).

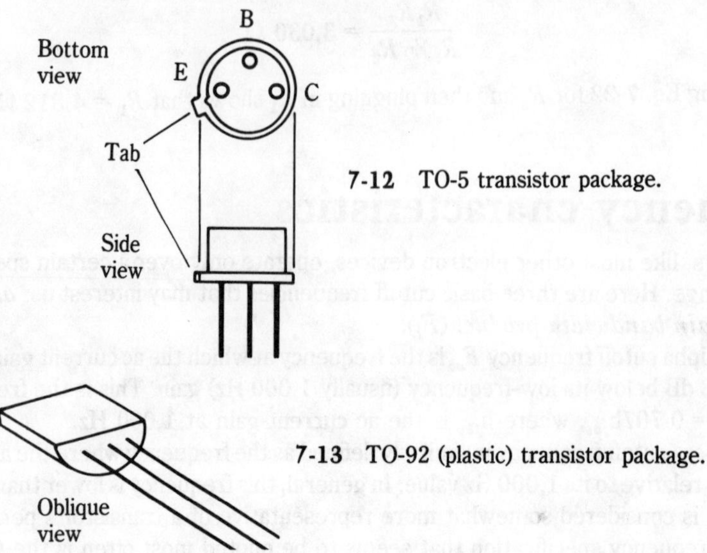

7-12 TO-5 transistor package.

7-13 TO-92 (plastic) transistor package.

Figure 7-14 shows several types of power transistor package. The TO-3 transistor in Fig. 7-14A is the so-called "standard" power transistor in a diamond-shaped package. A smaller diamond-shaped package is the TO-66, shown in Fig. 7-14B. There is also a Japanese "similar-to-TO66" package that looks on first blush like the TO-66, but has slightly different pin spacings. Finally, the big horse shown in Fig. 7-14C is the TO-7, or TO-36, depending on power level. This high-power transistor is used extensively in automotive audio applications, mobile two-way radio solid-state HV multivibrator dc power supply converters, and industrial electronics applications. Older tube-type mobile transmitters often used these transistors in the 13.6 Vdc-to-HV power supply.

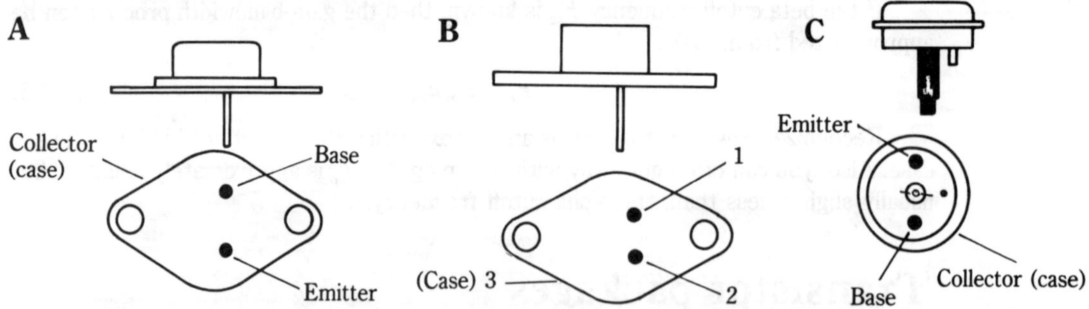

7-14 A. TO-3 transistor package; B. TO-66 transistor package; C. TO-36 transistor package.

Figure 7-15 shows several popular plastic power transistor packages. Some are listed as replacements for TO-3 or TO-66 diamond-shaped power transistors. The package in Fig. 7-15A is the TO-220, once also called "P-66", and is common in small, low- to medium-powered audio applications and most car radios. Two additional tab-mounted plastic power transistors are shown in Figs. 7-15B and C. Finally, the device shown in Fig. 7-15D is representative of a class of Motorola power transistor. These devices are not tab-mounted, but instead have a mounting hole through the body of the transistor.

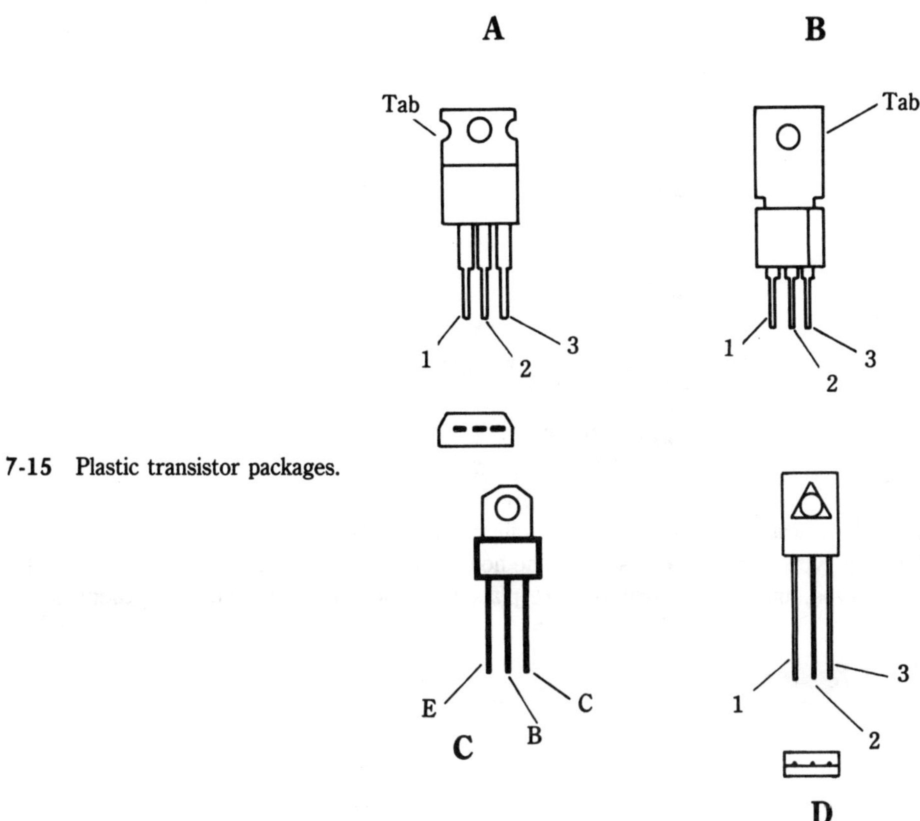

7-15   Plastic transistor packages.

Figure 7-16 shows how a plastic, tab-mounted TO-220 power transistor can replace a TO-3 power transistor. The center terminal (collector) is cut off the TO-220 — it won't be needed, the tab mount is also connected to the collector — and the base and emitter leads are bent down. The mounting screw is passed through the tab hole into the original mounting hole for the TO-3 transistor.

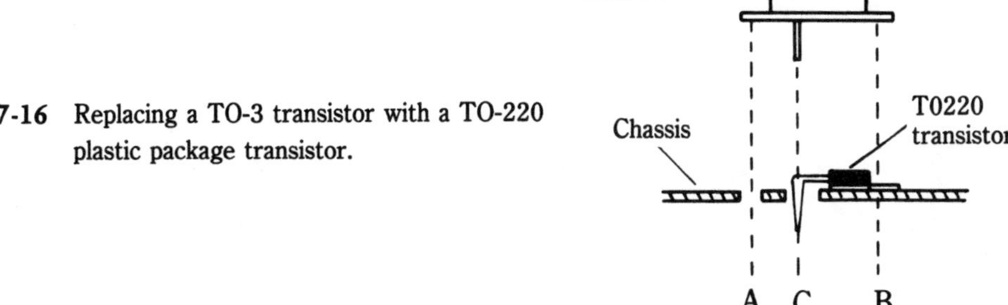

7-16   Replacing a TO-3 transistor with a TO-220 plastic package transistor.

Another transistor package is the RF power transistor in Fig. 7-17. This device uses thin, flat, low-inductance leads. Several sizes are available, and size doesn't always indicate relative power dissipation rating, although it usually does.

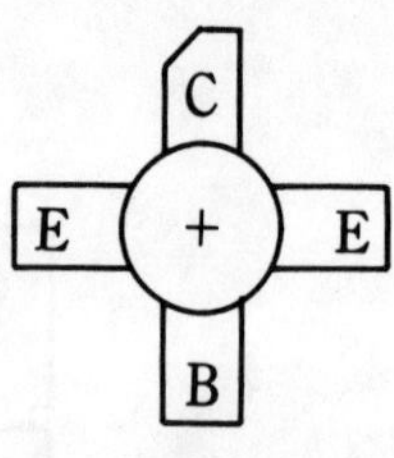

7-17   RF transistor.

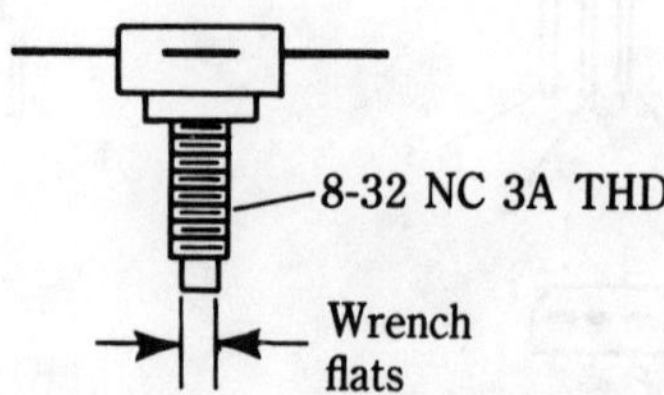

There was one 150-MHz FM mobile power amplifier that came with either of two different types of power transistor. The hole in the printed wiring board was cut for the larger type, and rubber O-rings were placed around the smaller to make them fit.

8

Bipolar transistor amplifier circuits

IN THIS CHAPTER I EXPAND ON THE SUBJECT OF TRANSISTOR AMPLIFIER circuits. You will find information on coupling methods (i.e., the method of sending signals from stage to stage) and power amplifier circuits.

Amplifier coupling methods

It is necessary to provide some method of coupling between amplifier stages that are connected in cascade. When you want to pass a signal from one stage to the next in a cascade chain, you must make sure that the proper conditions are met. For example, in some power amplifier applications, the coupling circuit must not just couple the signal, it must also match the impedances between the stages. This is true because the maximum power transfer occurs when the impedances of the source and load are matched.

In other cases, we are not interested so much in matching impedances but in keeping dc voltages from the first stage from interfering with the operation of the following stage. In all electronic amplifiers, it is a good rule to make the input impedance at least ten times the source impedance of the driver stages. In transistor amplifiers, the collector voltage may well be 10 to 50 Vdc, while the base-emitter voltage (V_{be}) will be in the 0 V to 5 V range. Clearly, when these stages are directly connected, it is likely that trouble will result. There is, however, a certain type of coupling circuit in which the transistor elements of succeeding stages are connected together: *direct-coupled amplifiers*.

Direct coupling

When direct coupling is used, the collector of the transistor in the input amplifier stage is connected directly to the base of the transistor in the following amplifier stage. It is necessary to design these stages (Fig. 8-1) such that the base voltage of Q2 is the same as the collector voltage of Q1.

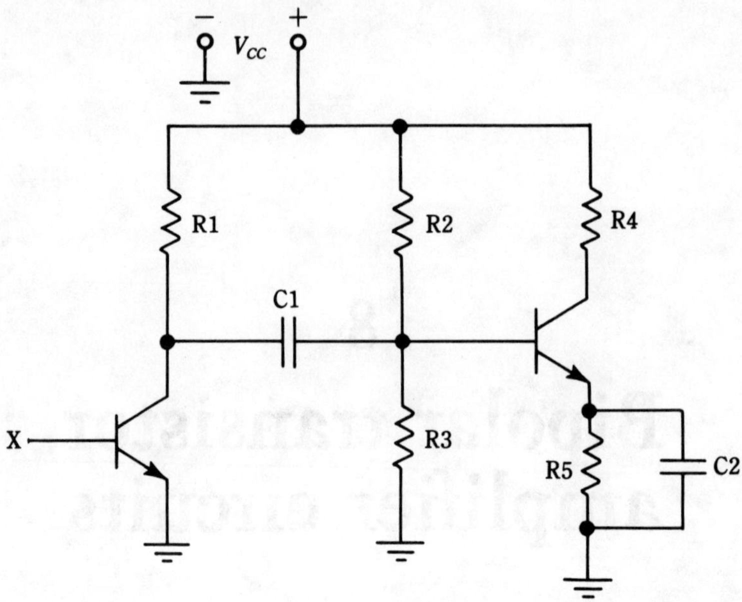

8-1   Capacitor coupling.

One advantage of the direct-coupled amplifier is that it will pass signals at all frequencies down to, and including, dc. These amplifiers are then sometimes called *dc amplifiers*, not after the idea that they are direct-coupled but from the fact that they will pass dc signals.

There theoretically could be almost any number of stages connected in this manner, but in most cases the amplifier will have only two to ten stages. Otherwise, you run into the supply voltage limit, and there is no longer any swing available for the signal.

Capacitor C1 in Fig. 8-1 is used to reduce the ac impedance seen by the emitter to a very low value relative to the value of the emitter resistor. The usual design protocol requires a capacitive reactance for C1 of less than one-tenth the emitter resistance at the lowest frequency of operation.

Transformer coupling

Figure 8-2 shows an example of transformer coupling. This particular circuit shows two transformers. T1 is an interstage transformer, and T2 is an output transformer. The principal reason for this type of coupling is impedance matching, so it is used mostly for power-amplifier applications. Another application for transformer coupling is to isolate a circuit from either the signal source or the load.

On any transformer, the impedance ratio is set by the turns ratio of the transformer's primary and secondary windings:

$$\frac{N_p}{N_s} = \sqrt{\frac{Z_p}{Z_s}} \tag{8-1}$$

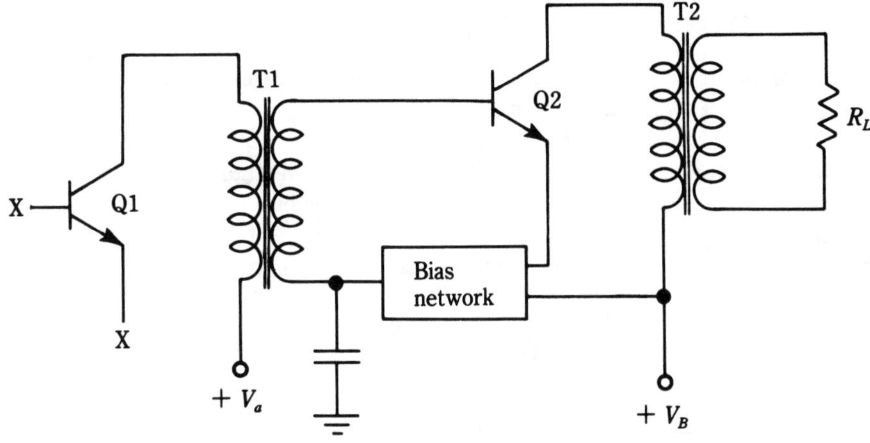

8-2 Transformer coupling.

Where: N_p is the number of turns in the primary winding
N_s is the number of turns in the secondary winding
Z_s is the impedance load connected to the secondary
Z_p is the impedance reflected to the primary

Capacitor coupling

Figure 8-3 shows an example of resistor-capacitor coupling. The idea here is to pass the signal from the input stage (Q1) to the output stage (Q2) without letting the dc voltage used to operate the input stage affect the bias state of the output stage. The low-frequency response of the circuit is partially dependent on the respective values of the resistances in the circuit and the coupling capacitor.

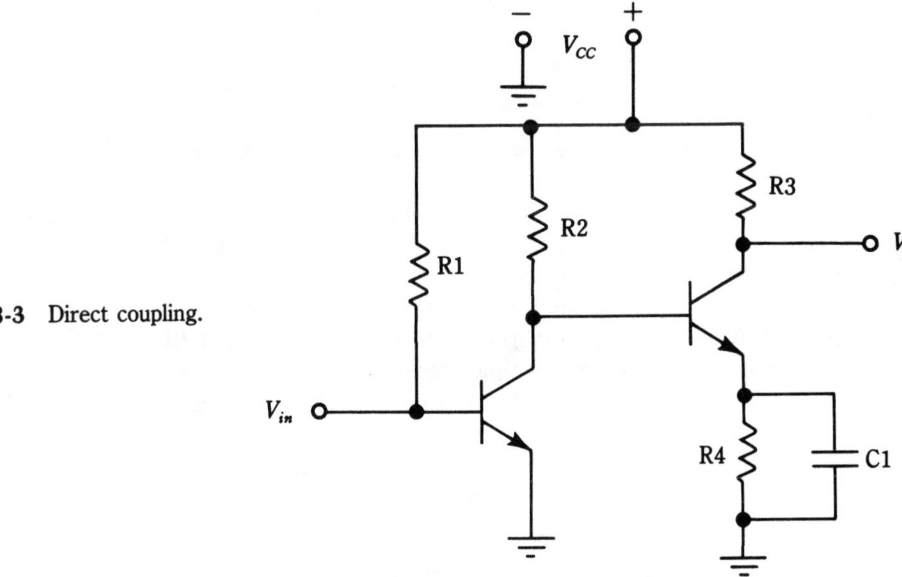

8-3 Direct coupling.

Power amplifier circuits

The power amplifier develops the audio signal into the strong current needed to drive the speaker voice coil. Power amplifiers are generally current amplifiers because of the low-impedance levels associated with speaker systems.

Figure 8-4 shows the schematic of a typical early solid-state power amplifier. This circuit uses a pnp germanium power transistor (Q1). The bias network contains a thermally sensitive resistor (thermistor) to provide some degree of temperature stability. In more modern designs, a solid-state diode is used in this function.

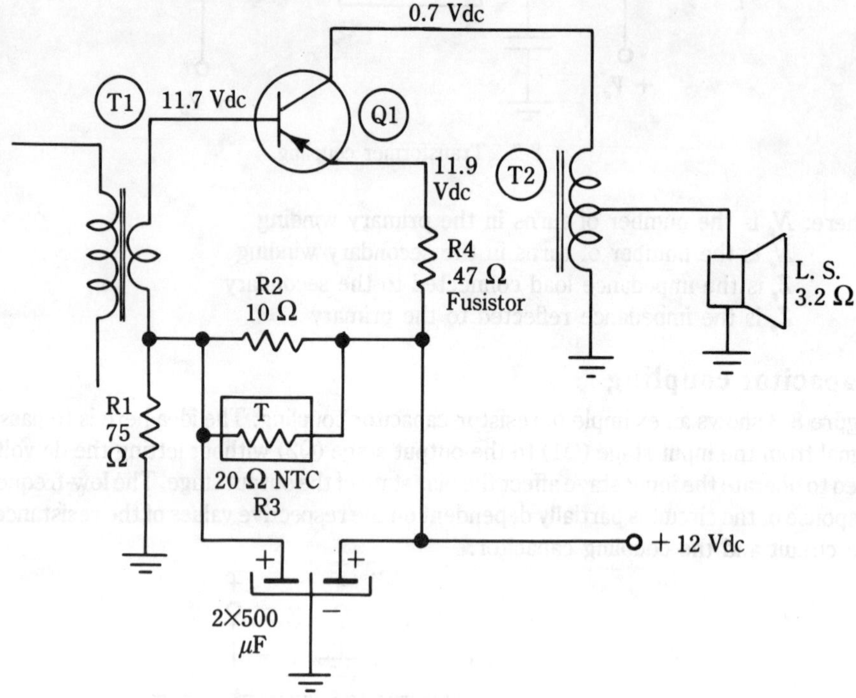

8-4   Transformer-coupled Class-A power amplifier.

The low-value resistor in the emitter circuit (R4) is called a *fusible resistor*, or *fusistor*. Its function is two-fold. First, since it is unbypassed, it offers a small amount of negative feedback. Because of its small value, this amount appears to be insignificant, but many circuits have been known to be particularly sensitive to changes in the value of the fusible resistor. Typical values for these parts run from approximately 0.20 to 1.2 Ω.

Although substitutions are sometimes permissible, do not use a replacement that is too far from the original value. It is a matter of percentages. If you replace a 0.22 Ω with a 0.47 Ω, you will change the value more than 100 percent.

The second function of the fusible resistor, as its name implies, is fusing. Should the output transistor become shorted, the fusible resistor is supposed to be destroyed by the excess current flow. Although destroying the resistor cannot help the transistor, it does offer some protection to the other power amplifier and the dc power supply parts. In cases where the fusible resistor fails to open, secondary damage to the radio can be extensive.

Performance improvements can be realized by the use of push-pull audio power amplifiers. These circuits come in several basic categories: *standard push-pull, complementary symmetry push-pull,* and *totem pole push-pull.* There were several other variations on the push-pull theme, but they had only a limited historical lifespan in the audio world.

Figure 8-5 is a standard push-pull amplifier. Once again we see the use of pnp germanium transistors. The interstage (input) transformer is of the center-tapped style usually associated with push-pull amplifier circuits. The output circuit, however, is a push-pull version of an output choke coil, which serves here as an autotransformer. However, in some radios it might be a transformer, rather than a choke.

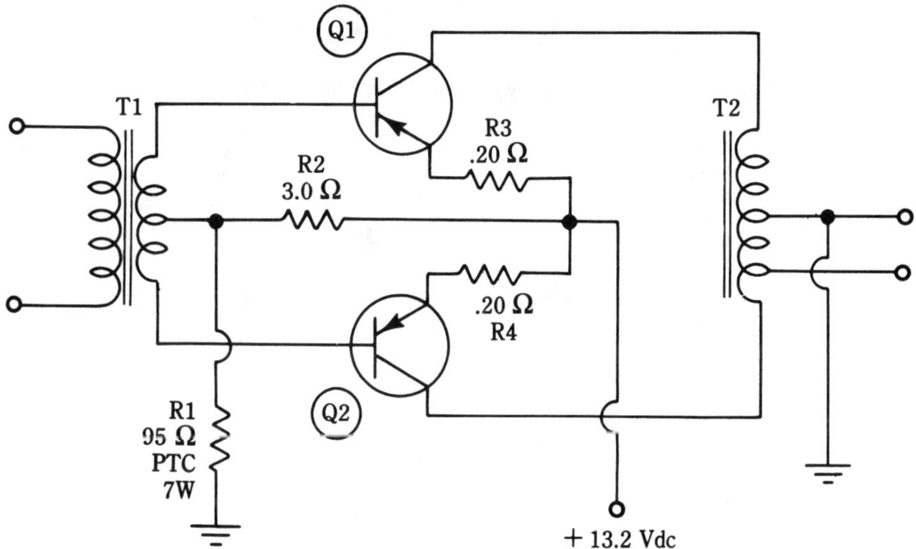

8-5 Transformer-coupled push-pull power amplifier.

In any solid-state push-pull amplifier, the transistors must be given a slight forward bias, even though the circuit is operated in Class-B or Class-AB, in order to prevent *crossover distortion.* This type of distortion occurs as the input signal approaches the zero axis. Crossover distortion is due to the nature of performance of transistors at low signal levels. The bias in Fig. 8-5 is provided by resistors R1 and R2.

Figures 8-6 and 8-7 show a variation on the normal push-pull theme. In these circuits, the loudspeaker voice coil is used as the load for the output transistors.

The circuit in Fig. 8-6 uses a 20 Ω center-tapped speaker. Although this speaker is now hard to locate, an adequate substitute can be fashioned from a universal multi-impedance speaker. These speakers usually have two 8 Ω to 10 Ω voice coils that can be used in series, parallel, or individually to nominally cover the 4 Ω to 20 Ω range. Phasing of the two windings, in this case, is crucial and must be determined experimentally by the cut-and-try method.

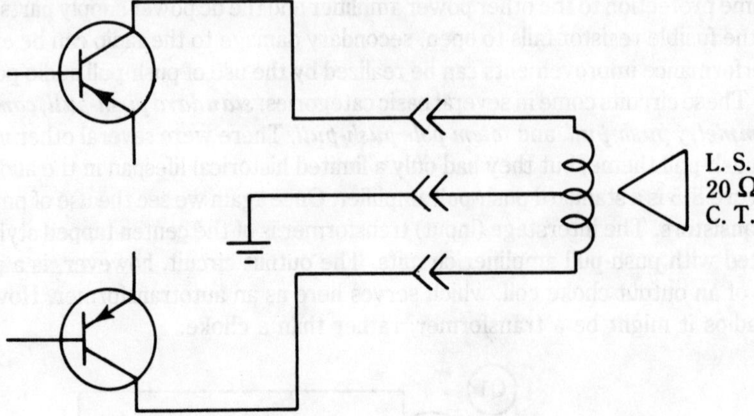

8-6   Use of a speaker voice coil for impedance matching.

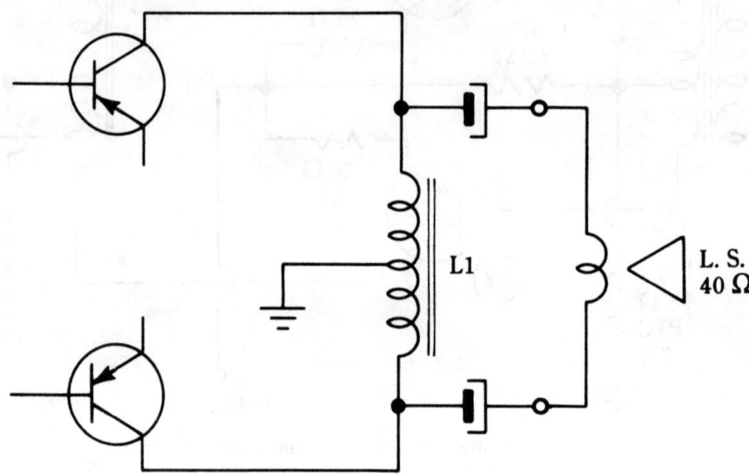

8-7   Choke-output push-pull amplifier.

Figure 8-7 shows a circuit that uses a 40 Ω speaker. In this case, it is actually necessary to order a 40 Ω speaker.

One of the requirements for push-pull amplification is that one transistor must operate over one-half of the input cycle, while the other transistor operates over the other half. In the standard push-pull circuit of Fig. 8-5, this phase inversion is accomplished by the use of a center-tapped transformer. If the center tap is considered common, the signal at the extreme end of one side of the secondary winding is 180 degrees out of phase with the signal at the opposite extreme. This situation will drive one transistor into conduction as the other is turned off.

The complementary-symmetry configuration in Fig. 8-8 uses the polarity differences between npn and pnp transistors to accomplish phase inversion. The npn transistor Q3 will conduct as the base voltage becomes more positive than the emitter. And since Q4 is a pnp transistor, it will conduct when the base voltage becomes less positive than the emit-

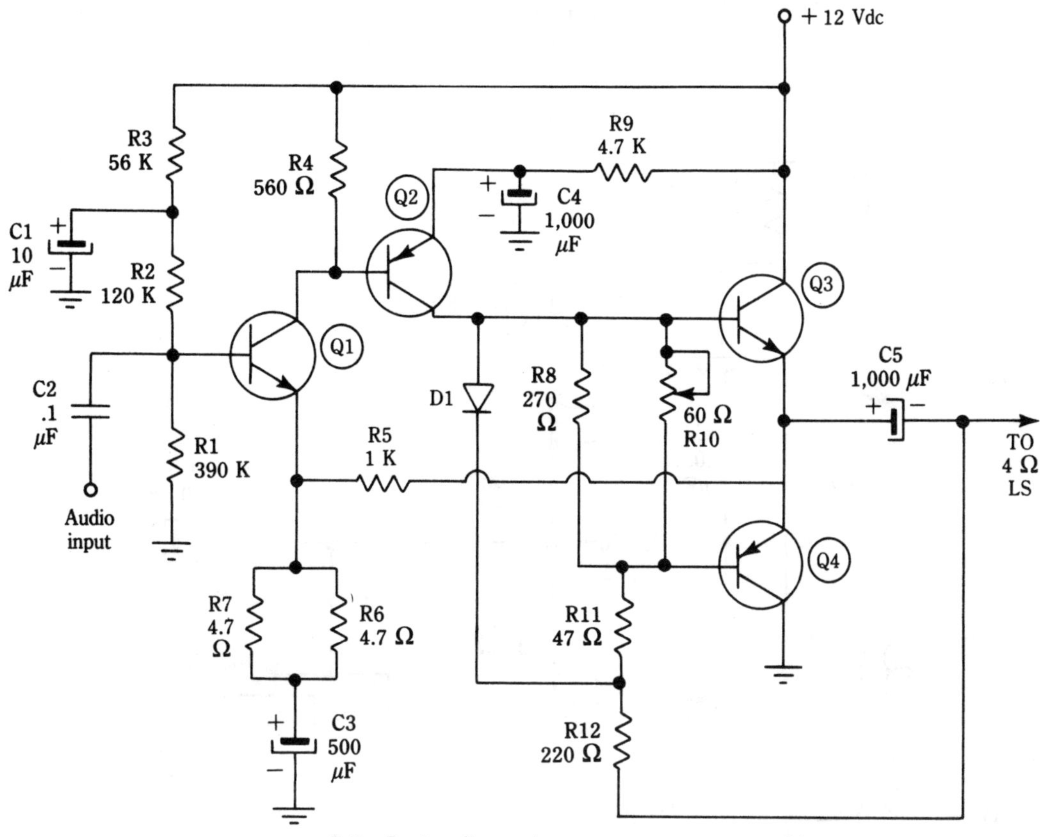

8-8 Push-pull complementary symmetry amplifier.

ter. Because of this alternate action, Q3 will operate (conduct) on the positive half-cycle of the input signal, while Q4 will operate on the negative half.

The signal is routed to the speaker through an electrolytic capacitor. Typical values for the speaker capacitor will normally be in the range of 300 to 2,000 μF. The diode used in this circuit gives the transistors a slight forward-bias in order to reduce crossover distortion.

Totem pole push-pull amplifiers have been very popular in modern amplifiers. An example of a typical transformer-coupled totem pole power amplifier is shown in Fig. 8-9. In this circuit, both output transistors are pnp. If a transistor should become defective, most manufacturers recommend replacement of both transistors so that a matched pair can be inserted.

The output coupling of the totem pole is almost identical to that of the complementary circuit. The input side of the amplifier is the distinguishing part. Phase inversion occurs by using a specially designed interstage transformer that has two secondary windings. The two windings are connected to their respective base terminals in opposite "sense" so that there will be a 180-degree difference between the two signals. Small disc ceramic capacitors are connected between the output transistor base and collector terminals to suppress RF oscillations that could occur.

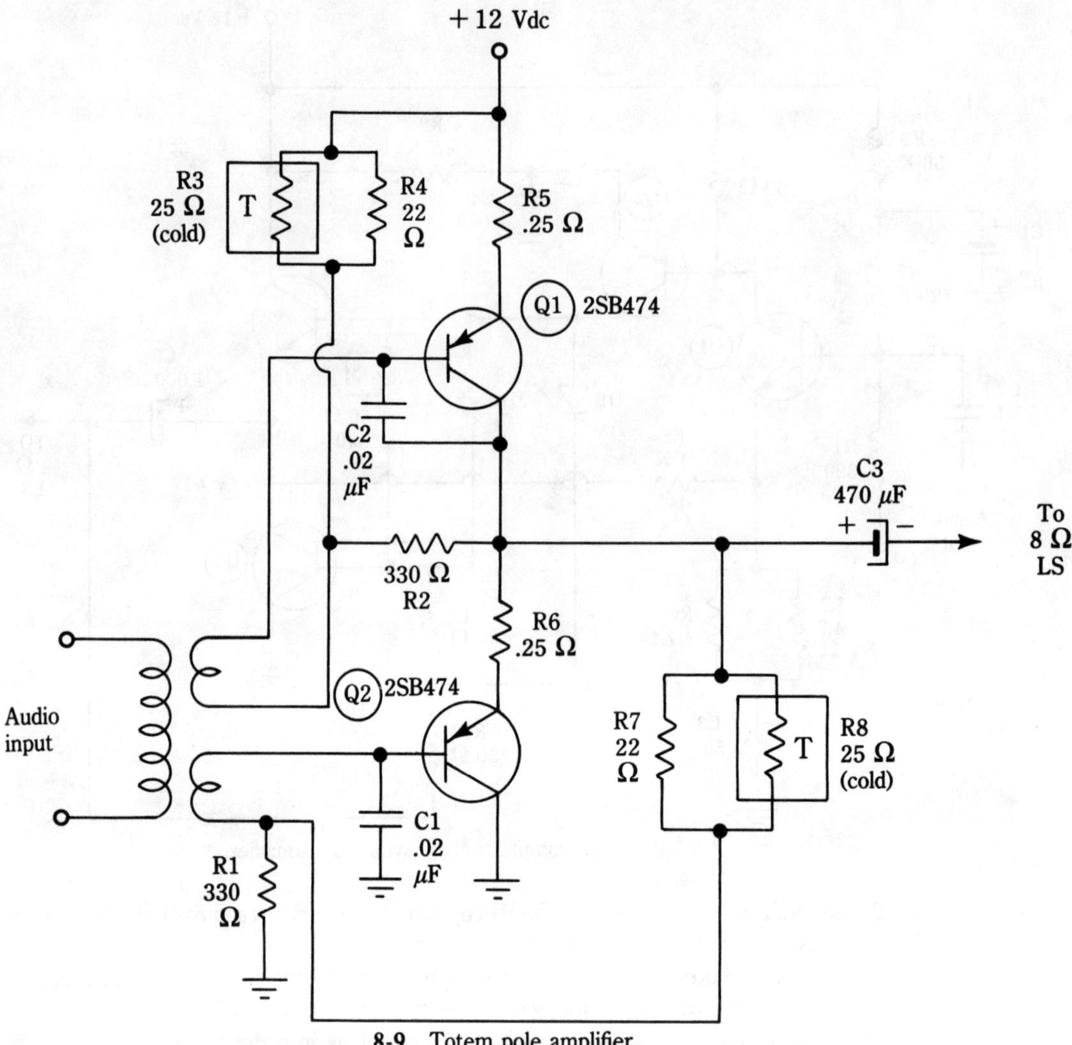

8-9   Totem pole amplifier.

Troubleshooting the amplifier

Figure 8-10 shows a complete audio section from a radio. Although this particular circuit is of more recent design, it is a direct descendent of earlier designs.

In troubleshooting these direct-coupled stages, remember that any one defect in the audio section can affect the dc conditions throughout the amplifier. For example, should Q2 become shorted (collector to emitter), the base voltage of the output transistor will reduce. Since the output transistor is a pnp, the forward bias on Q3 will increase by whatever voltage is removed from the base. Such a condition will drive Q3 into saturation and make it appear shorted.

Transistor Q1 also can affect the output transistor. If it ceases to conduct, its collector voltage will rise because of the decreased voltage drop across the 6.8K resistor. This

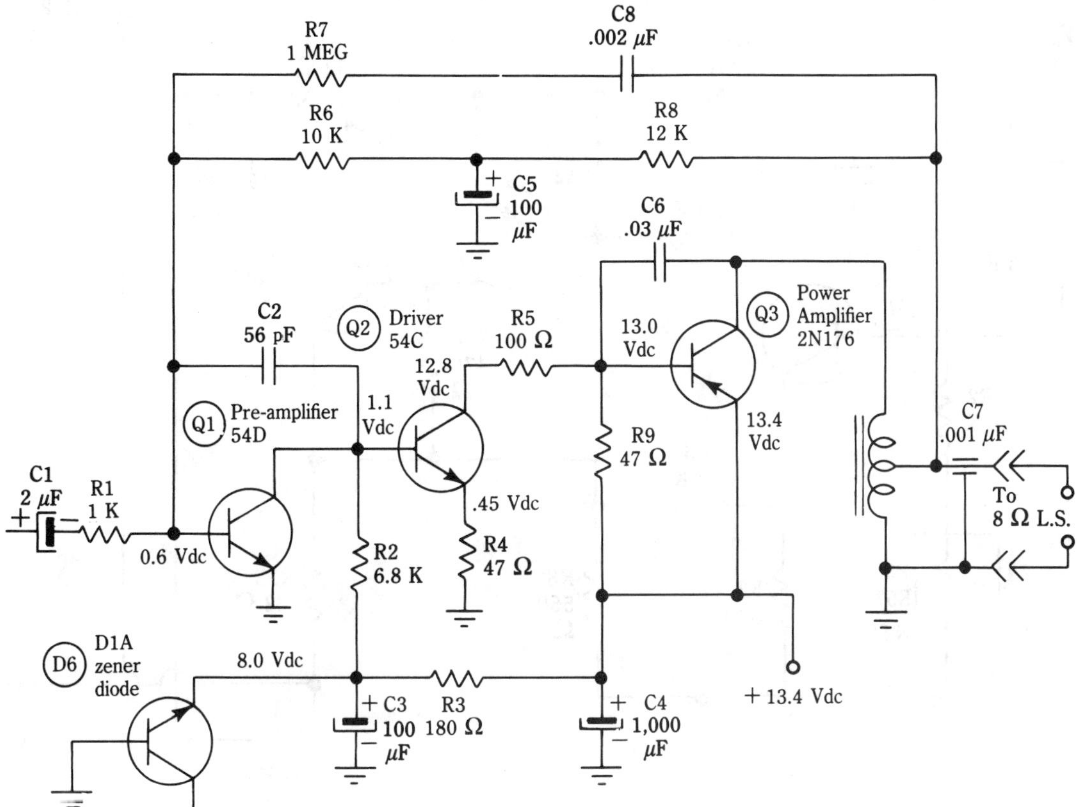

8-10   Audio amplifier section to a car radio.

increase in voltage saturates Q2, causing it to draw excessive collector current. With an increase in collector current, Q2 will have a considerable decrease in collector voltage, which reduces the Q3 base voltage (increasing forward bias) to a point where Q3 is saturated and appears shorted.

A collector-to-emitter short at Q1 also shorts the base of Q2 to ground, causing Q2, an npn transistor, to cut off. Since almost zero collector current will flow in Q2 under these conditions, the base voltage of Q3 will rise to a point where Q3 is cut off.

Figure 8-11 is another audio amplifier design. An increase in the collector current of transistor Q2 will increase the conduction of Q3 automatically. Since Q2 is a pnp device, its collector current increases when its base voltage becomes lower (less positive) than its emitter voltage. This voltage level for Q2 is a function of the collector current drawn by the preamplifier transistor Q1. As Q1 increases conduction, Q2 also will increase conduction. With increased current levels flowing in the Q2 collector circuit, a greater voltage will be impressed across the Q2 collector resistor. It is this voltage that is used to control the conduction of the npn output transistor Q3.

Notice that in this circuit the output incorporates a transformer, rather than a simple choke. This design is necessary in order to isolate the high positive voltage on the collector of Q3 from the grounded speaker circuit. A certain amount of negative feedback is gained by having both the collector and emitter windings on the output transformer.

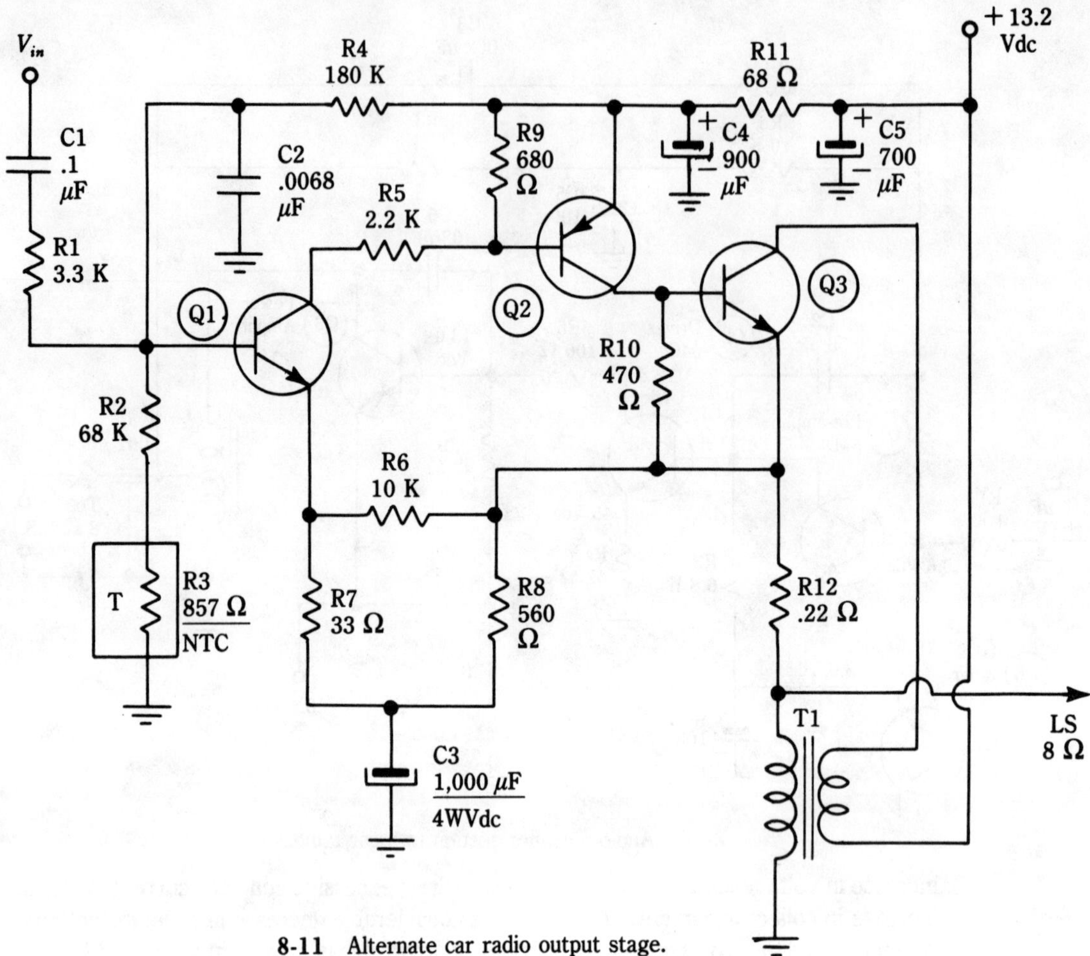

8-11 Alternate car radio output stage.

There have been two general areas of trouble in these circuits. One is failure of the output transistor. Most output transistors found defective are shorted. In audio amplifiers such as Fig. 8-11, however, the plastic-package npn output transistors are often victims of a base-to-emitter open condition. In these situations, the Q3 base voltage will be almost as high as the collector voltage.

Another area of fault is the 1,000 μF decoupling capacitor. The electrolytics will occasionally open, causing low volume and distortion. When the capacitor fails in this manner, the relative dc levels are not affected.

Motorola once took a different path in the design of the audio stage. Figure 8-12 shows a typical Motorola audio section using a tiny integrated circuit (MFC-4050). Preamplifier and driver stages are housed inside a plastic case less than 1/2 inch in its greatest dimension.

This circuit is the utmost in simplicity. It has only four terminals: input, output, common (signal and dc one terminal), and positive voltage source. The output terminal is directly coupled to the base of the power transistor. This power transistor is a pnp germa-

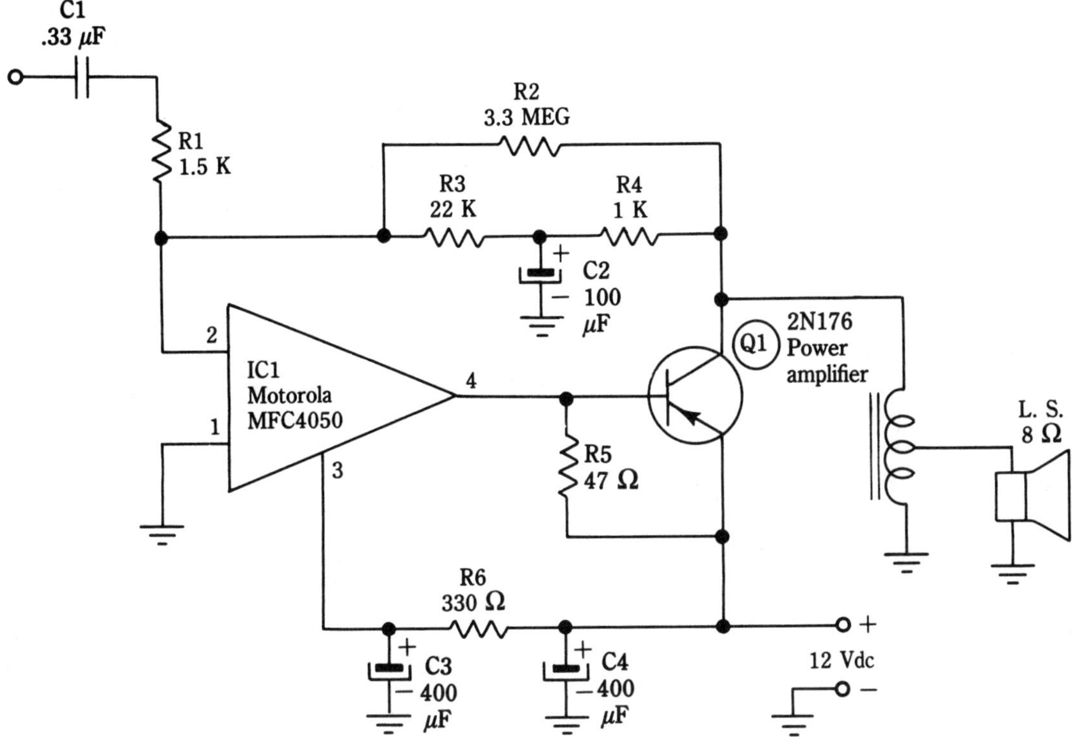

8-12   Audio amplifier with IC preamplifier/driver.

nium type that has enjoyed long popularity in Motorola car-radio designs. An external feedback network shapes the tonal characteristics of the overall circuit.

Failure of the MFC4050 can cause either saturation or cutoff of the output transistor. Should the output transistor be saturated (dc readings such as collector voltage and collector current will make it appear shorted), try shorting the output transistor base and emitter terminals together to determine the faulty component. If the collector current drops to zero, it proves that the transistor base junction is still capable of controlling the collector current. In this case, the finger of suspicion should fall on the MFC4050 IC. If, on the other hand, the collector current fails to drop appreciably when the base and emitter terminals are shorted together, you may safely assume that the transistor is shorted.

In cases where the output is cut off (almost zero collector current), connect a resistor (approximately 100 Ω) between the output transistor base terminal and ground. If the output transistor begins conducting, the IC is probably open. (IC terminal voltages offer a guide to some defects but cannot be relied upon to tell the final story in all situations.)

Today, most low-power, and some high-power, amplifier designs are based on integrated circuit technology. Some are covered in more detail in chapter 13.

9
Field-effect transistors

THE FIELD-EFFECT TRANSISTOR (FET) WAS THE SUBJECT OF SCIENTIFIC speculation in the late 1930s. A paper published in a physics journal during that decade gave details for the construction of the *junction field-effect transistor* (JFET), and that hypothetical transistor would have worked if the metallurgy of the era had been better. Following World War II, it was the bipolar transistor that received research attention. If the JFET had been built in 1939, then the semiconductor era would've seen a ten-year head start over what actually occurred — and it's likely that most transistors today would be field-effect devices, rather than bipolar devices.

FETs have a very high input impedance, and are transconductance amplifiers. They tend to have performance properties that are far more reminiscent of the pentode vacuum tube than of bipolar transistors.

There are two basic types of MOSFET devices: *depletion* and *enhancement.* These devices operate in somewhat different modes, and therefore will be discussed separately. Both types of MOSFET device are available in both p-channel and n-channel versions. Thus, the possible simple MOSFETs are:

Junction field-effect transistors

A basic JFET device is shown in Figs. 9-1 and 9-2. The JFET consists of a semiconductor channel and a gate structure. Two basic types of JFET are seen: *p-channel* and *n-channel.* The designations are based on the kind of semiconductor material used to form the channel section. There are three electrode elements on the JFET: *gate, source* and *drain.*

The gate structure is always made of the opposite type of semiconductor material from the channel structure. In normal operation, a voltage is applied across the channel in the polarity shown. Some JFETs, incidentally, are capable of operation when the drain and source ends of the channel are reversed.

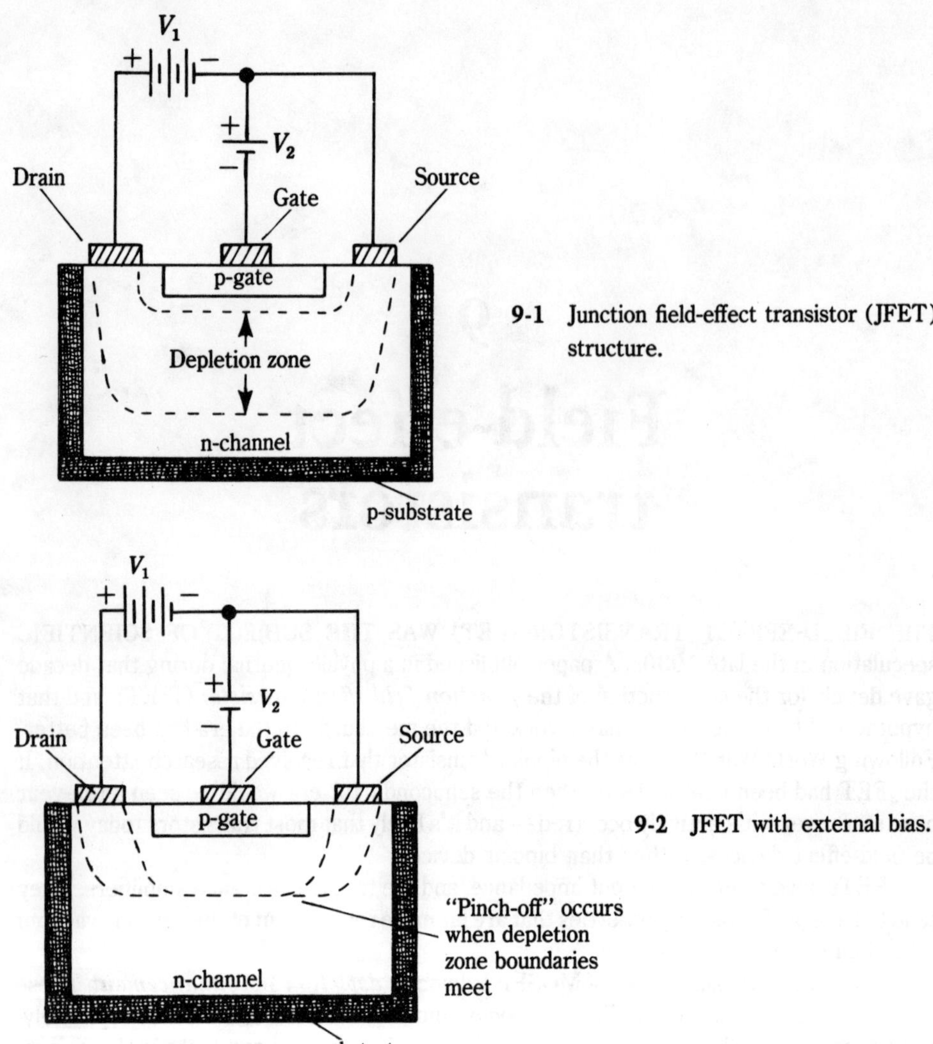

9-1 Junction field-effect transistor (JFET) structure.

9-2 JFET with external bias.

The channel has an electrical resistance less than infinity, so current can flow. When a potential is applied to the gate, however, it will create a *depletion zone* in the channel material. Very few electrical charge carriers (electrons or holes) can exist in this zone, so it has a very high electrical resistance. At 0 V gate potential, the depletion zone is at a minimum, so the channel resistance is the lowest. As the potential is increased, however, the depletion zone widens, thereby narrowing the portion of the channel in which charge carriers will flow. At some specific potential, called the *pinchoff voltage*, the depletion zone completely chokes off the channel and no current can flow from source to drain.

JFET circuits

The circuit of Fig. 9-1 is for an n-channel JFET. The gate structure, made of a different form of semiconductor material than the channel, is diffused into the channel in a manner resembling the emitter of a bipolar transistor. However, this junction is kept perpetually reverse-biased. In fact, if the junction were to become forward-biased, destruction of the device is a distinct possibility. The channel is formed such that it is placed between the p-type gate and a p^+-type substrate on which the transistor is built.

Electrons in the n-channel are encouraged to flow by the external potential source. If the depletion zone is wide, current will flow in the external circuit. By making the gate-substrate boundary negative with respect to the channels, electrons are repelled (like charges repel each other) in the channel, thereby widening the depletion zone. When the depletion zones from gate and substrate meet in the channel, pinchoff occurs. This effect is shown in Fig. 9-2.

Consider the channel resistance for a moment. When the gate voltage is 0, the depletion zones are narrow, so the channel resistance is low. When a high negative potential is applied to the gate and substrate, however, the depletion zones widen, causing the channel resistance to increase. When the gate potential reaches a certain critical level, the pinch-off voltage, the channel resistance becomes extremely high. The JFET, then, can be used as an electronic switch because it possesses an extremely high channel resistance under one voltage condition and a very low resistance under another. The CMOS digital device, type number 4066, is a series of six FETs used specifically as switches.

The p-channel device is exactly the same as the n-channel JFET, except for polarity. The channel structure will be made of p-channel semiconductor material, while the gate and substrate will be made from n-channel material. The circuits for p-channel devices are the same as for n-channel circuits, but with the external voltage polarities reversed. Symbols for the two types of device are shown in Fig. 9-3.

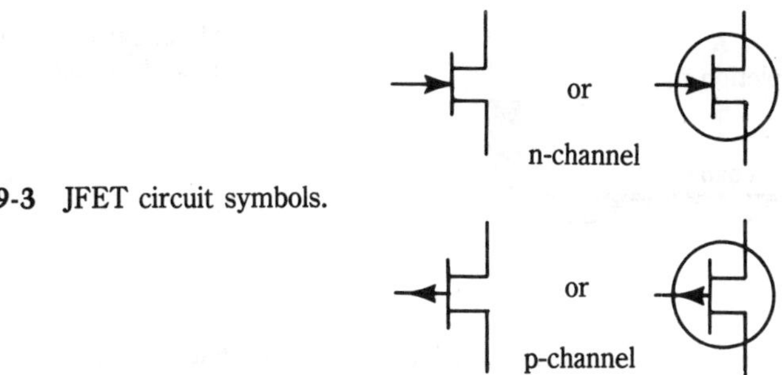

9-3   JFET circuit symbols.

MOSFET devices

MOSFET devices are a bit different from JFET devices. MOSFETs are capable of much higher input impedances than JFETs because there is no physical connection—i.e., an emitterlike pn junction — between the gate and the channel. In the JFET device, input (i.e., *gate*) impedance is limited by the reverse leakage current across the reverse-biased gate-channel junction.

There are two basic types of MOSFET devices: *depletion* and *enhancement*. These devices operate in somewhat different modes, and therefore will be discussed separately. Both types of MOSFET device are available in both p-channel and n-channel versions. Thus, the possible simple MOSFETs are:

- p-channel depletion MOSFET
- p-channel enhancement MOSFET
- n-channel depletion MOSFET
- n-channel enhancement MOSFET

MOSFET devices are sometimes called *insulated gate field-effect transistors* (IGFETs) because of their gate structure.

Depletion MOSFETs

The depletion type of MOSFET is shown schematically in Fig. 9-4. As in the JFET device, the channel will be of one of the two basic types of semiconductor material — n or p — and the substrate will be of the opposite type of material. It is always desirable to keep the substrate-channel pn junction either reverse-biased or zero-biased to prevent an excessive current flow.

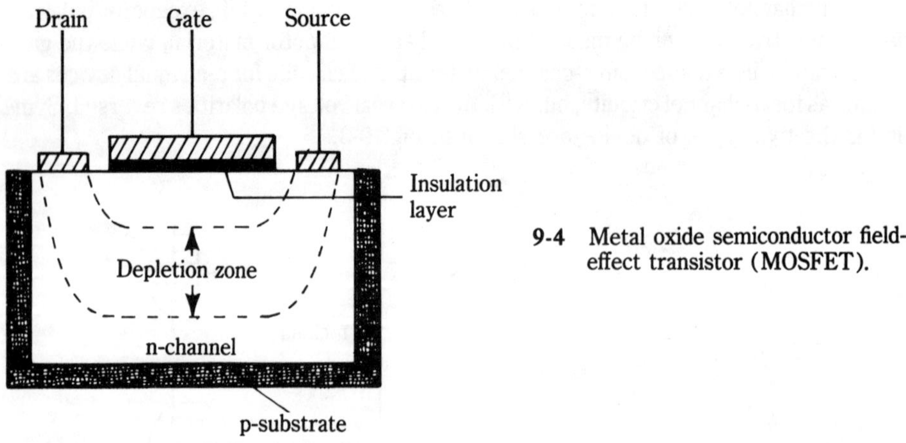

9-4   Metal oxide semiconductor field-effect transistor (MOSFET).

The gate is a metallic contact insulated from the channel material by an extremely thin metal oxide insulating material. This is the origin of the name, *insulated gate*. The depletion MOSFET is normally *biased on*, meaning that the channel resistance is low when the gate voltage is zero.

When an electrical potential is applied to the gate, however, an electrical field is created in the channel. If the voltage is positive with respect to the channel, then electrons will tend to draw closer to the gate structure. If the voltage is negative, however, then the field will repel electrons in the channel, creating a depletion zone. The higher the gate voltage, the wider the depletion zone. When the depletion zone is finally wide enough to pinch off the channel, current flow ceases. We can, therefore, control the channel resistance with a gate voltage.

Because channel resistance is controlled by the gate voltage, the output current flowing in the source-drain circuit is also under the control of the gate voltage (assuming constant voltage between drain and source).

Field-effect transistors are *transconductance amplifiers*. The definition of such amplifiers is that a change of input voltage (ΔV_{in}) causes a change of output current (ΔI_o). In the transfer function format that takes the ratio of output to input:

$$g_m = \frac{\Delta I_o}{\Delta V_{in}} \tag{9-1}$$

The unit of transconductance is the *siemen* (S); one Siemen equals one ampere per volt; i.e., $1\ S = 1\ A/1\ V$. For practical FET devices, the unit *millisiemen* (1/1,000 siemen) is typically used to express the gain of the device.

The symbol for a depletion type MOSFET is shown in Fig. 9-5. As in the case of JFET devices, an arrow pointing inward denotes an n-channel device, while an arrow pointing outward denotes a p-channel device. It is interesting to note that the manufacture of p-channel depletion transistors is somewhat more difficult than the manufacture of n-channel devices. As a result, n-channel devices predominate in the marketplace.

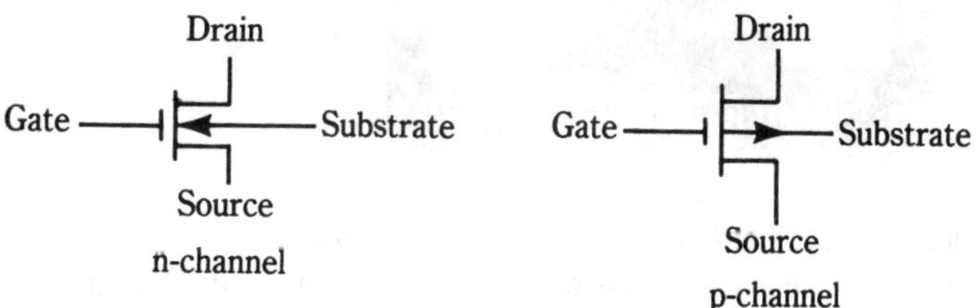

9-5   MOSFET circuit symbols.

Enhancement MOSFETs

The enhancement MOSFET is a normally off device (exactly the opposite of the depletion type MOSFET) in which it is necessary to apply a gate potential before channel conduction takes place. Again, both p- and n-channel devices are possible. The p-channel enhancement MOSFET requires a negative gate potential to begin conduction, while the n-channel device requires a positive gate potential.

Figure 9-6 shows the construction of the enhancement MOSFET. In this case, an n-channel device is shown.

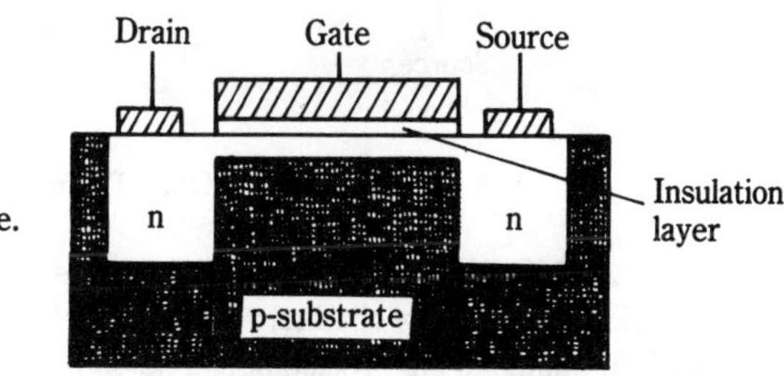

9-6   MOSFET structure.

In all MOSFETs there is an inherent tendency for electrons to cluster close to the interface between the metal oxide layer and the channel material. This phenomenon forms a thin n-type region in the p-type substrate, located immediately beneath the insulating layer. The drain and source regions are n-type semiconductor material diffused into the p-type channel material. There is ordinarily no conduction between them, except for the tiny n-type region close to the gate structure. When the gate voltage is zero, the electrons of the n-channel are prevented from migrating out of the drain and source regions. However, when a positive potential is applied (Fig. 9-7) to the gate, the electrons from these regions are attracted to the insulator-substrate interface, causing a conduction zone to appear between the source and drain. Current will flow under this circumstance.

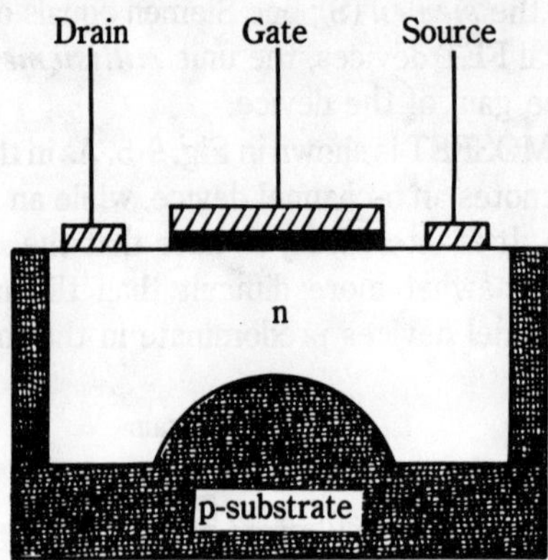

9-7 MOSFET structure.

The conduction of the enhancement MOSFET is controlled by varying the gate voltage. This action will increase or decrease the attraction of the drain and source electrons toward the gate region.

The circuit symbols for the enhancement-type MOSFET are shown in Fig. 9-8. These symbols are the same as the depletion MOSFET symbols, except that the channel bar is broken into sections to indicate the enhancement action taking place inside the device.

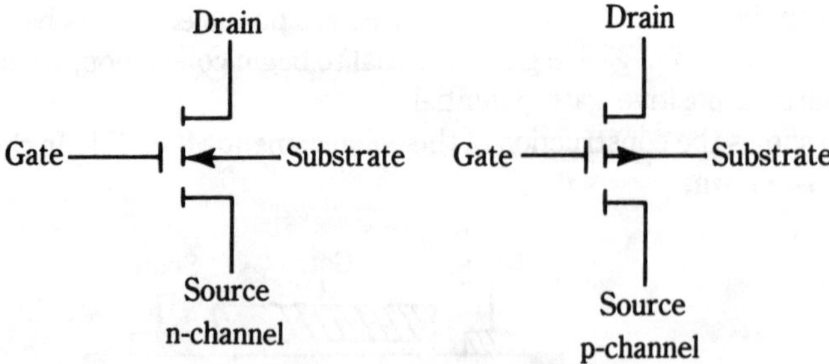

9-8   Enhancement MOSFET symbols.

In both types of MOSFETs, the activity can be used as an electronic switch. In both cases, one condition causes a low channel resistance, while the opposite condition causes an extremely high, "pinched off" resistance. The on resistance depends, in part, on the signal voltage applied, but it is typically between 10 and 2,000 Ω. Most devices will have a resistance in the 100 to 200 Ω range.

The variation of on resistance is something of a problem that can lead to distortion of the output signal, but manufacturers have been able to limit this problem by using a load resistance that is high compared with the channel resistance. Load resistances on the order of 10 kΩ to 500 kΩ often are seen.

The off resistance of the channel is usually many megohms. The specification is usually given in terms of leakage current, rather than resistance. In most switches, the leakage current under the off condition, called the *effective off resistance*, will be from 0.01 picoamperes (pA) to 100 pA. Assuming a $+10$ V applied signal, then, and a "typical" 10 pA leakage current, the resistance would be:

$$R_{off} = \frac{10 \text{ V}}{10^{-8} \text{ A}}$$

$$= 1,000 \text{ M}\Omega$$

(9-2)

The on resistance of the channel, on the other hand, drops to a much lower value, such as 100 Ω or so. This fact, incidentally, makes the FET almost ideal as an analog electronic switch. A control voltage can be applied to the gate in order to obtain an expected switching action from the difference between the on resistance and off resistance of the channel.

The primary use of the field-effect transistor is as an amplifier or as the amplifying element in an oscillator or active filter circuit. The input impedance of the FET is typically very high. For some JFETs, on the low end of the price scale, this figure might be less than 1 MΩ, but for most it is quite a bit higher. In some MOSFET devices, the input impedance exceeds *one teraohm* (10^{12} Ω).

In some circuit configurations, the high input impedance, coupled with a generally low output impedance, makes the device useful as an impedance converter. An amplifier for a signal source that has a high internal impedance can be made by using an FET in the "front-end" circuit.

Later sections will consider some aspects of FET circuit design and the types of circuit that might be selected. In Fig. 9-9, however, one elementary FET amplifier circuit is demonstrated.

The circuit of Fig. 9-9 shows a *common source amplifier*, which is analogous to the common emitter circuit seen earlier in the discussion of bipolar transistors. The input impedance of the common source amplifier is essentially the value of R1; the JFET input resistance is usually much higher than this resistor. The input signal voltage is applied across the gate and source terminals of the JFET. The source terminal must be bypassed for ac by capacitor C3. The value of this capacitor must follow the one-tenth rule: the capacitive reactance of C3 must be less than, or equal to, one-tenth the value of R3 at the lowest operating frequency.

The amplifier can operate as a voltage amplifier through a mechanism that is very similar to the scheme used to make a bipolar npn or pnp transistor operate as a voltage amplifier. In the case of the FET (a transconductance amplifier device), we use a resistance in series with the drain and the dc power supply. The channel of the FET operates as an electronically variable resistance when a signal voltage is applied to the gate. When the

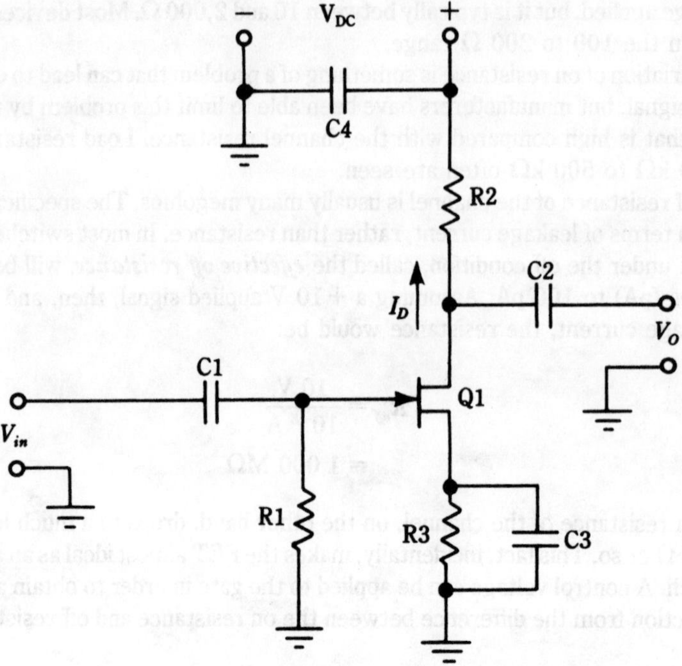

9-9 JFET common source circuit.

signal voltage increases in a positive direction, the channel resistance drops, so the voltage at the drain is reduced (the percentage of the terminal voltage dropped across R2 increases). Similarly, when the drain resistance increases, the voltage at the drain also increases. If the bias on the FET is set at a point where the drain voltage is (V+)/2, the output voltage will swing positively and negatively about this potential.

The bias is set by the voltage drop across resistor R3, which is analogous to self-bias in bipolar circuits. This voltage is equal to the product of the source-drain current and the resistance of R3. We normally want the gate to be negative with respect to the source (this is an n-channel device), so making the source more positive than the gate serves the same purpose.

Operating regions

There are two regions of operation for the field-effect transistor. In the *ohmic region*, which exists for low drain-source voltages (V_{DS}), the channel current is directly proportional to the voltage. In the *pinch-off region*, which is also called the *constant current region*, the current will not increase for further increase in V_{DS}.

It is this latter characteristic that allows the FET to operate as a constant current source (CCS). In this application, the source and gate are tied together, and the device operates as a diode with constant current characteristics. Several CCS devices available on the market are nothing more than internally connected JFETs.

A field-effect transistor can be biased in any of several ways. Self-bias can be used, as can external bias and certain combinations of self- and external bias. Figures 9-10 and 9-11 show two methods of FET biasing. Figure 9-10 shows an example of external bias. A reverse-bias dc potential is applied to the gate of the FET through resistor R1. The resistor is used to limit current flow across the junction and to provide a load for the input signal (if needed). It also allows the discharge of electrons that build up on the gate structure. The level of bias is set by varying V1.

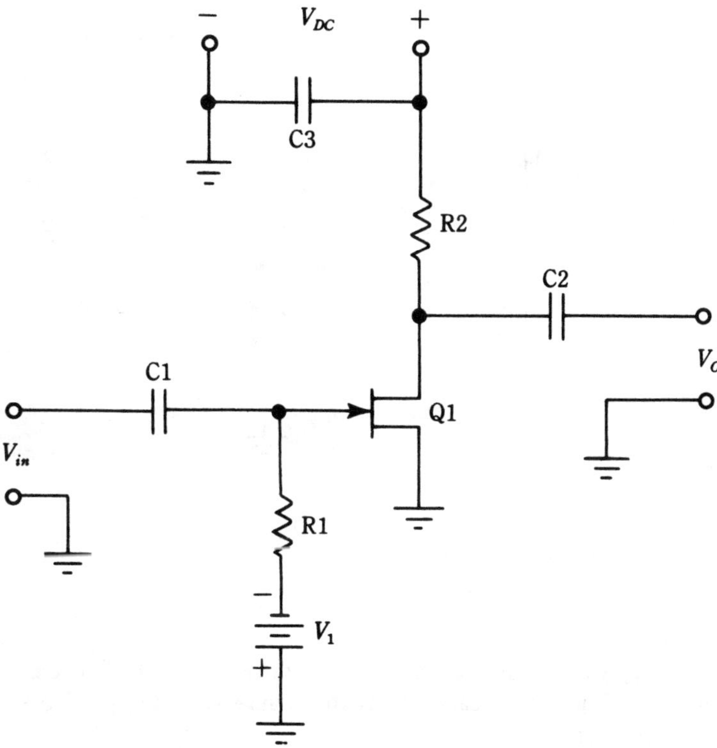

9-10   JFET circuit with a bias voltage.

An example of self-bias, which is by far the most common form, is shown in Fig. 9-11. In this case, the voltage drop across source resistor R2 is the bias voltage. The bias must make the gate more negative (or less positive) than the source. A negative voltage was applied in the previous example, but in this circuit the source is made more positive than the gate (which is conceptually the same thing).

The value of the resistor must take into consideration the source-drain current of the FET and the desired level of bias. Ohm's law is applied. We know that the gate-source voltage is given by:

$$V_{GS} = -I_D R_S \tag{9-3}$$

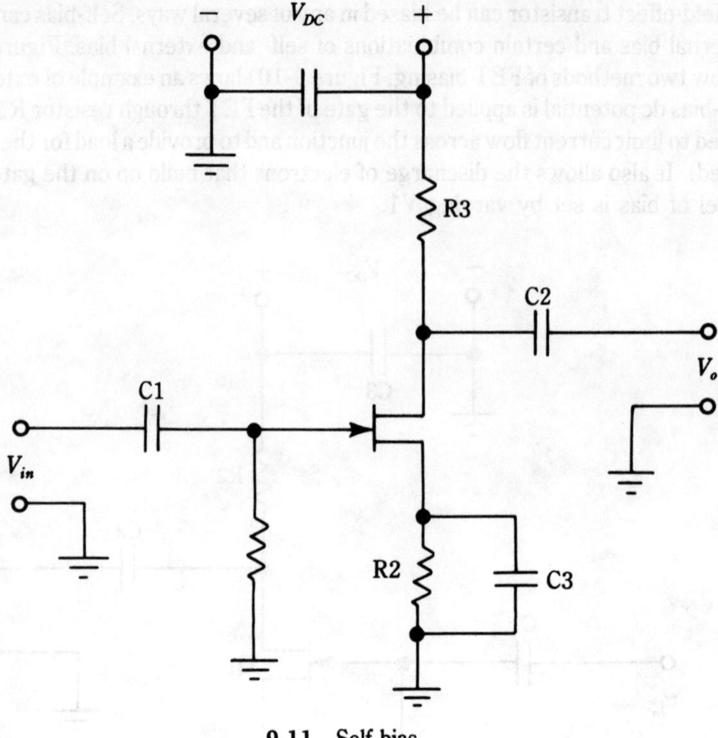

9-11   Self-bias.

Where: V_{GS} is the gate-source voltage
I_D is the drain-source current
R_S is the source resistor

A power supply potential must be supplied to the FET (i.e., V_{DC}), but only a portion of the power supply voltage appears across the drain-source (V_{DS}). The voltage drops around the circuit are:

$$V_{DD} = I_D(R_D + R_S) + V_{DS} \tag{9-4}$$

Where: V_{DD} is the dc power supply voltage
I_D is the drain-source current
R_S is the source resistor
R_D is the drain resistor
V_{DS} is the drain-source voltage

We could select a value for the collector current (see the data sheet for any specific device) and a drain-source voltage. In most cases, the value of V_{DS} will be ½ of V_{DD}. For example, if we applied 28 Vdc to the circuit (V_{DD}) in Fig. 9-12, and want a 2 mA current to flow in the drain circuit, we would select values for the resistors consistent with this goal. In general, we will make the source resistor one-tenth to one-fifth the drain resistor.

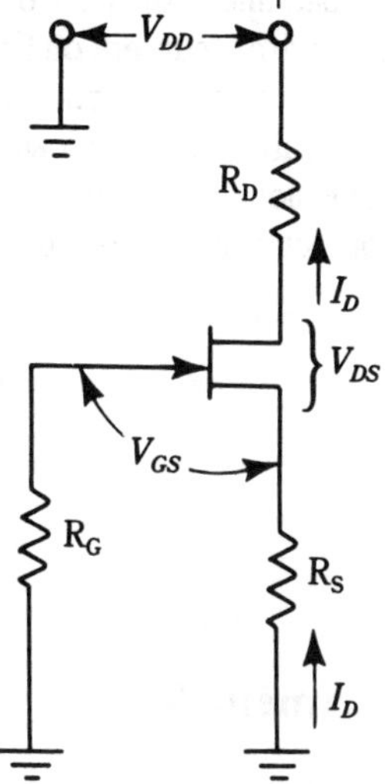

9-12 JFET with voltage and current relationships.

Graphical method

One of the methods for determining the proper values for the source resistor is to use the graph of I_D-vs.-V_{GS} characteristic of the particular device under consideration. These graphs are printed in the manufacturer's data sheets. An example in shown in Fig. 9-13.

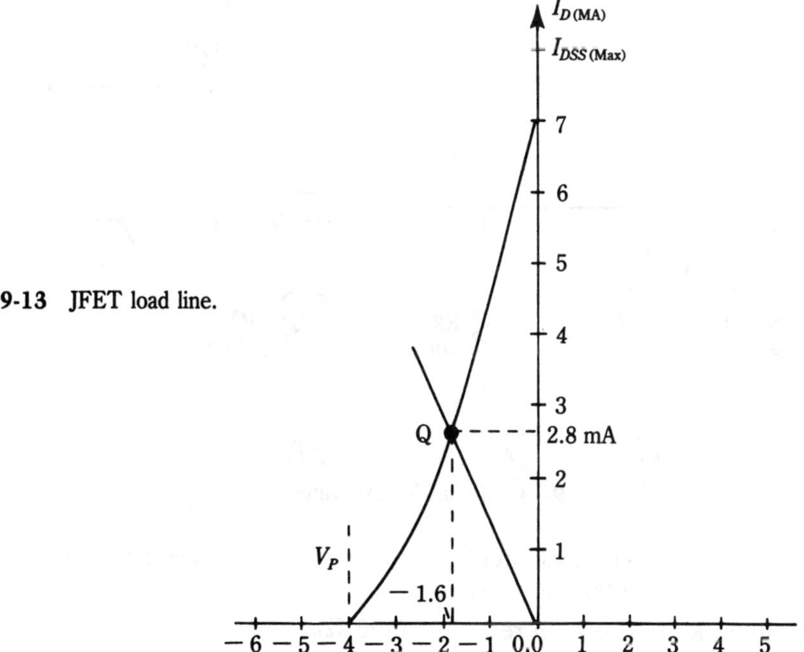

9-13 JFET load line.

A load line is drawn from the pinch-off voltage region on the horizontal axis and the maximum drain current on the vertical axis. A proper drain current must be chosen. Then draw a line from the vertical axis at that point to the load line. The point at which this current line intersects the load line is directly over the proper gate-source voltage (V_{GS}).

The slope of the line from the origin (0,0) to the *Q point* represents the resistance of the correct source resistor. In this case, the slope is:

$$9S = \frac{0.028 \text{ A}}{1.6 \text{ V}} \tag{9-5}$$
$$= 0.0175 \text{ siemens}$$

$$R = \frac{1}{G_m} = \frac{1}{0.0175 \text{ s}} \tag{9-6}$$
$$= 57 \text{ ohms}$$

Experiment 9-1

1. Connect the circuit in Fig. 9-14.

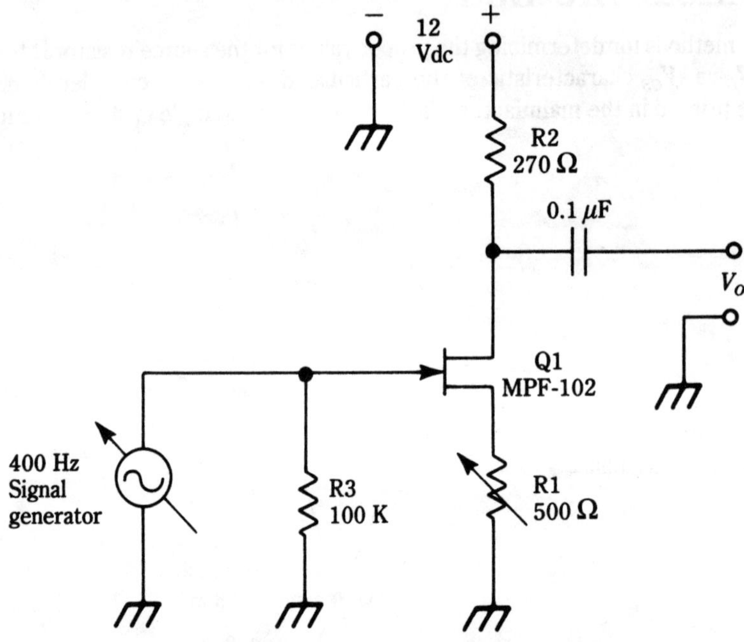

9-14 Circuit for experiment 1.

2. Set an audio signal generator for a frequency of 400 Hz and set the output signal level to 0. Use the sine waveform.

3. Adjust R1 to maximum resistance (500 Ω).

4. Connect the output of this circuit (V_o) to the vertical input of an oscilloscope.

5. Advance the level control on the signal generator until a signal waveform is seen on the scope (adjust the scope controls for a full presentation).

6. Examine the waveform shown on the oscilloscope while adjusting R1. Note the effect of the setting of R1 on the amplitude and waveshape.

7. Measure the dc voltages at the source, drain, and gate of Q1.

8. Reverse the positions of the 270 Ω resistor and 500 Ω potentiometer, and then repeat this experiment.

10

IC amplifier devices

WHEN I FIRST BECAME INTERESTED IN ELECTRONICS AS A HIGH SCHOOL
student in the late 1950s, one of my teachers predicted that semiconductor manufacturers were working on technology that would permit putting an entire audio amplifier chain, or even an entire radio, on a single semiconductor crystal not larger than a thumbnail. Although that prediction seemed a little on the wild side in 1959, by 1969 it was a fact in production home, portable, and car radio receivers.

The prediction that my electronics teacher mentioned was the invention of the integrated circuit (IC), or *chip*. These devices integrate onto a single monolithic crystal of semiconductor material all of the transistors, diodes, and resistors needed to make a circuit. Some modern circuits also include inductances and capacitances, but only at very high frequencies. These components, built onto a small crystal, are very dense, so the size of electronic devices can be reduced considerably.

Digital electronics IC devices also progressed at a tremendous pace. The modern computer would not be practical for most users unless there was a semiconductor industry that makes integrated circuits. The first digital devices in large-scale production were the resistor transistor logic (RTL) and diode transistor logic (DTL). In the 1960s, the now ubiquitous transistor transistor logic (TTL) devices were produced. The TTL devices were followed soon thereafter with the complementary metal oxide semiconductor (CMOS) devices, which required such low power that many portable devices that were previously impossible now became easy.

One of the startling aspects of integrated circuit electronics that has been repeated endless times is the well-known "learning curve" phenomenon. The price of IC devices typically starts out very high when the technology is new, and then drops rapidly. For example, the μA-709 operational amplifier started out over $120 in the early 1960s (when dollars were bigger!), but now costs less than 50¢. The μA-703 RF/i-f amplifier cost $15 to replace in the middle to late 1960s, when I was servicing electronic equipment, but now costs two bits, when you can find one. Similarly, I paid $10 for a 741 operational amplifier in the mid-1960s, but now they cost ten for $2 in commercial plastic packages. The low cost of these devices makes it reasonable for hobbyists and students to experiment with ICs and use them in actual, very practical, projects.

Today, electronics includes a bewildering variety of functions from radio broadcasting, communications of many types, control systems, instrumentation, and many thousands of other applications. Even your automobile is highly dependent on electronics. Although older cars had only a radio as an electronics device, modern cars are run on central computers and have dozens of other electronics devices on board. Even the ignition system is now an electronics circuit. In nearly all of these cases, the electronics are part of IC technology.

A portion of this book is devoted to the operational amplifier. One reason is the fact that the "op amp" is very popular. It is used in a wide variety of different applications and is, perhaps, the easiest IC device that is both flexible and sufficiently well behaved for the newcomer to use. The other reason is that the device really defines the field of linear ICs. Even many supposed "non-op amps" are actually special-purpose op amps in disguise (e.g., internal connections or components that make the device work in a special way).

However, op amps do not define the entire universe of IC devices, so we will also take a look at certain nonoperational amplifier devices such as the current difference amplifier (CDA), also called the *Norton amplifier*, and the operational transconductance amplifier (OTA).

Common IC package styles

The integrated circuit is formed on a tiny "chip" of silicon material by a photolithographic process. Typical chip "die" sizes range around 100 mils (0.100 inch), with some being larger and others being smaller. The die is typically mounted inside of a package and connected to the package pins by fine wires.

Figure 10-1A shows a die with wire attached, while Fig. 10-1B shows a packaged die with a see-through window for illustration purposes. The connecting wires between package pins and "solder" pads on the die are around 10 mils (0.010 inch) diameter, and are made of either gold or aluminum in most cases. Either an electric current or a thermosonic process is used to melt the end of the wire onto, and bond it with, the connecting pad on the semiconductor die.

The particular package style selected for any given IC device depends in part on the intended application and the number of pins required. For many IC devices, several different packages are available. The earliest IC packages were the 6-, 8-, 10- and 110-lead metal can devices (Fig. 10-2A). These packages were redesigns of (and similar to) the TO-5 metal transistor package.

When viewed from the bottom of the package, the keyway marks the highest number lead or "pin" (e.g., pin no. 8 in Fig. 10-2A), and pin no. 1 is the next pin clockwise from the keyway. You must be careful when looking at IC base diagrams to know whether a top or bottom view is depicted.

Perhaps the largest number of IC devices on the market today are sold in *dual inline packages* (DIPs), examples of which are shown in Fig. 10-2B. DIP packs are available in a wide variety of sizes, from 4 to more than 48 pins. Although many devices are available in other size DIP packs, most linear devices are found in 8-, 14-, or 16-pin DIP packs.

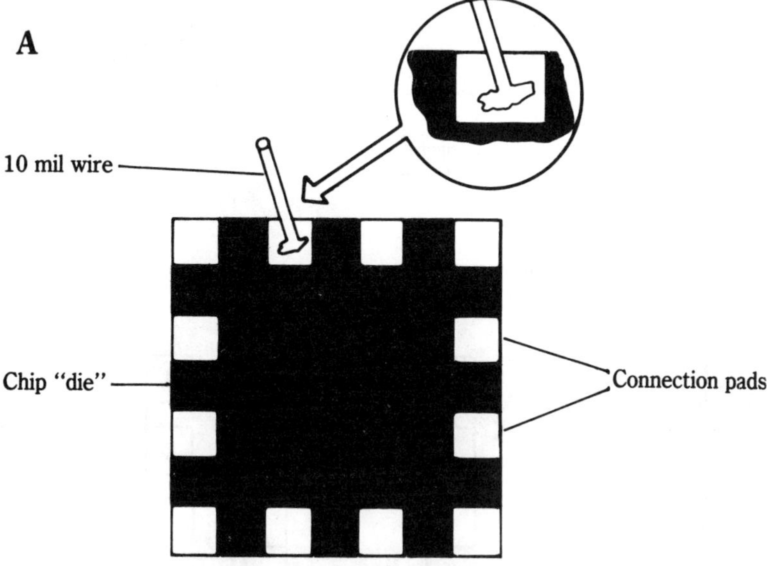

10-1   A. Integrated circuit die; B. Integrated circuit.

The DIP pack is symmetrical with respect to pin count, regardless of the package size, so some other means is needed to designate pin no. 1. Figure 10-2B shows several common methods. In all cases the IC DIP pack is viewed from the top. In some devices a paint dot, "pimple," or "dimple" will mark pin no. 1. In other cases, a semicircle or square notch marks the end where pin no. 1 is located. When viewed from the top, with the notch pointed away from you, pin no. 1 is to the left of the notch, while the highest numbered pin is to the right of the notch.

Both plastic and ceramic materials are used in DIP pack construction. In general, the plastic packages are used in consumer and noncrucial commercial (or industrial) equipment, while the ceramic packs are used in military and crucial commercial equipment. The principal difference between the plastic and ceramic packages is the intended temperature range. Although exceptions exist, typical temperature range specifications are 0° to +70°C for commercial plastic devices, and −10° to +80°C for ceramic. Military and some crucial commercial ceramic devices are rated from −55° to +125°C. The principal

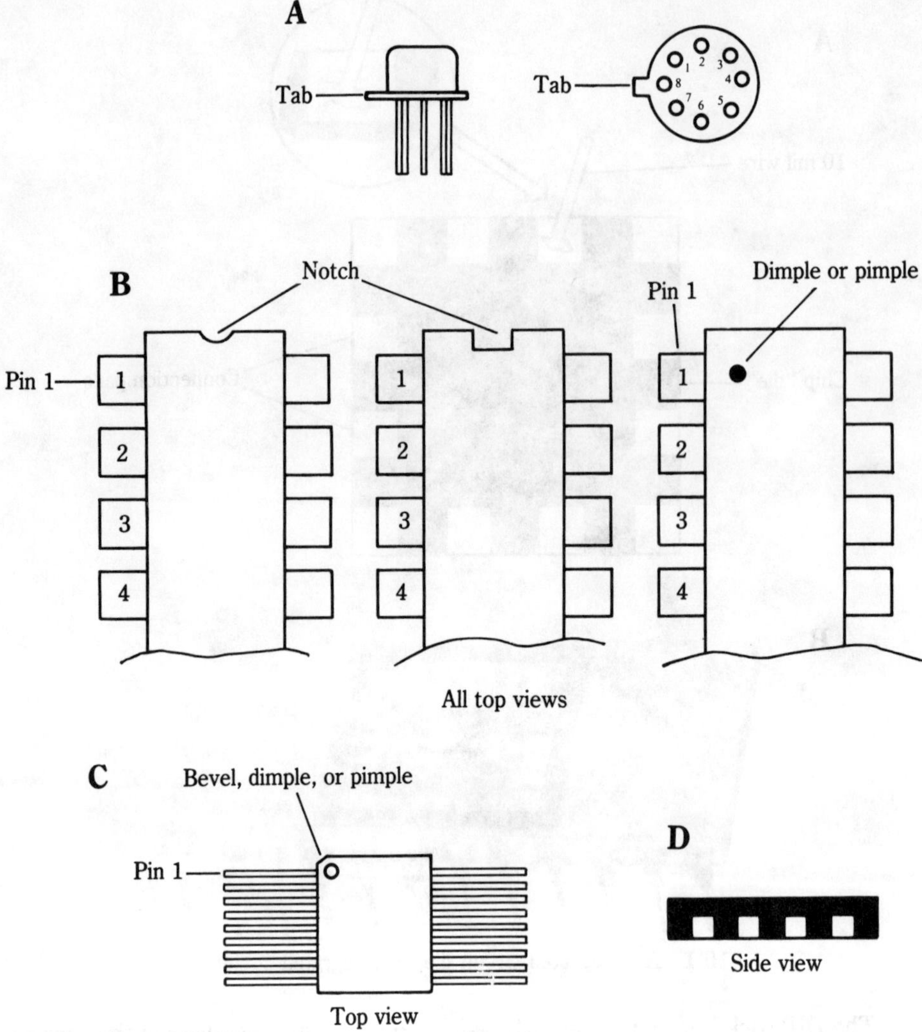

10-2   A. Metal can IC; B. Dual in-line package (DIP); C. Flat package; D. Surface mount.

difference between military chips and the highest grade commercial or industrial chips is the amount of testing, "burn-in," and documentation that accompanies each device.

An example of an IC flat-pack is shown in Fig. 10-2C. This type of package is typically used where very high component density is required. Most flat-pack devices are digital ICs, although a few linear devices are also offered. DIP-package IC devices are mounted either in sockets or by insertion of the pins through holes drilled in a printed circuit board (PCB). Flat-packs, on the other hand, are mounted on the surface by direct soldering to the conductive track of the PCB.

A relatively new style of package is the *surface-mounted device* (SMD), an example of which is shown in Fig. 10-2D. The SMD technology represents a significant improvement in packaging density. SMD components can be mounted closer together than other types of package, and are amenable to automatic PCB production methods. It is expected that the SMD package eventually will overtake and replace other forms, especially in VLSI applications.

Scales of integration

Several different scales of integration are found among IC devices. The ordinary *small-scale integration* (SSI) device consists of single gates, small amplifiers, and other smaller circuits. The number of components on each chip is on the order of twenty of less.

Medium-scale integration (MSI) devices have a slightly higher degree of complexity, and may have about 100 or so components on the chip. Devices such as operational amplifiers, shift registers, counters and so forth are usually classed as MSI devices.

Large-scale integration (LSI) devices are mostly digital ICs and include functions such as calculators and microprocessors. Typical LSI devices contain from about 100 to 1,000 components. Some newer devices are called *Very large scale integration* (VLSI) devices and include some of the latest computer chips.

The numbers and descriptions listed for SSI, MSI, and LSI devices are approximate only, but serve to provide guidelines. Most linear IC devices are either SSI or MSI, with the latter predominating.

Skills experiment

Integrated circuit electronics requires the ability to solder small pins that are close together. This experiment is really just a skills exercise, but it is very important to learn.

1. Buy a grab bag of miscellaneous integrated circuits and a perforated circuit board. The board should be designed especially for DIP IC devices. Radio Shack® has a number of these for sale, as well as souvenir bags of (probably useless) "miscellaneous" ICs (sold "as is").

2. Install a DIP IC on the board and practice soldering it in place. Use a 25- to 75-watt "pencil" soldering iron.

3. Examine your work with a magnifying glass or eyeloop. The joints should be bright, shiny, and clear of flux, and adjacent connections should not be shorted together.

4. Solder a dozen or more chips into place and repeat the inspection. Do it until you get it right.

11

Operational amplifiers

IN THIS CHAPTER, THE BASICS OF OPERATIONAL AMPLIFIERS ARE DISCUSSED. The role of these devices in analog electronics is so great that the material in this chapter is very important to any discussion of amplifiers. Topics to be considered include operational amplifiers and certain other forms of amplifier that are based on IC operational amplifiers.

One of the early texts on operational amplifiers tells us that the IC operational amplifier has made ". . . the contriving of contrivances a game for all." A physics professor at my college was struggling with the design of a transistor amplifier for his undergraduate students to use in a classroom experiment. It seemed very difficult to build a circuit that both had a gain of precisely 100 without the need for adjustment, and was replicable for the thirty or so amplifiers required for the class (using "on-hand" components of mixed and doubtful ancestry). After he learned about the 741 IC operational amplifier, his problems were all but over!

The little 741, although today not even considered for many jobs because of the newer types now on the market, then cost just over a dollar, yet it obeyed very simple design equations. We can design op-amp voltage amplifiers by using only the ratio of two resistors forming a voltage-divider feedback network.

The operational amplifier originally was designed to perform mathematical operations in analog computers (hence the name "operational" amplifiers). The first op amps were vacuum-tube models and only approximated the behavior of the ideal mathematical model of the device. Later, transistor operational amplifiers were available, and finally integrated circuit types came on the market. One of the first linear ICs made commercially was the μA-709 operational amplifier. Although now considered primitive, that device sold for $109 in one mid-sixties catalog. The current price is about five for a dollar — where they are still available.

Figure 11-1A shows the usual circuit symbol of the operational amplifier. An alternate symbol shown in Fig. 11-1B is preferred by some people. The second symbol is tech-

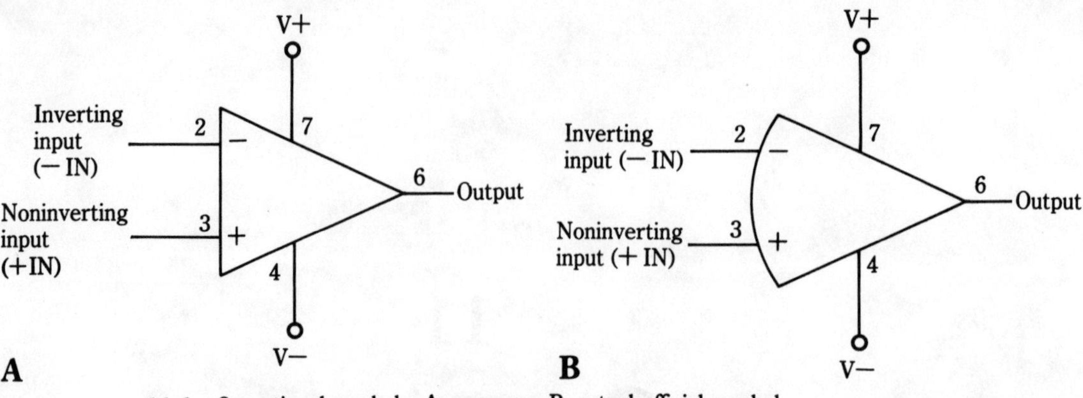

11-1   Operational symbols: A. common, B. actual official symbol.

nically the correct symbol and uses a curved back to which the input leads are attached. The version shown in Fig. 11-1A is used almost universally, however, even though it technically denotes any amplifier — including, but not limited to, operational amplifiers. We will maintain the de facto standard of Fig. 11-1A.

Note the pin-outs for the amplifiers in Fig. 11-1. The pin numbers given are for the 741 device, but they have become something of an industry standard. There are two input connections, two power supply connections, and one output connection. There is no ground or common connection. The ground and signal common are taken from the power supply common line (of which, more in a moment).

The two power-supply connections are V+ and V−. The V+ supply is positive with respect to common, while the V− is negative with respect to common. The range for these voltages is typically ±4 to ±18 V, although a number of examples exist with wider (or slightly different) voltage ranges. An RCA CA-3140 BiMOS device, for example, operates at potentials up to ±22 V for V− and V+, while a certain "low-power" operational amplifiers operate down to ±1.5 Vdc.

In addition to the absolute voltage limits, there are also sometimes relative limitations. For example, other 741 devices had a 30 volt limit for the voltage defined by the expression $[(V+) - (V-)]$, even though each $V-$ and $V+$ can be as high as 18 V. Thus, if $V+$ is +18 V, then $V-$ must be not greater than −12 V in order that the differential not be greater than 30 V $[(+18) - (-12) = -30 \text{ V}]$.

The selection of power supply voltages also might depend somewhat on the maximum anticipated output voltage. If the amplifier is being designed for use with an analog-to-digital converter that has a range of −10 to +10 V input, then we certainly want the output of the amplifier to be capable of achieving those limits.

However, there is a limit to how high the output voltage can reach, and that limit is a function of the power supply voltage. In general, the limitation is based on the number of pn junctions between the output terminal on the IC and the two power-supply terminals. Each pn junction has a 0.7 V drop that must be accounted for. If there are four pn junctions between the output terminal and the V+ power supply terminal, for example, the maximum allowable output voltage will be $[(V+) - (4 \times 0.7)]$ volts, or 2.8 volts lower than V+. Thus, when we want the output terminal to swing to +10 V, the absolute minimum

V+ power supply voltage will be 10 + 2.8 V, or +12.8 Vdc. Obviously, a +12 Vdc power supply will not work in this case.

In general, ordinary bipolar transistor operational amplifiers require a supply voltage of 2 to 4.5 volts higher than the maximum required output voltage, yet must remain within the V+ and V− constraints of the device. Some BiMOS and BiFET devices are available in which the maximum output signal voltage can be as close as 0.5 volt to the dc power supply potential.

The two inputs for the operational amplifier form a so-called *differential pair* because they operate 180 degrees out of phase with each other. The inverting input produces a 180-degree phase shift between input signal and output signal. (In other words, a positive-going input signal produces a negative-going output signal, and vice versa.) The noninverting input produces a 0-degree phase shift in the output signal. Since one input produces an in-phase output and the other produces an out-of-phase output, application of the same voltage to both inputs simultaneously produces a zero net output potential. We will use this fact in a later section to form the immensely useful differential amplifier.

The two inputs on the operational amplifier have a very high impedance (which is infinite in the ideal model). Thus, they are a perfect voltage amplifier input.

The output of the operational amplifier is also suited to use as a voltage amplifier circuit. The output impedance of the typical op amp is usually quite low (50 to 200 Ω), so forms a nearly perfect voltage source.

Operational amplifier dc power supplies

Figure 11-2 shows a model of the typical operational amplifier power supply. Although batteries are shown for sake of simplicity, electronic power supplies operated from the ac power lines are commonly used. Recall that we have two different voltages in the op-amp power supply: V+ and V−. Voltage V+ is supplied by battery B1, while V− is supplied by battery B2. The common (or ground) connection is the junction between the two batteries. Normally, B1 and B2 will have the same voltage rating, although that is not strictly a requirement unless other circuit considerations apply.

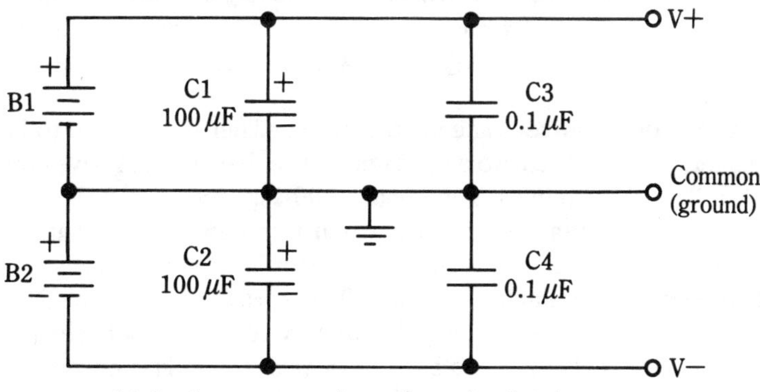

11-2   dc power supply configuration for the op amp.

The capacitors shown in Fig. 11-2 are used for decoupling, especially when multiple stages are fed from the same power supply. Capacitors C1 and C2 are normally 50 to 200 μF and are used for decoupling low-frequency signals. Capacitors C3 and C4 are generally 0.1 μF and are used for decoupling higher frequency signals. We cannot normally use the higher value C1 and C2 for high-frequency signals because they are normally electrolytic capacitors, which are ineffective at high frequencies.

The signal common connection is used as the zero reference for input and output signals on the operational amplifier. Whether it is actually grounded or not depends upon circuit design considerations. In most cases, however, it is grounded for the sake of simplicity.

In most applications, electronic power supplies used for B1 and B2 must be voltage-regulated. Although there are certainly numerous applications where voltage-regulated dc power supplies are not strictly required, they are almost always "good engineering practice."

The ideal operational amplifier

Before getting further into operational amplifier circuits, let's set the stage for our simplistic circuit analysis by discussing the properties of the so-called "ideal operational amplifier." This ideal device has the following properties:

1. Infinite open-loop gain
2. Zero output impedance
3. Infinite input impedance
4. Zero noise contribution
5. Infinite bandwidth
6. Differential inputs that "stick together"

What do these statements mean, and how do they compare with real, practical IC operational amplifiers?

Infinite open-loop gain   *Infinite open-loop gain* means that the voltage gain of the ideal operational amplifier in the open-loop (i.e., no feedback) configuration is infinite. Real operational amplifiers do not even approach the ideal, but are still good enough approximations to make the device behave properly. The ability of the practical operational amplifiers to approach the ideal is dependent upon having an extremely high open-loop gain (otherwise, the equations behave badly). In practical devices, we find that the open-loop voltage gain — A_{vol} — will be 20,000 in low-cost devices, and well over 1 million in premium devices.

Zero output impedance   The operational amplifier is supposed to be a perfect voltage amplifier, so requires a zero output impedance. Real devices have output impedances of 50 to 200 Ω, with most being under 100 Ω.

Infinite input impedance   *Infinite input impedance* means that the input will neither sink nor source electrical current. Recall that input impedance is $Z_{in} = V_{in}/I_{in}$, so for input impedance to be infinite, I_{in} must be 0. In real operational amplifiers, of course, we find this value is nonzero and is one of the primary differences between premium and low-cost devices. Low-cost amplifiers have input (transistor) bias currents to contend with of up to 1 or 2 mA, but certain other devices measure the input current in picoamperes or nanoamperes. The RCA BiMOS operational amplifiers (CA-3140, etc.) use MOSFET input transistors to produce an input impedance of more than 10^{12} Ω.

Zero noise contribution   The *noise* referred to is internal noise added to the signal. This is one area of primary difference between the low-cost and premium devices. The low-cost amplifiers add considerable "hiss" noise, so are unusable on low-signal applications.

Infinite bandwidth   *Infinite bandwidth* means that there is no limit to the operating frequency of the device, which in real operational amplifiers is patently absurd. Unconditionally stable, frequency-compensated devices like the 741 might have an upper frequency limit of only a few kilohertz, while other operational amplifiers operate to several megahertz. Only a few high HF or low VHF devices are available, and they are usually labeled "video operational amplifiers," or something similar.

Differential inputs that stick together   This property is essential to the simplified circuit analysis which we use. The implication of this property is that a voltage applied to one input also will appear on the other input. We must treat both inputs alike mathematically in this regard. Thus, if we apply a voltage to the noninverting input, we must treat the inverting input as if it also sees that voltage. This is not just some theoretical device used to make equations work. If you apply a real voltage to a real noninverting input, and then use a real voltmeter at the inverting input, you will measure a real voltage at that point.

We will use these properties in the circuit descriptions that follow.

Standard operational amplifier parameters

Designing circuits with operational amplifiers requires understanding the various parameters given in the specification sheets. The following list is not intended to be exhaustive, but rather it represents the most commonly needed parameters.

Open-loop voltage gain (A_{vol})   *Voltage gain* is defined as the ratio of output signal voltage to input signal voltage (V_o/V_{in}), which is a dimensionless quantity. The open-loop voltage gain is the gain of the circuit without feedback (i.e., with the op amp's external feedback loop open-circuited). In an ideal operational amplifier, A_{vol} is infinite, but in practical devices it will range from about 20,000 for low-cost devices to over 1 million in premium devices.

Large signal voltage gain   *Large signal voltage gain* is defined as the ratio of the maximum allowable output voltage swing (usually several volts less than V− and V+) to the input signal required to produce a swing of ±10 V (or some other standard).

Slew rate   *Slew rate* specifies the ability of the amplifier to transition from one output voltage extreme to the other extreme, while delivering full rated output current to the external load. The slew rate is measured in terms of voltage change per unit of time. The 741 operational amplifier, for example, is rated for a slew rate of 0.5 volt per microsecond (0.5 V/μS). Slew rate is usually measured in the unity gain noninverting follower configuration (of which, more later).

Common-mode rejection ratio   A *common-mode voltage* is one that is presented simultaneously to both inverting and noninverting inputs (voltage V3 in Fig. 11-3). In an ideal operational amplifier, the output resulting from the common-mode voltage is zero, but in real devices it is nonzero. The common-mode rejection ratio (CMRR) is the

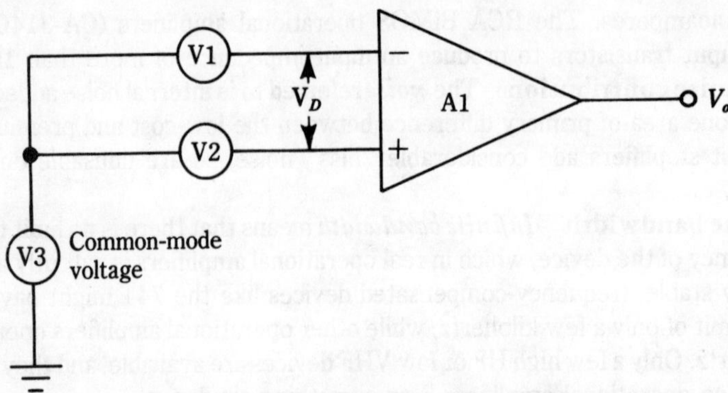

11-3   Input voltages to a differential op amp.

measure of the device's ability to reject common-mode signals. It is expressed as the ratio of the differential gain to the common-mode gain. The CMRR is usually expressed in decibels (dB), with common devices having ratings between 60 dB and 120 dB (the higher the number, the better the device).

Power supply rejection ratio   Also called *power supply sensitivity*, the power supply rejection ratio (PSRR) is a measure of the operational amplifier's insensitivity to changes in the power supply potentials. The PSRR is defined as the change of the input offset voltage (see following) to a 1-volt change in one power supply potential, while the other is held constant. Typical values are in microvolts, or millivolts, per volt of power supply potential change.

Input offset voltage   The voltage required at the input to force the output voltage to zero when the signal voltage is zero is called the *input offset voltage*. The output voltage of an ideal operational amplifier is 0 when $V_{in} = 0$.

Input bias current   *Input bias current* is the current flowing into or out of the operational amplifier inputs. In some sources, this current is defined as the average difference between currents flowing in the inverting and noninverting inputs.

Input offset (bias) current   *Input offset current* is defined as the difference between inverting and noninverting input bias current when the output voltage is held at zero.

Input signal voltage range   The *input signal voltage range* is the range of permissable input voltages as measured in the common-mode configuration; i.e., the maximum allowable value of V3 in Fig. 11-3.

Input impedance   The *input impedance* is the resistance between the inverting and noninverting inputs (Z_{in} in Fig. 11-4). This value is typically very high: 1 mΩ in low-cost bipolar operational amplifiers and over 10^{12} Ω in premium BiMOS devices.

Output impedance   The *output impedance* refers to the "resistance looking back" (to borrow a concept from Thevinin's theorem) into the amplifier's output terminal and is usually modeled as a resistance in series with the output signal and ground (despite the lack of a ground terminal on the device). Typically the output impedance is less than 100 Ω.

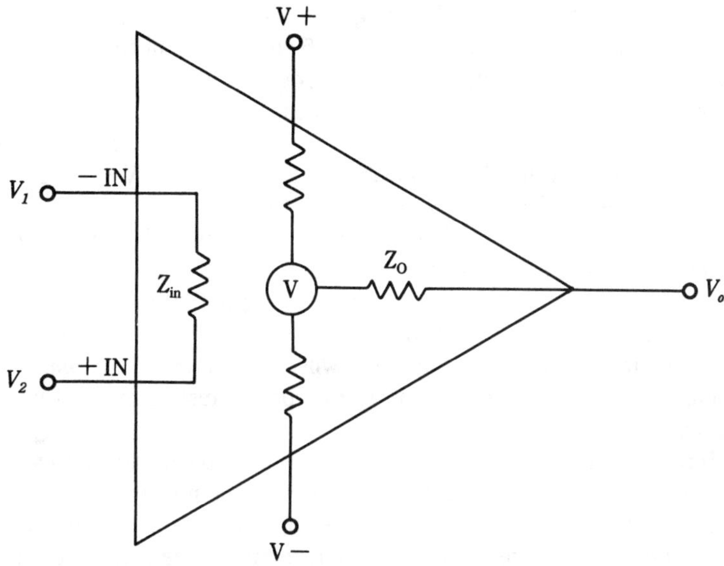

11-4   Internal equivalent circuit.

Output short-circuit current   The current that will flow in the output terminal when the output load resistance external to the amplifier is $0\ \Omega$ (i.e., a short to ground) is called the *output short-circuit current.*

Channel separation   This parameter is used on multiple operational amplifier integrated circuits; i.e., two or more operational amplifiers sharing the same package with common power supply terminals. The separation specification tells us something of the isolation between the op amps inside the same package and is measured in decibels. The 747 dual operational amplifier, for example, offers 120 dB of channel separation. From this specification we may imply that a 1 microvolt (μV) change will occur in the output of one of the amplifiers when the other amplifier output changes by 1 volt (1 V/1μV = 120 dB).

Minimum and maximum parameter ratings

Operational amplifiers, like all electronic components, are subject to certain maximum ratings. If these ratings are exceeded, the user can expect premature — often immediate — failure, or unpredictable operation. The following ratings are the most commonly used.

Maximum supply voltage   The *maximum supply voltage* is the maximum that can be applied to the operational amplifier without damaging the device. The operational amplifier uses V+ and V− dc power supplies that are typically ± 15 Vdc, although some exist with much higher maximum potentials. The maximum rating for either V− or V+ often depends upon the value of the other.

Maximum differential supply voltage   This potential is the algebraic sum of V− and V+, namely $[(V+) - (V-)]$. It is often the case that this rating is not the same as the summation of the maximum supply voltage ratings. For example, one 741 operational amplifier specification sheet lists V− and V+ at 15 volts each, but the maximum differential supply voltage is only 28 volts. Thus, when both V− and V+ are at maximum (i.e., 15 Vdc each), the actual differential supply voltage is [(+ 15 V) − (− 15 V)] = 30, which is 2 volts over the maximum rating. Therefore, when either V− or V+ is at maximum value, the other must be proportionally lower. For example, when V+ is + 15 V, the maximum allowable value of V− is [28 V − 15 V] = 13 V.

Power dissipation   This rating is the maximum power dissipation of the operational amplifier in the normal ambient temperature range (80°C in commercial devices, and 125°C in military-grade devices). A typical rating is 500 mW (0.5 W).

Maximum power consumption   The maximum power required, usually under output short circuit conditions, that the device will survive is called the *maximum power consumption*. This rating includes both internal power dissipation and device power requirements.

Maximum input voltage   This potential is the maximum that can be applied simultaneously to both inputs. Thus, it is also the maximum common-mode voltage. In most bipolar operational amplifiers, the maximum input voltage is equal to the power supply voltage, or nearly so. There is also a maximum input voltage that can be applied to either input when the other input is grounded.

Differential input voltage   The *differential input voltage* rating is the maximum differential-mode voltage that can be applied across the inverting (−) and noninverting (+) inputs.

Maximum operating temperature   The maximum operating temperature is the highest ambient temperature at which the device will operate according to specifications with reasonable reliability. The usual rating for commercial devices is 70° or 80°C, while military components must operate to 125°C.

Minimum operating temperature   There is a minimum operating temperature; i.e., the lowest temperature at which the device operates within specifications. Commercial devices operate down to either 0° or − 10°C, while military components operate down to − 55°C.

Output short-circuit duration   The length of time the operational amplifier will safely sustain a short circuit of the output terminal is called the *output short-circuit duration*. Many modern operational amplifiers are rated for indefinite output short-circuit duration.

Maximum output voltage   The output potential of the operational is related to the dc power supply voltages. Most operational amplifiers have several bipolar pn junctions between the output terminal and either V− or V+ terminals, and the voltage drop across these junctions reduces the maximum output voltage. For example, if there are three pn junctions between the output and power supply terminals, then the maximum output voltage is [(V+) − (3 × 0.7)], or [(V+) − 2.1] volts. If the maximum V+ voltage permitted is 15 volts, then the maximum allowable output voltage is [(15 V) − (2.1 V)], or 12.9 volts.

It is not always true that the maximum negative output voltage is equal to the maximum positive output voltage. A related rating is the maximum output voltage swing, which is the absolute value of the voltage swing from maximum negative to maximum positive.

Inverting follower circuits

Figure 11-5 shows the inverting follower circuit. In this circuit, the noninverting input is grounded, so we must treat the inverting input as if it were also grounded (recall ideal property number 6). This fact gives rise to a somewhat confusing concept: *virtual ground.* The inverting input is not actually grounded, but since it is at zero potential because the other input is grounded, we say that it is "virtually grounded." The concept is simple, only the semantics are confounding.

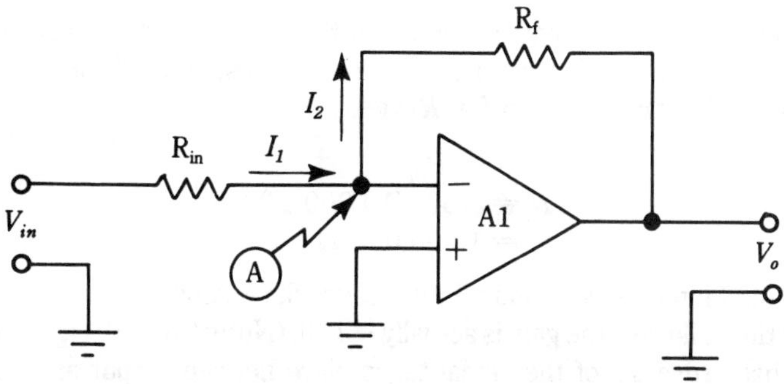

11-5   Inverting follower circuit.

Let's consider the currents appearing in node "A" of Fig. 11-5. We know from property no. 3 that I3, the input bias current, is zero. We also know from Kirchoff's Current Law (KCL) that all currents into and out of a junction algebraically sum to zero. Thus,

$$I_1 = -I_2 \tag{11-1}$$

We also know from Ohm's law that:

$$I_1 = \frac{V_{in}}{V_o} \tag{11-2}$$

and,

$$I_2 = \frac{V_o}{R_{in}} \tag{11-3}$$

Thus, when we substitute these two equations into KCL:

$$\frac{V_{in}}{R_{in}} = \frac{-V_o}{R_f} \tag{11-4}$$

We know that a voltage amplifier's transfer function is V_o/V_{in}, so solving this equation for the transfer function yields:

$$\frac{V_o}{V_{in}} = \frac{-R_f}{R_f} \tag{11-5}$$

Thus, the voltage gain A_v of the inverting follower is given by the ratio of two resistors:

$$A_v = \frac{-R_f}{R_{in}} \tag{11-6}$$

We may, then, design the inverting amplifier simply by manipulating the values of these two resistors (see Fig. 11-5).

There is sometimes a constraint on the minimum allowable value of R_{in}. Point "A" is essentially grounded, so the input impedance of this circuit is simply the resistance of R_{in}. There is a rule of thumb in voltage amplifier design that says the input impedance of a stage must be five times (most prefer ten times) the source impedance of the signal source.

Experiment 11-1

Design and build a gain-of-100 inverting amplifier that has an input impedance of 10kΩ or more. Because the input impedance must be 10kΩ, R_{in} must be 10kΩ or more. Let's set it at 10kΩ. Solving the gain equation for R_f yields:

$$R_f = A_v R_{in}$$
$$R_f = (100)\ (10,000\ \Omega)$$
$$= 1,000,000\ \Omega$$

Thus, a 10kΩ input resistor and a 1 mΩ feedback resistor yields a gain of 100. Since this is an inverting follower, the gain is actually −100. (Note: the "−" sign indicates that a 180-degree phase reversal of the signal takes place between input and output.)

Now that you've designed the circuit, build it and test it to compare the results with the theory.

Would you like to test the "virtual ground" idea? Set the signal generator output to 0. Use a dc voltmeter to measure the potential at the inverting and noninverting input pins of the op amp IC.

Noninverting followers

The noninverting follower applies a signal to the noninverting input. There are two basic configurations for the noninverting follower:

1. Unity gain noninverting follower
2. Noninverting follower with gain

Unity gain noninverting follower

Figure 11-6 shows the unity gain noninverting follower. The output terminal is connected to the inverting input, producing 100 percent feedback. The output voltage is equal to the input voltage. So what use is a unity gain (i.e., gain of 1) voltage amplifier? There are three uses: *buffering, impedance transformation,* and *power amplification.*

Buffering means using an amplifier to isolate a circuit from its load. Some transducers and most oscillators or astable multivibrators will change frequency if the load impedance changes, so a unity gain noninverting follower will help "buffer" the circuit.

Impedance transformation occurs because the input impedance is terribly high, while the output impedance is very low. We can use this circuit, for example, when acquiring a signal from biological or chemical sensor sources where the natural impedance is extremely high. A pH electrode, for example, has source impedances ranging from 10 mΩ to 100 mΩ.

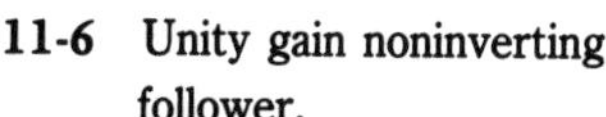

11-6 Unity gain noninverting follower.

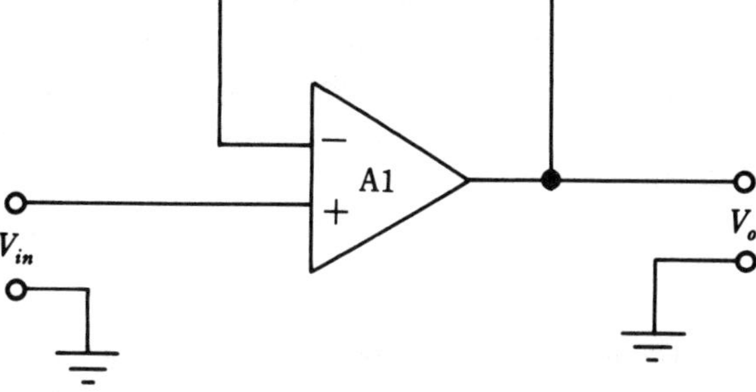

The power amplification is small, but is illustrated by the fact that the voltage remains constant ($V_o = V_{in}$), but the impedances are unequal. Obviously, since power is defined by $P = V^2/R$, reducing R while keeping V the same results in higher power (P).

Experiment 11-2

1. Build the circuit of Fig. 11-7.

2. Using either an oscilloscope of ac voltmeter, measure the input and output signal voltages.

3. Compare the input and output voltages by taking the ratio V_o/V_{in}.

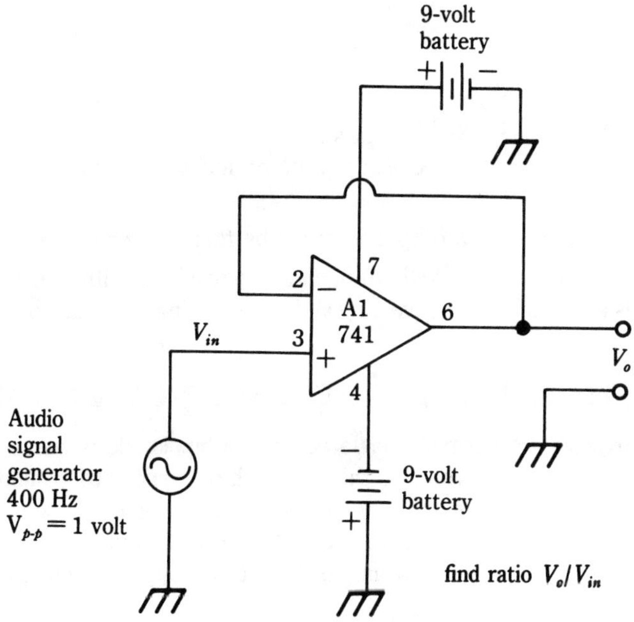

11-7 Circuit for experiment 2.

Noninverting follower with gain

The noninverting follower with gain circuit of Fig. 11-8 retains the properties of the unity gain circuit, but produces voltage gain as well. Keeping in mind that the inverting input (point "A") is at V_{in}, analysis similar to the previously used method produces a voltage gain of:

$$A_v = \frac{R_f}{R_{in}} + 1 \qquad\qquad (11\text{-}7)$$

The noninverting follower circuits are used wherever extremely high input impedance is needed or where no phase reversal can be tolerated.

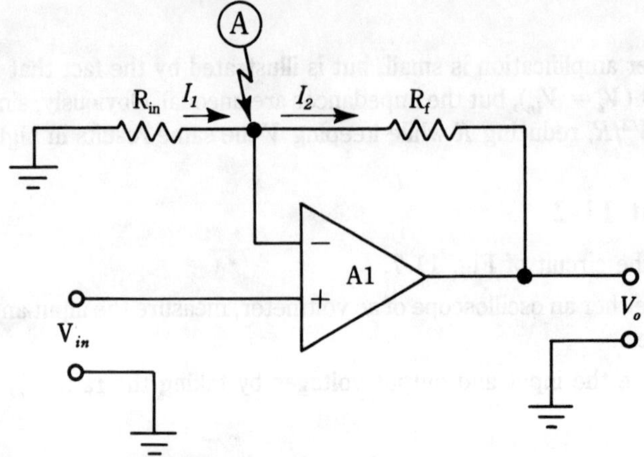

11-8   Noninverting follower with gain.

Experiment 11-3

1. Build the circuit of Fig. 11-9.

2. Using either an oscilloscope of ac voltmeter, measure the input and output signal voltages.

3. Compare the input and output voltages by taking the ratio V_o/V_{in}.

4. Set the signal generator level to 0, and then use a dc voltmeter to measure the voltages appearing at the inverting and noninverting input pins of the op amp IC.

Operation from a single dc power supply

Operational amplifiers are normally powered from a bipolar dc power supply. Such a power supply has V+ and V− voltages that are each referenced to ground or common. This system essentially requires two semi-independent dc supplies. In some cases, either ultimate use or other design constraints force the use of a single, monopolar dc power supply. In this section we will discuss simple methods for operating the amplifier from a single dc power supply.

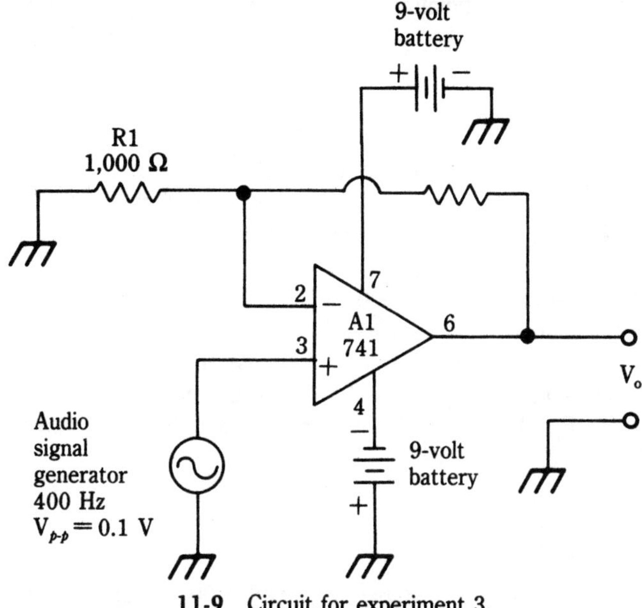

11-9   Circuit for experiment 3.

Some schemes exist for creating a split power supply from a monopolar supply in order to mimic bipolar power supply operation. One such scheme connects two zener diodes in series across the single supply, along with the necessary current-limiting resistors. The junction between the two zener diodes becomes the signal common. A limitation of this method is that the dc supply cannot be chassis-referenced.

Another scheme is to use the regular monopolar dc power supply for V+ and then use a dc-to-dc converter circuit for V−. Such a circuit is little more than an ac oscillator in the 20 to 500 KHz range, with its output signal rectified and filtered to produce the V− voltage.

Figure 11-10 shows the method for biasing the operational amplifier inputs to permit single-supply operation. This technique is based on the simple resistor voltage divider cir-

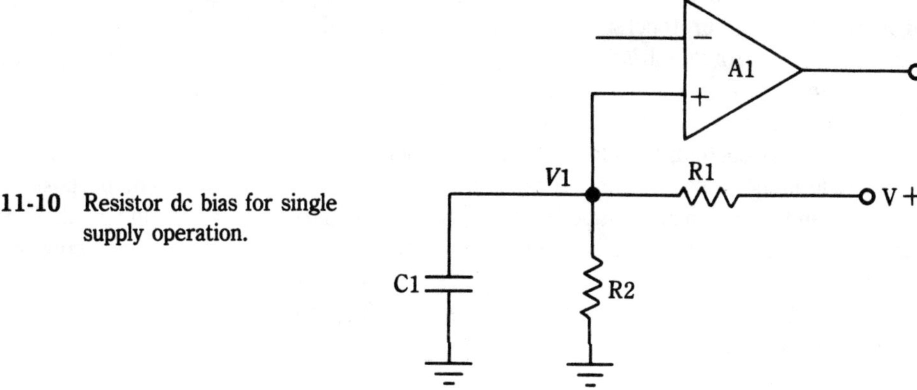

11-10   Resistor dc bias for single supply operation.

cuit of Fig. 11-10. The output voltage (V_1) is given by the standard voltage divider equation:

$$V_1 = \frac{R_2\,(V+)}{R_1 + R_2} \tag{11-8}$$

In most cases, the value of V_1 will be ½ V+, so that the operational amplifier has a quiescent output point that is midway between extremes. This bias level is achieved by making $R_1 = R_2$. The value of R_1 and R_2 is usually selected such that it falls between 1 kΩ and 100 kΩ.

The capacitor shunting resistor R2 (i.e., C1) is used to decouple ac variations. The value of capacitance is selected for a reactance value of one-tenth R_2 (i.e., $R_2/10$) at the lowest frequency of operation. For example, suppose $R_2 = 10$kΩ, and the lowest frequency of operation is 10 Hz. If R_2 is 10kΩ, then the capacitive reactance of the shunt capacitor should be: $R_2/10 = (10$k$\Omega)/10 = 1$kΩ. Solving the usual capacitive reactance equation for C_1 give us:

$$C1_{\mu F} = \frac{10^6}{2\,\pi\,FX_c} \tag{11-9}$$

$$= \frac{1{,}000{,}000}{(2)\,(3.14)\,(10\ \text{Hz})\,(1{,}000\ \text{ohms})}$$

$$= \frac{1{,}000{,}000}{62{,}800}$$

$$= 15.9\ \mu F$$

The value 15.9 μF is nonstandard, so a 20 or 22 μF unit normally would be selected as a practical matter.

Figure 11-11 shows the method for biasing an operational amplifier in the inverting follower configuration. In the bipolar supply version of this circuit (Fig. 11-5), the noninverting input is grounded (i.e., set to zero volts). In single-supply operation, however, we apply bias voltage V1 to the noninverting input. This voltage (V1) will also appear on the inverting input (hence the need for dc blocking capacitor C2). The output terminal will be biased up according to the value of V1, so might require a dc blocking capacitor also (shown in Fig. 11-12) if such a voltage adversely affects the following stage.

The value of capacitor C2 is selected to have a low impedance at the lowest frequency of operation, using a protocol similar to that discussed for the voltage divider shunt capacitor. A general rule of thumb is to regard $R_{in}C1$ as a high-pass filter with a cut-off frequency, F_c, equal to $1/(2\pi R_{in}C_2)$. The object is to select a value of C2, given a value for R_{in}, that results in a value of F_c lower than the lowest operating frequency.

The circuit configuration for noninverting follower circuits is shown in Fig. 11-12. This circuit is the same as for inverting followers except for resistor R3. The purpose of R3 is to maintain a high input impedance to signals applied to the noninverting input. The minimum value of R3 is at least ten times the output resistance of the driving stage. In practical cases, however, the source impedance is usually low enough that it is possible to set R3 to 100 or 1,000 times the source impedance. Typical values range from 10kΩ to 1 mΩ, with 100kΩ predominating.

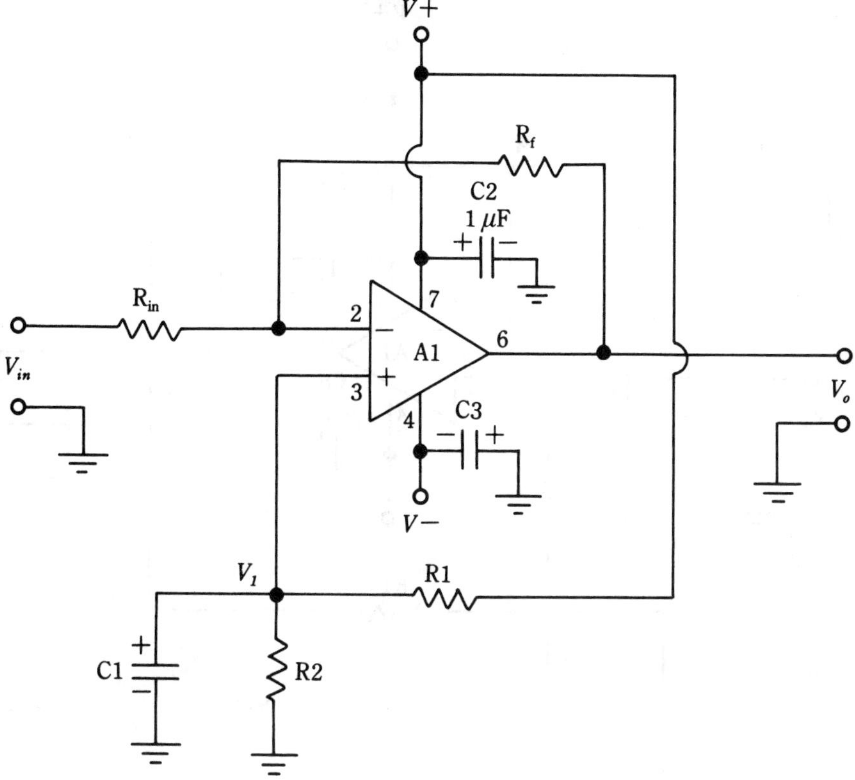

11-11   Inverting follower operated from a single dc supply.

The value of C1 is selected such that the cut-off frequency of the filter formed by C1R3 is lower than the lowest operating frequency. The same equation applies here as previously.

Capacitor C4 and resistor R4 are used when the (V+)/2 bias on the output terminal will adversely affect a following stage or instrument. Again, the "lowest frequency of operation" rule is invoked when setting the value of C1, with the resistance being the input resistance of the stage following.

Experiment 11-4

1. Build the circuit of Fig. 11-10, using a 741 operational amplifier (pinouts shown in preceding experiments). Use a value of resistance between 1 kΩ and 20 kΩ for the voltage divider resistors. Use 9 Vdc batteries or ±12 Vdc electronic power supplies.

2. Measure the dc voltages on the noninverting input, the inverting input, and the output pins of the operational amplifier.

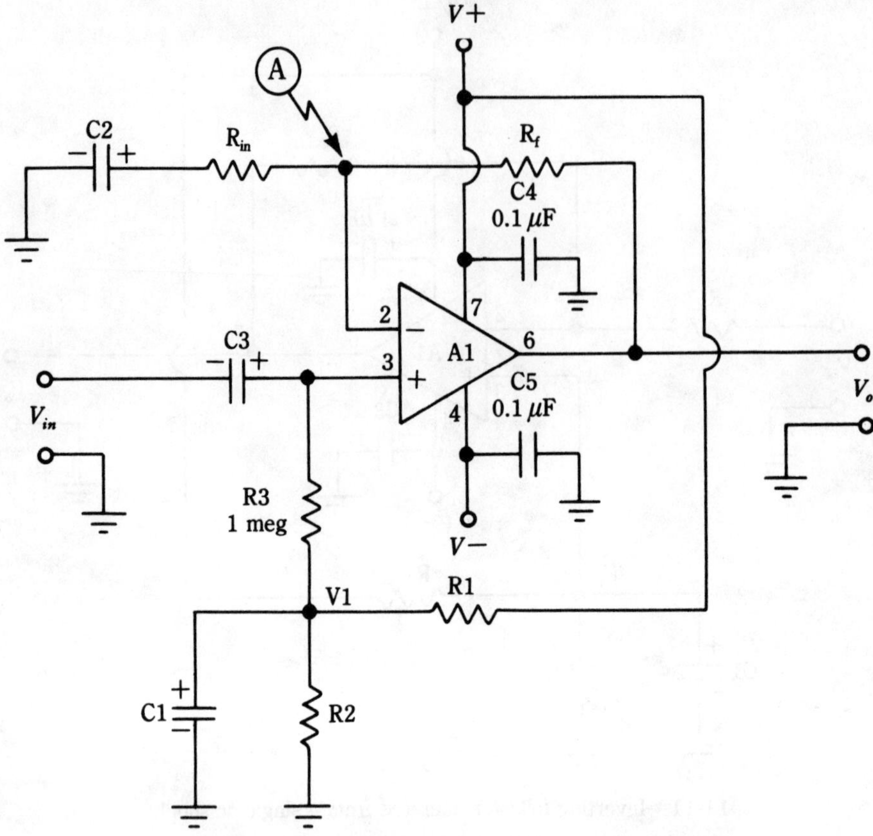

11-12   Noninverting follower operated from a single dc supply.

3. Replace the resistor voltage divider with a potentiometer. Connect one end of the potentiometer to ground, one end to V+, and the wiper to the resistor leading to the noninverting input of the op amp.

4. Repeat the measurements above.

Op amp problems and their solutions

Earlier in this chapter, we discussed the concept of the ideal operational amplifier. Such a hypothetical device is merely a learning tool and doesn't really exist. Although it makes our analysis easier, it cannot be purchased and used in practical circuits. All real operational amplifiers depart somewhat from the ideal, and the quality of the device is sometimes measured by the degree of that departure. We find, for example, that open-loop gain is not really infinite, but rather it is a value from 20,000 to over 1 million. Similarly, real operational amplifiers don't have infinite bandwidth, and in fact some are heavily bandwidth-limited. The latter statement is especially true of "unconditionally stable" or "frequency compensated" operational amplifiers such as the 741 device.

While the stability is highly desirable, it is obtained at the expense of frequency response. In this section we will deal with some of the more common problems in real devices, and their solutions.

Offset cancellation

The ideal operational amplifier will produce a zero output voltage when the two inputs are at the same potential. However, real operational amplifiers often produce an offset potential at the output. These voltages are those that exist on the output terminal when it should be zero.

There are several sources. One is the input bias currents used to operate the transistors in the input stage of the operational amplifier. Typically, these currents are nonzero. When the current flows into or out of the inverting input, it will create a voltage drop equal to the product of the current and the parallel combination of R_{in} and R_f. How do we deal with this problem? The voltage drop produced by the input bias currents is amplified and appears at the output as an offset potential.

Figure 11-13 shows the use of a compensation resistor, R_c. This resistor has a value equal to the parallel combination of the other two resistors. Since the same bias current flows in both inputs, this resistor will produce the same voltage drop at the noninverting input as appears at the inverting input. Since the inputs are differential, the net output voltage is zero.

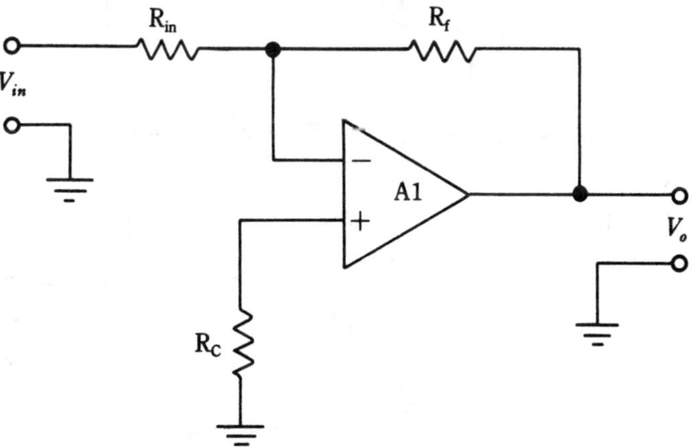

11-13   Use of a compensation resistor.

Figure 11-14 shows two methods for nulling output offsets, regardless of the source. Figure 11-14A shows the use of offset null terminals that are found on some operational amplifiers. A potentiometer is placed between the terminals, while the wiper is connected to the V— power supply. This potentiometer is adjusted to produce the null required. The input terminals are shorted together, and the pot is adjusted to produce 0 V output.

Figure 11-14B shows a circuit that can be used on any operational amplifier, inverting or noninverting, except the unity gain noninverting follower. A counter-current (I3) is injected into the summing junction (point "A") of a magnitude and polarity to cancel the

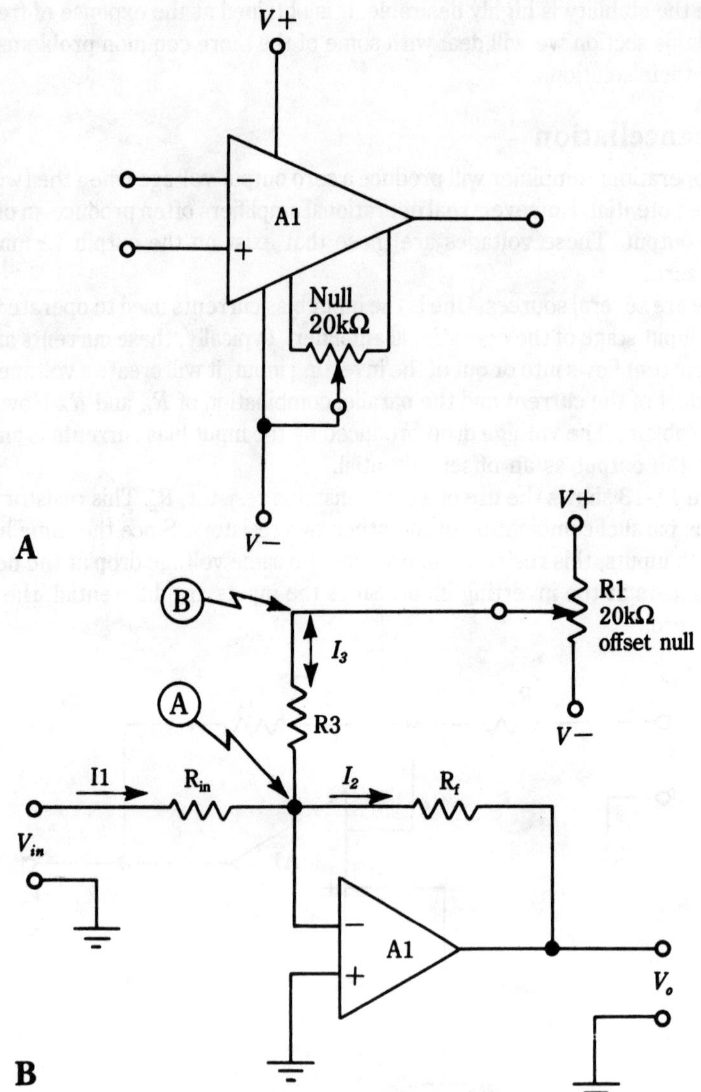

11-14   Offset compensation methods: A. Use of offset null terminals; B. Use of countercurrent at inverting input.

output offset voltage. The voltage at point "B" is set to produce a null offset at the output. The output voltage component due to this voltage is:

$$V_{o'} = \frac{-V_B R_f}{R_2} \tag{11-10}$$

If greater (i.e., finer) control over the output offset is required, then one of the two circuits of Fig. 11-15 can be used to replace the potentiometer of Fig. 11-14B.

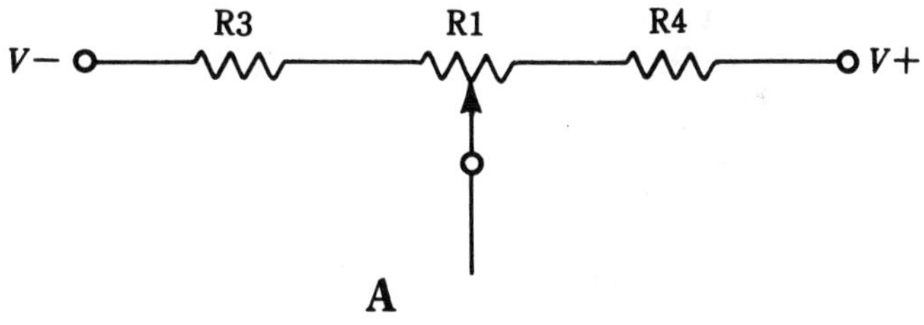

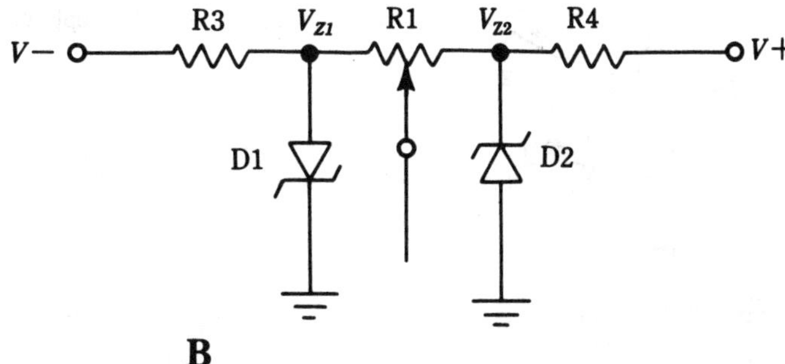

11-15   Countercurrent networks with better resolution: A. Resistive; B. Zener diode.

Power supply decoupling

Operational amplifiers, like all other active electronic components, are somewhat affected by variations in power-supply voltage and noise signals riding on the power-supply potentials. It is also possible that signal from one operational amplifier is coupled to others over the power-supply lines.

The cure for these problems, and certain stability problems which we will discuss in due course, is the decoupling scheme shown in Fig. 11-16. The V− and V+ power supply lines are each decoupled with two capacitors. C1 and C2 are 0.1 μF capacitors and are used for high-frequency signals; C3 and C4, on the other hand, are higher valued and are for low-frequency signals. The reason for using two capacitors is that electrolytic capacitors used for C3 and C4 are ineffective at high frequencies.

The decoupling capacitors should be mounted as close as possible to the body of the operational amplifier. This constraint is especially true when the operational amplifier is not frequency-compensated and has a high gain-bandwidth product.

Frequency stability

Operational amplifiers that are not internally frequency-compensated are susceptible to oscillations. Figure 11-17 shows a plot of the open-loop phase shift vs. frequency for a typical operational amplifier. From dc to a certain frequency there is essentially zero phase-shift error, but above that breakpoint the phase error increases rapidly. This change is a result of the internal resistances and capacitances of the amplifier acting as a phase-shift network. At some frequency, F?? , the phase-shift error reaches 180 degrees, which when added to the 180-degree inversion normal to inverting follower amplifiers adds up to the

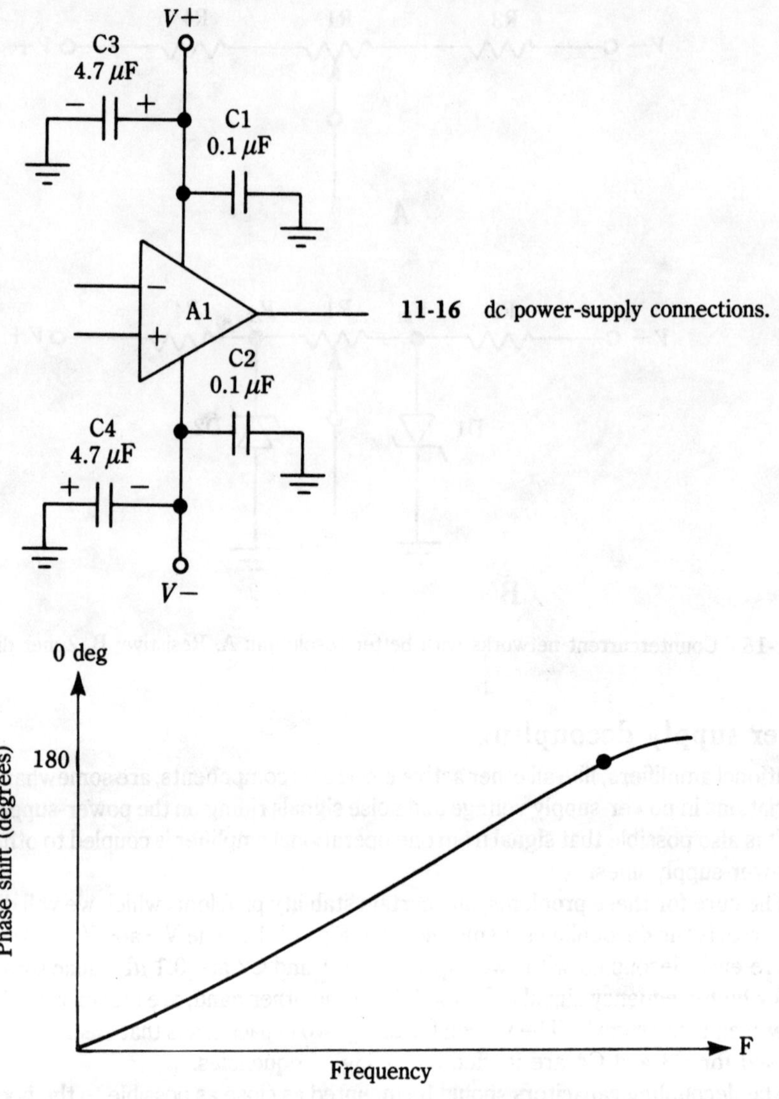

11-16 dc power-supply connections.

11-17 Phase shift vs. frequency plot.

360-degree phase shift that satisfies Barkhausen's criterion for oscillation. At this frequency the amplifier will become an oscillator.

The use of power-supply decoupling helps somewhat for this problem, and it is considered poor engineering practice to use an uncompensated operational amplifier without those decoupling capacitors. In other cases, you might need to use a variant of the methods shown in Fig. 11-18.

In Fig. 11-18A we see lead compensation. If the operational amplifier is equipped with compensation terminals (usually pins 1 and 8 on "standard" packages), then connect a small-value capacitor (20 to 100 pF) as shown. An alternate scheme is to connect the capacitor from a compensation terminal to the output terminal.

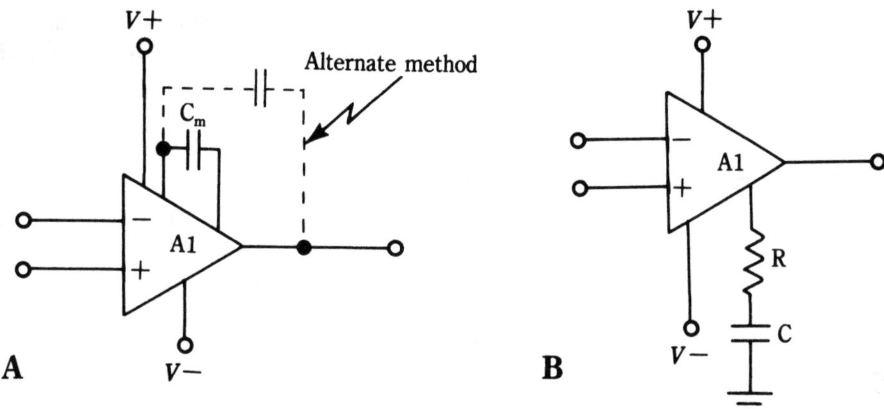

11-18   A. Lead compensation; B. RC lag compensation.

The recommended capacitance in manufacturer's specification sheets is for the unity gain noninverting follower configuration. For a gain follower, the capacitance is reduced by the feedback factor, β:

$$C = C_m\beta \qquad\qquad (11\text{-}11)$$

Where: C is the required capacitance
C_m is the recommended unity gain capacitance
β is the feedback factor $R_{in}/(R_{in} + R_f)$

Lag compensation is shown in Fig. 11-18B. In this case we connect either a single capacitor or a resistor-capacitor network from the compensation terminal to ground. A related method places the resistor-capacitor series network between the inverting and noninverting input terminals.

The object of these methods (see Fig. 11-19) is to reduce the high-frequency loop gain of the circuit to a point where the total loop gain is less than unity at the frequency

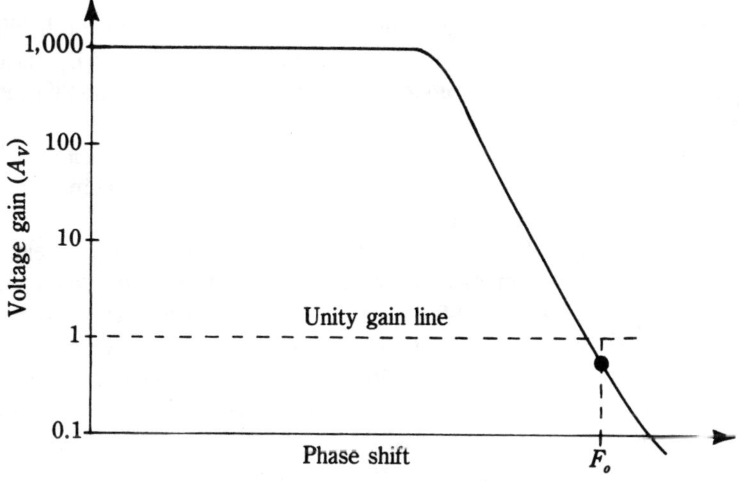

11-19   Gain-vs-phase shift.

where the 180-degree phase shift occurs (F_o). The amount of compensation required to accomplish this goal determines the maximum amount of feedback that can be used without violating the stability requirement.

dc differential amplifiers

A *differential amplifier* is one that produces an output voltage that is the product of the gain and the difference between the two input voltages. Figure 11-20 shows a simple dc differential amplifier that is based on a single operational amplifier. The gain is:

$$A_{vd} = \frac{R_3}{R_1} = \frac{R_4}{R_2} \tag{11-12}$$

Provided that $R_1 = R_2$ and $R_3 = R_4$.

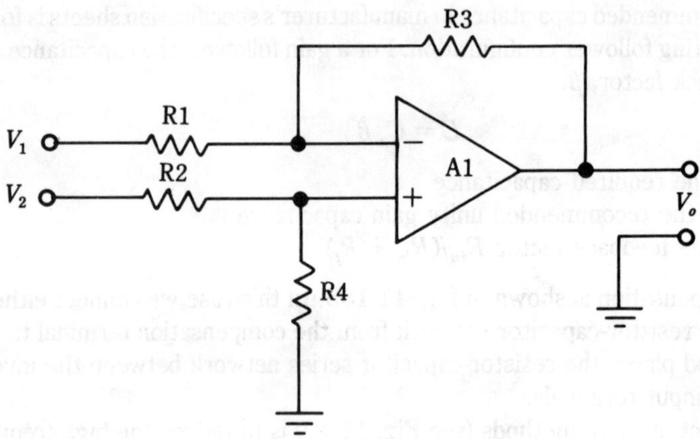

11-20   dc differential amplifier.

The output voltage from the dc differential amplifier is the product of the differential voltage gain and the difference between the two input voltages ($V_2 - V_1$). This difference potential is known as a *differential mode signal*, and the differential amplifier ideally responds only to such signals.

A common-mode signal is one that is applied to both inputs simultaneously. In the ideal differential amplifier, the common-mode gain is zero, so the common-mode signal causes no change in the output signal.

Instrumentation applications, including many in the data acquisition field, are ideal for differential amplifiers. Consider a case where a low-level signal must pass over wires in an environment that is intense with 60 Hz ac fields from the power lines. In that case, you normally would expect the signal to be obscured with 60 Hz noise. If you use a differential amplifier, however, you can make the design such that the signal voltage is a differential mode potential. The lines from the signal source can be made equal length and pass through the same environment. In that case, both lines would receive the same 60 Hz signal levels, in phase, and that noise signal is therefore common mode. The differential amplifier will reject the interfering signal.

Good common-mode rejection is possible in the circuit of Fig. 11-20, but only if the resistors are matched (R1 = R2 and R3 = R4 to a close tolerance). An improved circuit is shown in Fig. 11-21. Here resistor R4 is replaced with a series combination of a fixed resistor and a potentiometer (or just a potentiometer, in some cases). Potentiometer R5 becomes a common mode adjust control. It is adjusted with V1 = V2 (which means shorted together and applied to a single signal source) for minimum output voltage.

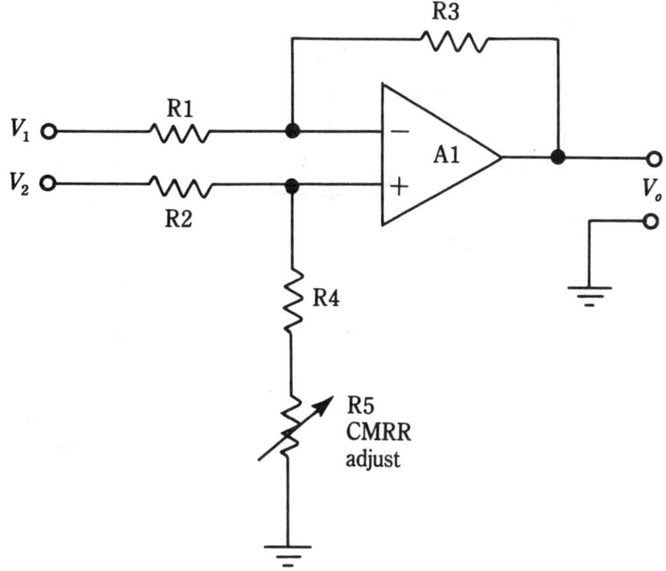

11-21   CMRR ADJUST control.

The basic differential amplifier presented in Figs. 11-20 and 11-21 suffers from the same defects as the inverting follower circuit. If you want to make a differential amplifier with a very high input impedance, you must use the three-device instrumentation amplifier circuit of Fig. 11-22.

Where would such a circuit be used? Wherever there is a requirement for extremely high impedances, of course. Such applications might include biopotential amplifiers (EEG, ECG, etc.), chemical electrodes (pH, oxygen, CO_2, etc.), and certain physics instruments.

The circuit in Fig. 11-21 consists of three operational amplifiers. It is preferred if amplifiers A1 and A2 be in the same IC package, to improve thermal stability, but they can be separate if needed. (In a moment we will examine the IC instrumentation amplifier.)

Operational amplifier A3 is used in the same dc differential amplifier circuit as seen earlier, but its inputs are driven from a balanced circuit consisting of two noninverting followers with gain (A1 and A2) that share a common "input" resistor, R1. The overall voltage gain of this amplifier is given by:

$$A_v = \left(\frac{2\,R_2}{R_1} + 1 \right) \left(\frac{R_6}{R_5} \right) \qquad (11\text{-}13)$$

Assuming that $R_2 = R_3$, $R_4 = R_5$, and $R_6 = R_7$.

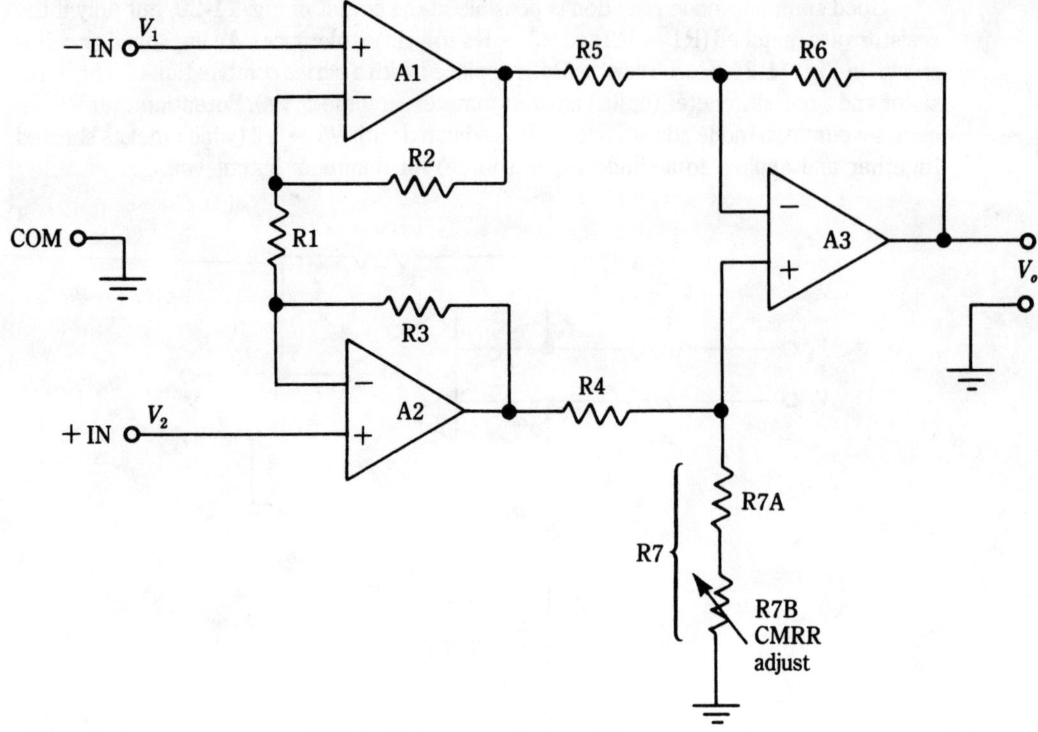

11-22   Instrumentation amplifier.

In some cases, you will want to make a variable gain control for the differential amplifier. This neat job is accomplished by making R1 adjustable, although you must be cautious in this regard because the term *R1* appears in the denominator of the gain equation. If you place a potentiometer in this slot, then it will attain zero toward one end of its range, and that makes the gain go very high — perhaps higher than the circuit can tolerate. In those cases, it is prudent to place a fixed resistor in series with the potentiometer, which has a value that produces the highest gain desired.

Resistor R7 is used to adjust the common-mode rejection ratio (CMRR) of the differential amplifier. In an ideal differential amplifier, common-mode signals will null each other to zero, so no output occurs. In a practical amplifier, however, there will be some small differential gain error, so some signal will appear on the output in response to common-mode signals.

One source of differential-gain errors is the normal tolerance differences of the resistors. Minimizing these errors (e.g., using 1 percent or better, low-temperature coefficient resistors) helps. If you make R7 adjustable, then it is possible to adjust out the error.

The CMRR adjustment is made by shorting the two inputs together and then connecting them to a signal source (1 V at 100 Hz, for example). Monitor the output of the amplifier on an oscilloscope, and adjust R7 for minimum output. You will have to continuously adjust the gain of the oscilloscope to more sensitive levels in order to "fine-tune" the CMRR adjustment.

Instrumentation amplifiers

The three-device instrumentation amplifier (IA) has been made available in IC form. There are several advantages to making the instrumentation amplifier in integrated form (let's call such a device an *ICIA*). For one thing, drift is better controlled because all three operational amplifiers and all of the resistors except R1 share a common silicon substrate, hence a common thermal environment. All stages drift in the same manner and often cancel each other's effects. Drift is not zero, but with proper design the ICIA drift is reduced significantly compared with all but the best discrete circuit designs.

A second advantage is that of component density. An IA made from discrete components will require at least 2 to 3 square inches of printed circuit board space. In ICIA form, however, the same amplifier might require less than 1 square inch!

Still another advantage for many users is cost. Although the unit cost of the ICIA, especially in the higher grades, is often large compared with simple discrete op amps, the aggregate cost of the final product is often lower because the design is simpler, assembly easier, printed circuit board layout simpler, and general overall "hassle factor" for production people less. The advantages of the ICIA are sufficiently interesting that these chips, in all probability, will someday eclipse, if not replace, the conventional operational amplifier.

The size advantages of the ICIA are not always of primary importance. In some cases, however, size is of concern, or it can make an existing product better. In the case of a transducer or electrode, in a noisy environment especially, it becomes possible to install the amplifier on or in the transducer. This design allows you to send a larger signal down the line to the receiving instrument or data acquisition system.

In one example, a pressure transducer had a built-in X100 ICIA so that it could produce an output of 100 mV to 1 V, instead of 100 to 1,000 mV. Obviously, if you expect to pick up some 60 Hz line noise, you are better off sending a 1,000 mA signal than a 1 mA signal. In the section to follow we will examine several commercial ICIA devices.

Commercial ICIA devices

The typical ICIA device will contain a circuit similar to the one shown in Fig. 11-21. An exception is that resistor R1 will be external to the IC package. This external resistor is sometimes designated R_g, and the pins that connect it to the ICIA are labeled *gain set* or something similar.

Typical gain equations for the ICIA devices will be of the form:

$$A_v = \left(\frac{50 \text{ k}\Omega}{R_g} + 1\right) \tag{11-14}$$

which, of course, means that for circuits such as Fig. 11-21, resistors R2 and R3 are 25 kΩ each, and that $R_4 = R_5 = R_6 = R_7$.

As is common in too many texts, the form of the equation given is not what is actually needed. For most of us, the gain is known or otherwise determinable, and what we need to know is the resistor that will yield that required gain. Thus, I will help you avoid the minor algebra needed to convert Eq. 11-14 to find the resistance required:

$$R_g = \left(\frac{50 \text{ k}\Omega}{A_v - 1}\right) \tag{11-15}$$

Example

Select a resistor that will produce a gain of X100 in the commercial ICIA just described.

$$
\begin{aligned}
R_g &= \frac{50 \text{ k}\Omega}{A_v - 1} \\
&= \frac{50 \text{ k}\Omega}{100 - 1} \\
&= \frac{50 \text{ k}\Omega}{99} \\
&= 0.51 \text{ k}\Omega \\
&= 510 \ \Omega
\end{aligned}
$$

The ICIA may come in any one of several packages. There are several examples in 8- or 10-pin round metal cans (the original form of IC package!), while others are available in 8-, 14-, or 16-pin DIP packages. There are also several hybrid instrumentation amplifiers that are similar enough in function to the ICIA to be considered under the same rubric.

12
Other IC
linear amplifiers

THE OPERATIONAL AMPLIFIER HAS LONG BEEN RECOGNIZED AS THE PRIMARY linear integrated circuit amplifier on the market. The growth in the number of different op amps available testifies to that claim. There are also some other types of IC amplifiers, however, and they are also very important in the overall scheme of things electronics. Two are the operational transconductance amplifier (OTA) and the current difference amplifier (CDA), which are discussed in this chapter.

Operational transconductance IC amplifiers

The operational transconductance amplifier is based on a transfer function that relates an output current to an input voltage. In other words:

$$G_m = \frac{\Delta I_o}{\Delta V_{in}}$$

(12-1)

Where: G_m is the transconductance in siemens or microsiemens
I_o is the output current
V_{in} is the input voltage

The OTA-equivalent circuit is shown in Fig. 12-1. The differential input circuit is similar to the input circuit of the operational amplifier because each are differential voltage inputs ($-$IN and $+$IN). The input voltages are differential signal, V_d (i.e., $V_2 - V_1$), and common mode signal, V_{cm}. The output side of the amplifier, however, is a current source that produces an output current, I_o, which is proportional to the gain and the input

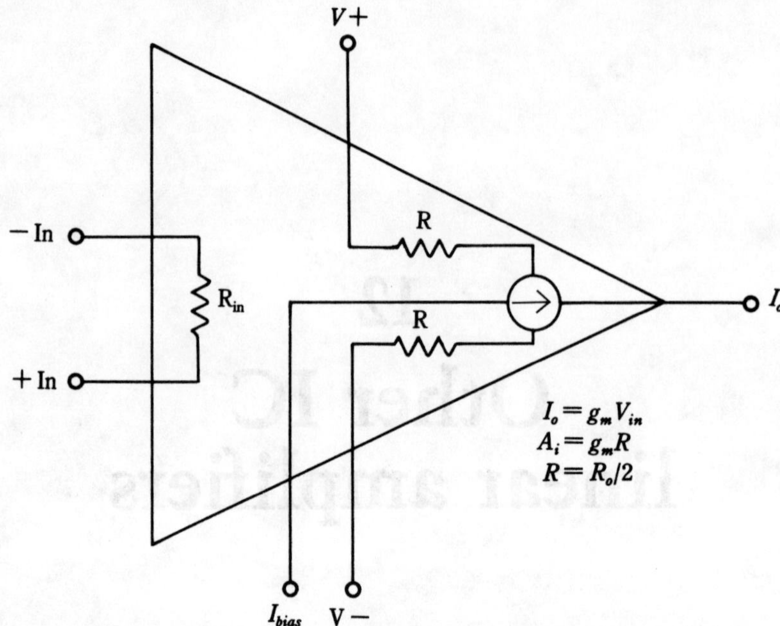

12-1   Operational transconductance amplifier (OTA) equivalent circuit.

voltage. The current gain (A_{gm}) of this circuit is a function of the transconductance (I_o/V_{in}) and the load resistance, R:

$$A_{gm} = G_m R \tag{12-2}$$

Where: A_{gm} is the gain
G_m is the transconductance (I_o/V_{in})
R is the load resistance (one-half the output resistance R_o).

Note: G_m and R must be expressed in equivalent reciprocal units. In other words, when G_m is in Siemens then R is in ohms. Likewise, millisiemens–milliohms and microsiemens–micro-ohms are also paired.

Perhaps the most common commercial versions of the operational transconductance amplifier are the RCA CA-3080, CA-3080A, and CA-3060 devices. The CA-3080 devices are available in the eight-pin metal IC package using the pinouts shown in Fig. 12-2. The CA-3080 devices will operate over dc power-supply voltages from ±2 to ±15 V, with adjustable power consumption of 10 μW to 30 mW. The gain is 0 to the product $G_m R$. The input voltage spread is ±5 V. The bias current can be set to as high as 2 mA.

Note that the pinouts for the CA-3080 device are "industry standard" operational amplifier pinouts, except for the bias current applied to pin number 5:

- V− on pin 4
- V+ on pin 7
- Inverting input (−IN) on pin 2
- Noninverting input (+IN) on pin 3
- Output on pin 6

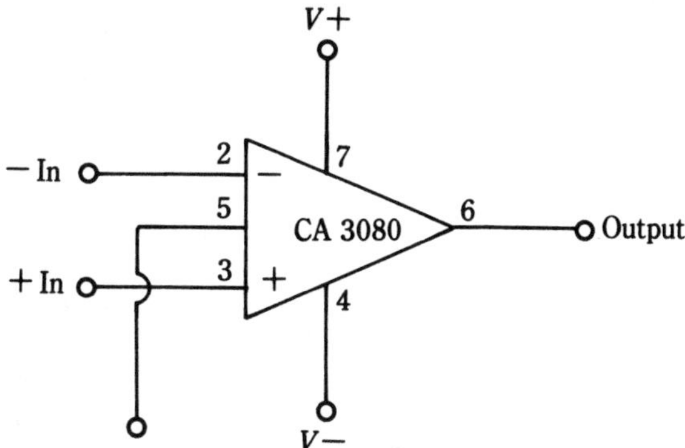

12-2 CA-3080 circuit symbol.

The operating parameters of the operational transconductance amplifier are set by the bias current (I_{bias}). For example, on the CA-3080 device the transconductance is 19.2 times higher than the bias current:

$$G_m = 19.2\ I_{bias} \tag{12-3}$$

Where: G_m is in millisiemens
I_{bias} is in milliamperes

In many actual design cases you will know the required value of G_m from knowledge of I_o/V_{in}, so the G_m required can be set by adjusting the bias current. In those cases, the I_{bias} is found by rewriting the above expression:

$$I_{bias} = \frac{G_m}{19.2} \tag{12-4}$$

The CA-3080 output resistance of the device is also a function of the bias current:

$$R_o = \frac{7.5}{I_{bias}} \tag{12-5}$$

Where: R_o is the output resistance in megohms
I_{bias} is the bias current in milliamperes

Voltage amplifier from the IC OTA

The OTA is a current-output device, but it can be used as a voltage amplifier when one of the circuit strategies of Fig. 12-3 are used. The simplest method is the resistor load shown in Fig. 12-3A. Because the output of the OTA is a current (I_o), you can pass this current through a resistor (R1) to create a voltage drop. The value of the voltage drop (and the output voltage, V_o) is found from Ohm's law:

$$V_o = I_o\ R_1 \tag{12-6}$$

A problem with this circuit is that the source impedance is very high, being equal to the value of R1. In the example shown in Fig. 12-3A, the output impedance is 10 kΩ. This problem can be overcome by adding a unity gain noninverting operational amplifier such

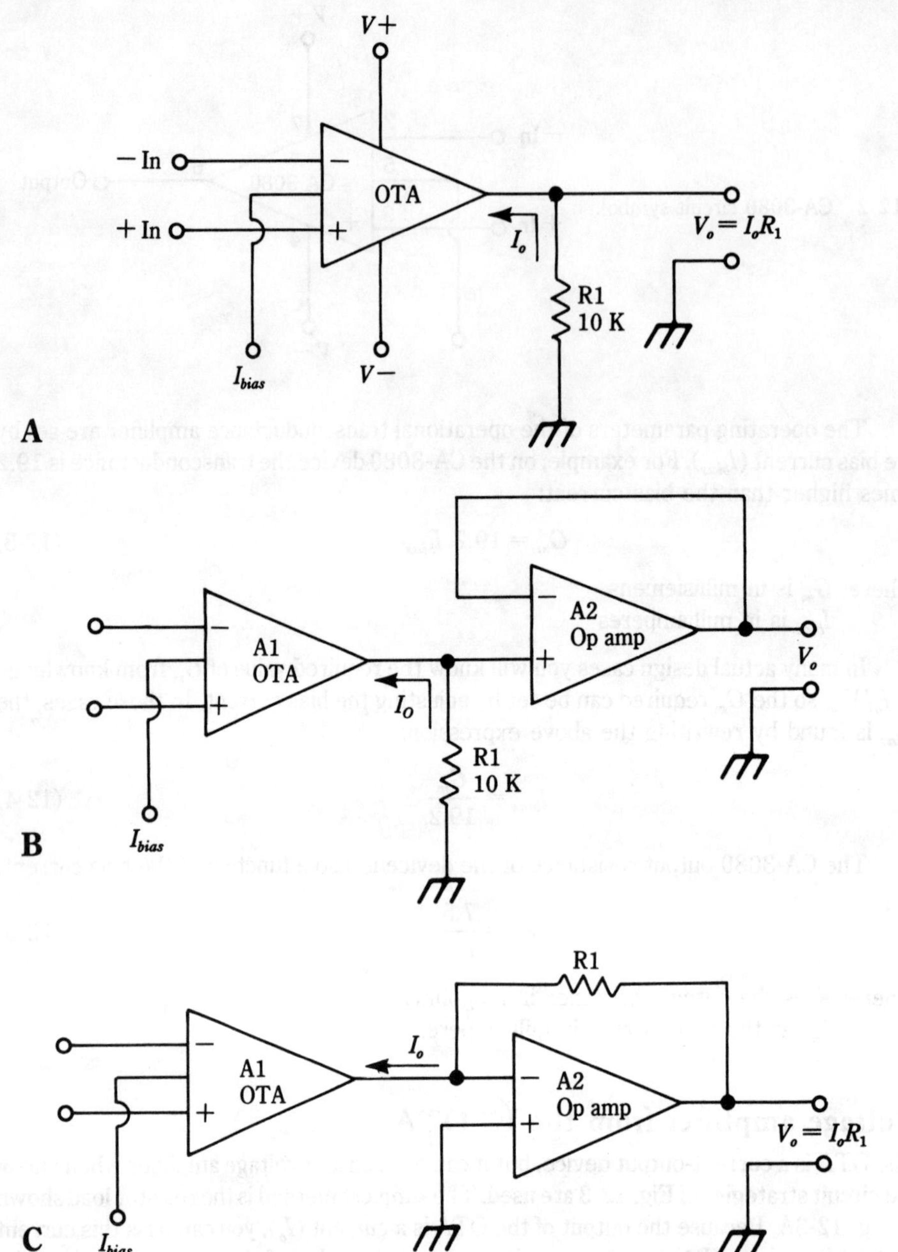

12-3 A. OTA circuit connections; B. Circuit for producing voltage output; C. Alternate method.

as A2 in Fig. 12-3B. The output voltage in this case is the same as for the nonamplified version: $I_o R_1$, although the output impedance is very low.

Another form of low-impedance output circuit is shown in Fig. 12-3C. This form uses the inverting follower configuration of the operational amplifier (A2). The output voltage is the product of the OTA output current (I_o) and the op amp feedback resistor (R_1). See Eq. 12-6.

In Fig. 12-3B the output impedance is equal to the operational amplifier output impedance, which is typically something less than 100 Ω.

OTA applications

An *analog multiplier* (Fig. 12-4) is a circuit that produces a voltage that is the product of two input voltages:

$$V_o = rV_x V_y \qquad (12\text{-}7)$$

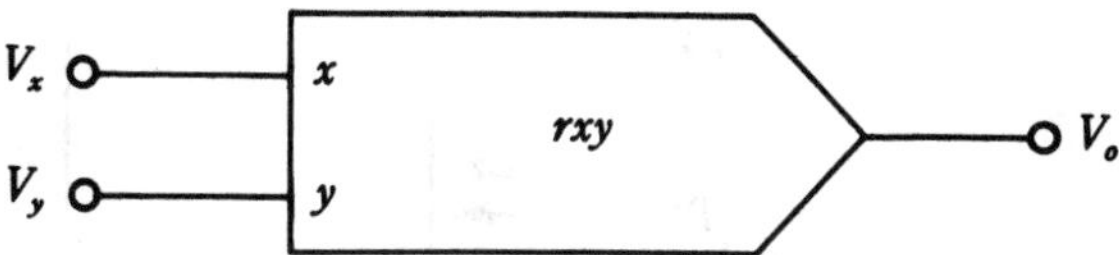

12-4 XY analog multiplier.

Where: V_o is the output voltage
V_x is the voltage applied to the X input
V_y is the voltage applied to the Y input
r is a proportionality constant

There are many applications for the multiplier circuit, even though some of them are now generally performed in a digital computer or processor. Immediately one can see instrumentation applications, even in this era of computers. Also possible are amplitude modulation and demodulation tasks for analog multipliers.

Operational transconductance amplifiers can be used to make two-quadrant and four-quadrant analog multipliers. Consider Fig. 12-5, which is an analog XY multiplier based on the CA-3060 quad OTA. Recall from Eq. 12-1 that:

$$G_m = \frac{I_o}{V_{in}} \qquad (12\text{-}8)$$

Therefore:

$$I_{o1} = -V_x G_{M1} \qquad (12\text{-}9)$$

and,

$$I_{o2} = +V_x G_{M2} \qquad (12\text{-}10)$$

Because the value of R_o for each OTA is very large compared with the load, we can simply sum the two output currents:

$$I_o = I_{o2} + I_{o1} \qquad (12\text{-}11)$$

by $V_o = I_o R_L$:

$$V_o = (I_{o2} + I_{o1})\,R_L \qquad (12\text{-}12)$$

$$V_o = [(+V_x)(G_{M2}) + (-V_x)(G_{M1})]R_L \qquad (12\text{-}13)$$

$$V_o = (G_{M2} - G_{M1})V_x R_L \qquad (12\text{-}14)$$

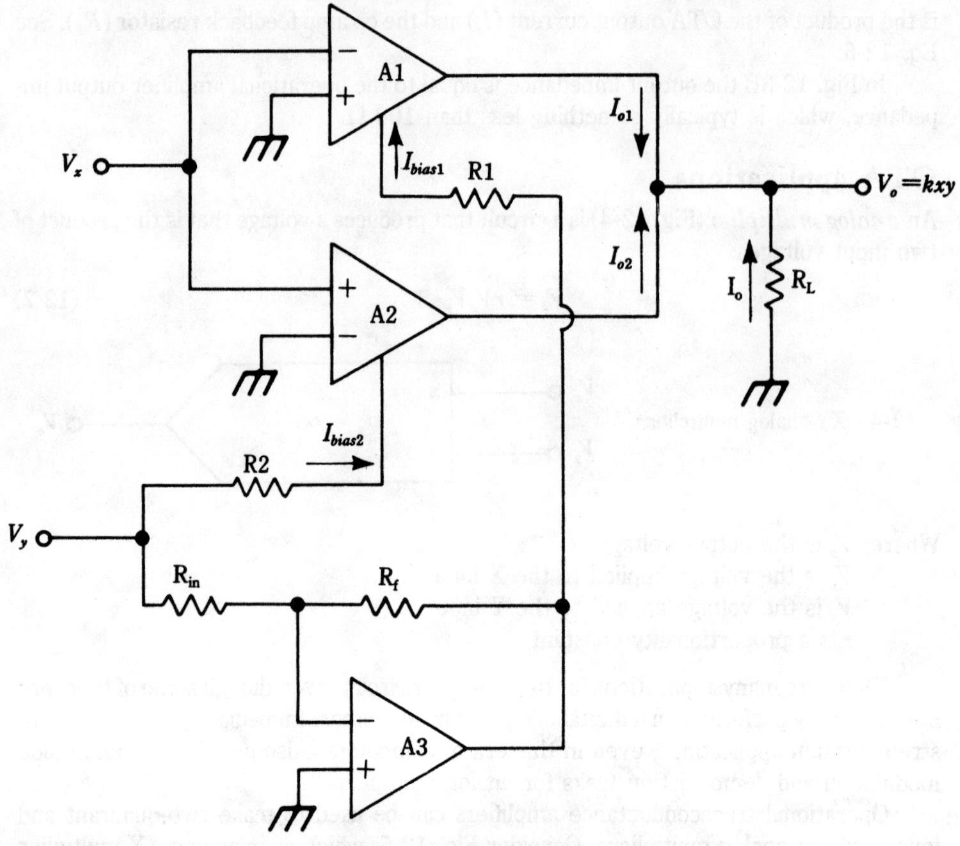

12-5   OTA XY analog multiplier circuit.

Recall that $G_m = kI_{bias}$. For amplifier A2 we know that:

$$I_{bias} = \frac{(V-) + (V_x)}{R_1} \tag{12-15}$$

hence,

$$G_{M2} = K[(V-) + (V_x)] \tag{12-16}$$

Using a similar line of reasoning, we arrive at:

$$G_{M1} = K[(V-) + (V_y)] \tag{12-17}$$

Combining Eqs. 12-15, 12-16, and 12-17 yields:

$$V_o = V_x KR_L[((V-) + V_x] - [(V-) - (V_y))] \tag{12-18}$$

or, after simplifying terms:

$$V_o = 2KR_L V_x V_y \tag{12-19}$$

Equation 12-19 is the transfer equation for Fig. 12-5. It has the same form as Eq. 12-7, in which $r = 2KR_L$.

Project 12-1   20 dB gain OTA voltage amplifier

Figure 12-6 shows the circuit for a gain-of-ten voltage amplifier that is based on the CA-3080 operational transconductance amplifier. Gain is set by the combination of the resistor feedback network (R_4/R_1) and the bias current created by connecting resistor R3 from the bias terminal of the OTA to the $+12$ Vdc power supply. The basic configuration of this amplifier is the inverting follower, and it uses an ac-coupled input circuit.

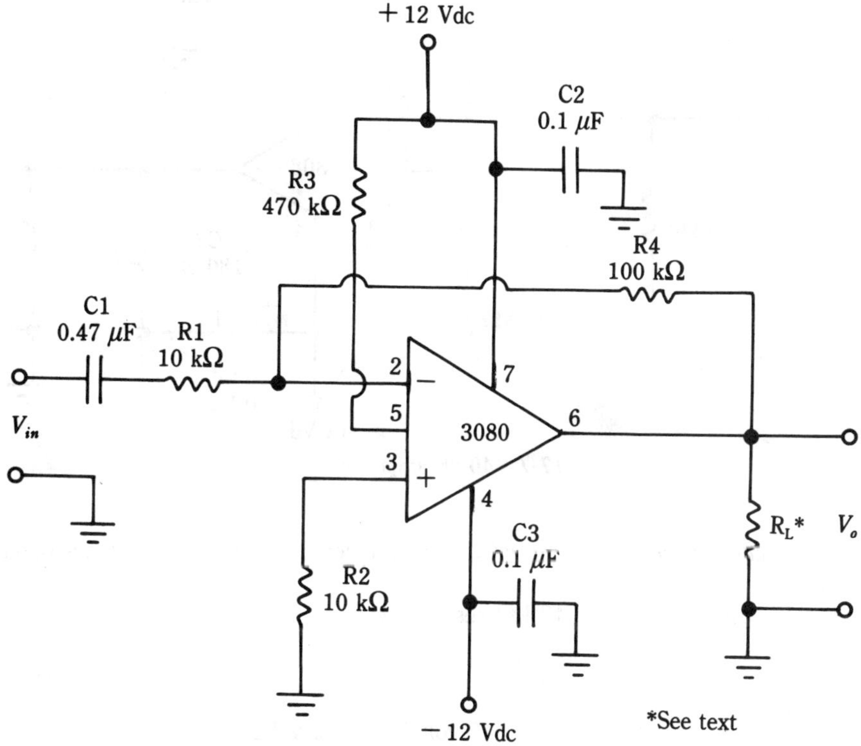

12-6   Gain-of-ten amplifier.

Output current of the OTA is converted to a voltage by the action of optional load resistor R_L and feedback resistor R4. The feedback resistor is seen effectively as a load resistor because the inverting input is seen as a virtual ground (the same concept as found in operational amplifier circuits). The maximum output voltage permitted is determined by the value of R_L. When R_L is infinite — i.e., when there is no resistor present — then the maximum output voltage is set by R4, and is on the order of 10 volts. The maximum output voltage and the gain are reduced by shunting R4 with a load resistor from pin no. 6 and ground. A minimum value of 10 kΩ reduces the output voltage to 0.5 volt.

Project 12-2   40 dB gain OTA voltage amplifier

A 40 dB gain amplifier has a voltage gain of 100. Unlike the previous project, this amplifier's gain is set entirely by the bias resistor and the output load resistor (see Fig. 12-7). This mode of operation is normal to operational transconductance amplifiers. Like the

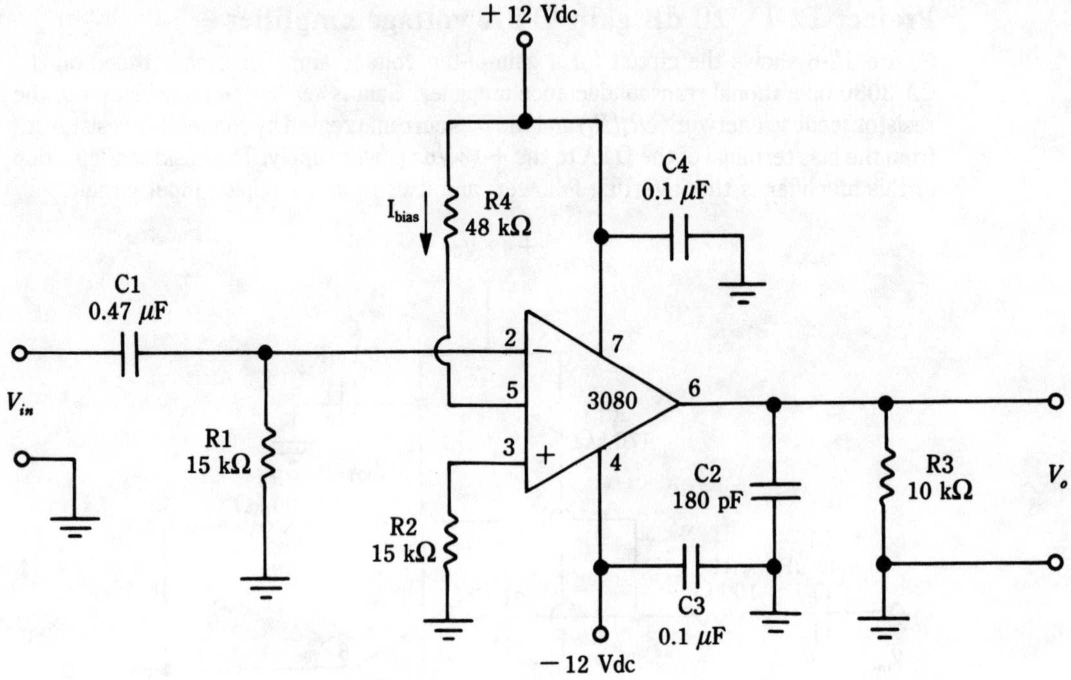

12-7   40 dB amplifier.

previous amplifier, however, this circuit is also an inverting follower with an ac-coupled input circuit.

The bias current is set by 48 kΩ resistor R4:

$$I_{bias} = \frac{(V+) - (V-)}{R4} \tag{12-20}$$

$$= \frac{(+12 \text{ volts}) - (-12 \text{ volts})}{48,000 \ \Omega}$$

$$= \frac{24}{48,000}$$

$$= 5 \times 10^{-4} \text{ amperes}$$

The transconductance is found from:

$$G_m = 19.2 \ I_{bias} \tag{12-21}$$

$$= (19.2)(5 \times 10^{-4} \text{ amperes})$$

$$= 9.6 \times 10^{-3} \text{ mhos}$$

And the gain A_{gm} is:

$$A_{gm} = G_m R3 \tag{12-22}$$

$$= (9.6 \times 10^{-3})(10,000 \ \Omega)$$

$$= 96$$

$$\approx 100$$

If you want to make the gain exactly 100, then make R4 adjustable and set it to produce the exact gain of 100. This technique is used in the adjustable gain amplifier in Project 12-3.

The output terminal of this project has a capacitor to ground for slew rate/frequency compensation. This 180 pF capacitor will set the slew rate to:

$$S_r = \frac{I_{bias}}{C_L}$$

$$= \frac{500\ \mu A}{180\ pF}$$

$$= 2.78\ V/\mu S$$

(12-23)

The bias current can be varied to produce other gains as required.

Project 12-3   Variable OTA voltage gain amplifier

Figure 12-8 shows the circuit for a variable-gain voltage amplifier. This circuit will offer gain from 14 to 40 dB, depending on the setting of the 500 kΩ potentiometer, R5. Also provided is a means for nulling the dc output offset voltage. The circuit is essentially the same as previous circuits, except that the bias current is adjustable. Therefore, we will not go through the calculations again.

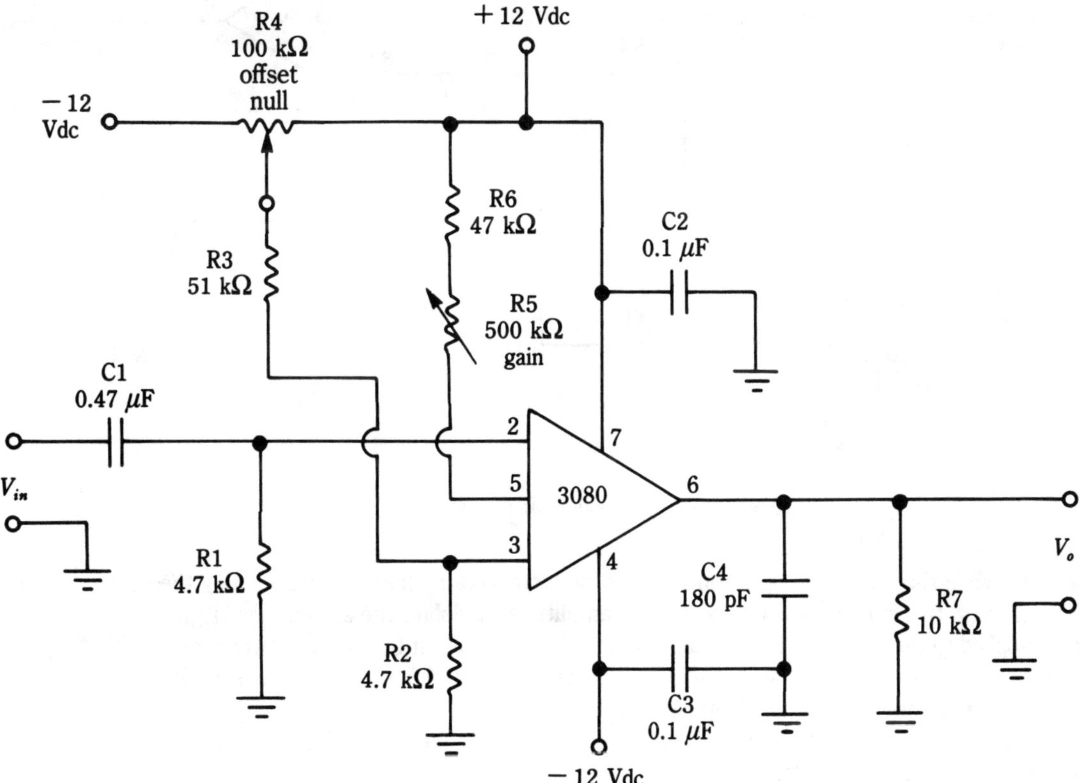

12-8   Variable-gain OTA amplifier.

Adjustment of the dc output offset voltage is done by setting R5 to its minimum value (i.e., 0 Ω), and then adjusting potentiometer R4, the *offset null* control, for 0 Vdc output when the input ac signal is 0. In order to make this adjustment, either short the input to ground or connect it to an audio signal generator that is turned off.

Project 12-4 Voltage-controlled gain OTA amplifier

An amplitude modulator circuit, such as Fig. 12-9, can be used to make CW signal into an AM signal, as in a signal generator or radio transmitter. It can also be used to make a voltage-controlled variable gain amplifier (VCGA).

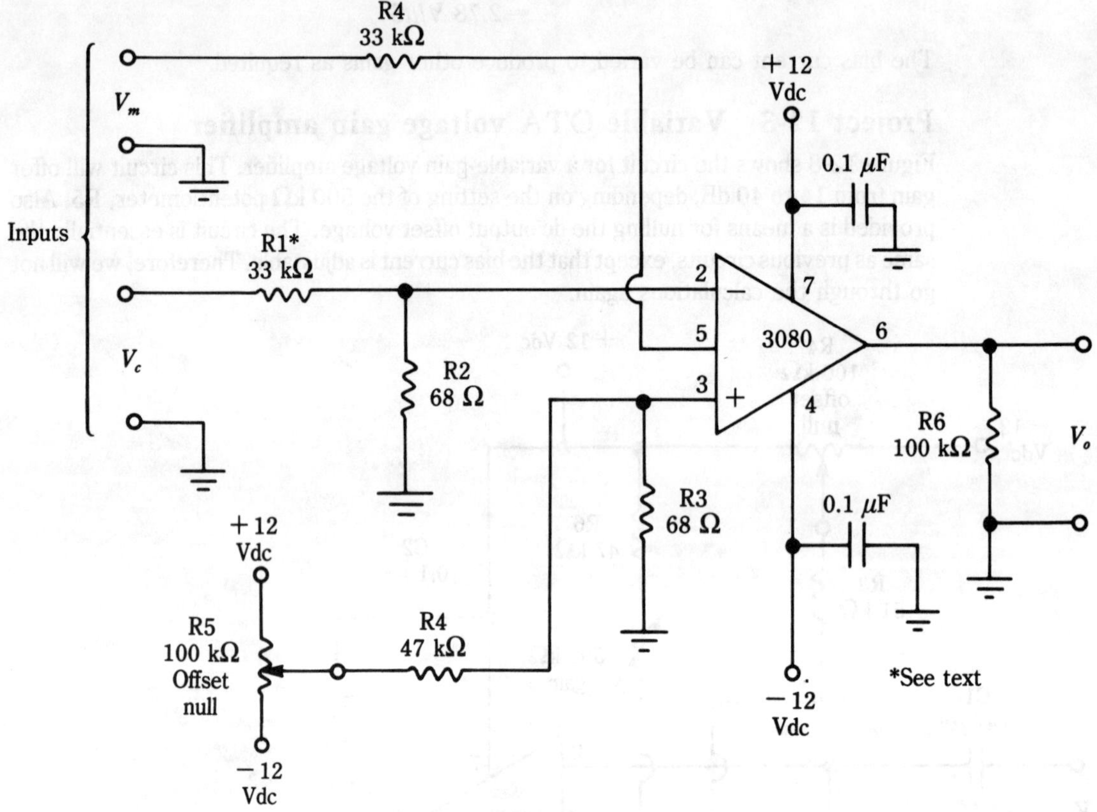

12-9 Amplitude modulator circuit.

Because the gain of an operational transconductance amplifier is controlled by the bias current, this current also can be used to amplitude-modulate the ac "carrier" signal (V_c) applied to the inverting input. You can vary the bias current by either of two means. You can vary resistor R1 (as was done in the variable gain amplifier), or you can leave R1 fixed, but vary the modulating voltage, V_m.

The dc output offset control is adjusted as in the case above, but with the modulating voltage V_m set to +12 Vdc, and V_c set to 0.

Project 12-5   40 dB gain OTA dc differential amplifier

A differential amplifier is one that has two inputs, each of which has identical gain but opposite polarity. The inverting input ($-$ IN) produces an output signal that is 180 degrees out of phase with the input signal, while the noninverting input produces an output signal that is in phase (0 degrees) with the input signal.

Differential amplifiers are used for two main reasons: first because the input signal source has a natural output configuration that is either differential or push-pull; second, when there is a high probability of receiving a noise level from external electrical or magnetic fields. For example, when an input line is subjected to the ubiquitous 60 Hz ac fields from the power lines, a differential amplifier is needed.

The voltage gain of the circuit in Fig. 12-10 is set by the bias resistor, R3, and is therefore similar to that of the previous 40 dB single-ended amplifier.

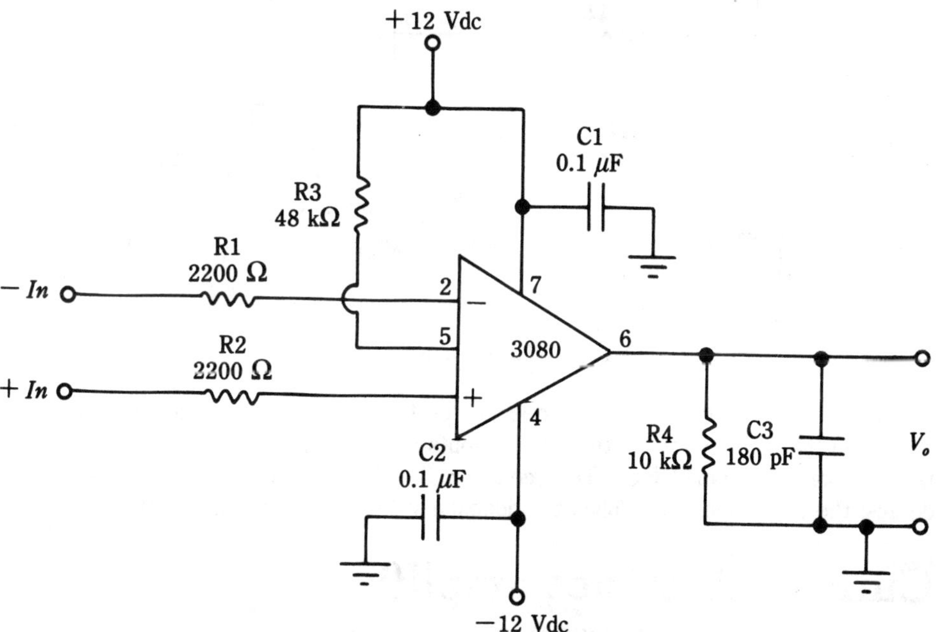

12-10   40 dB differential amplifier.

Project 12-6   Wideband OTA ac amplifier

A wideband amplifier has a very wide bandwidth and is used either for high-fidelity reproduction of complex waveforms or to pass video signals. Figure 12-11 shows the circuit for a single-ended OTA amplifier, based on the CA-3080, that will serve as a wideband amplifier.

The input impedance of this circuit is set to approximately 50 Ω by resistor R1. The reason for this impedance is that 50 Ω is the standard system impedance for RF systems (75 Ω in television antenna systems). In wideband and RF circuits, the system impedances should be matched in order to reduce the standing wave ratio (SWR) errors that can occur.

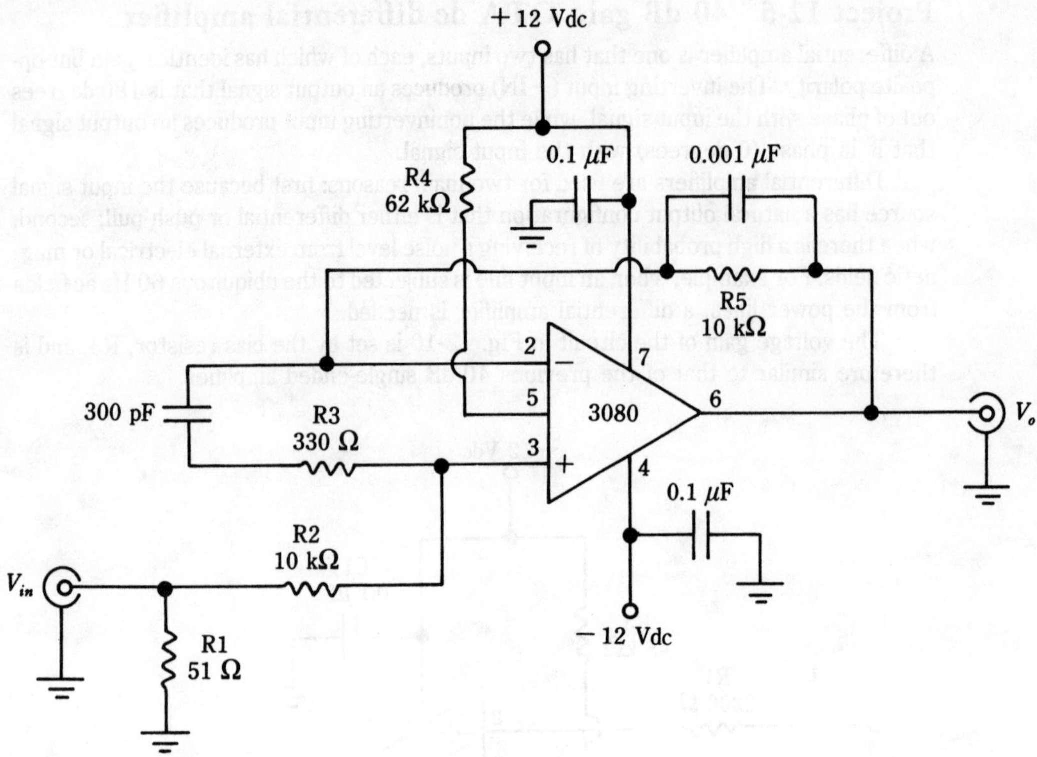

12-11   Wideband amplifier circuit.

Be careful when laying out this project on perf board. RF and video circuits can oscillate when the layout places outputs near inputs or where stray capacitance and inductances resonate at some HF or VHF frequency. See the RF section of this book for more details; the discussion also applies to wideband video amplifiers such as this one.

Current difference amplifiers

The current difference amplifier (CDA), also called the *Norton amplifier,* is another nonoperational linear IC amplifier that performs similarly to the op amp, but not exactly the same. The CDA has certain features that make it uniquely useful for certain applications. One place where the CDA is more useful than the operational amplifier is in circuits that process ac signals but are limited to a single-polarity dc power supply, for example, in automotive electronics equipment limited to a single 12 to 14.4 Vdc battery power supply that uses the car chassis for negative common return.

There are other cases where the linear IC amplifier is but a minor feature of the circuit, most of which operates from a single dc power supply. It might be wasteful in such circuits to use the operational amplifier. You would either have to bias the operational amplifier with an external resistor network or provide a second dc power supply.

The normal circuit symbol for the CDA is shown in Fig. 12-12. This symbol looks much like the regular op amp symbol, except that a "current source" is placed along the

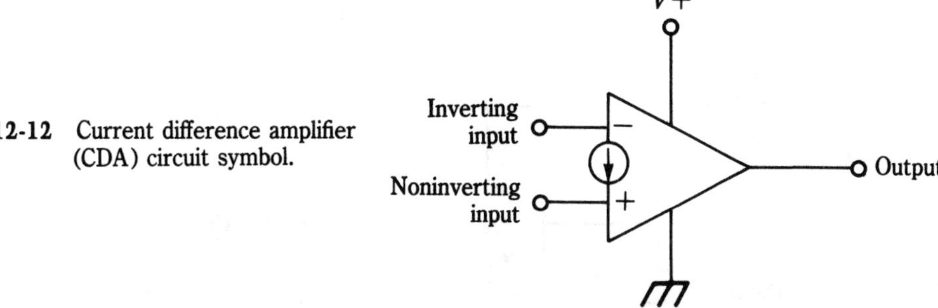

12-12   Current difference amplifier (CDA) circuit symbol.

side opposite the apex. This symbol is typically used for several products such as the LM-3900 device, which is a quad Norton amplifier. You sometimes might find schematics where the op amp symbol is used for the CDA, but that is technically incorrect usage.

CDA circuit configuration

The input circuit of the CDA differs radically from the operational amplifier. Recall that the op amp used a differential input amplifier driven from a constant current source supplying the collector-emitter current. The CDA is quite different, however, as can be seen in Fig. 12-13. The overall circuit of a typical CDA is shown in Fig. 12-13A, while an alternate form of the input circuit is shown in Fig. 12-13B.

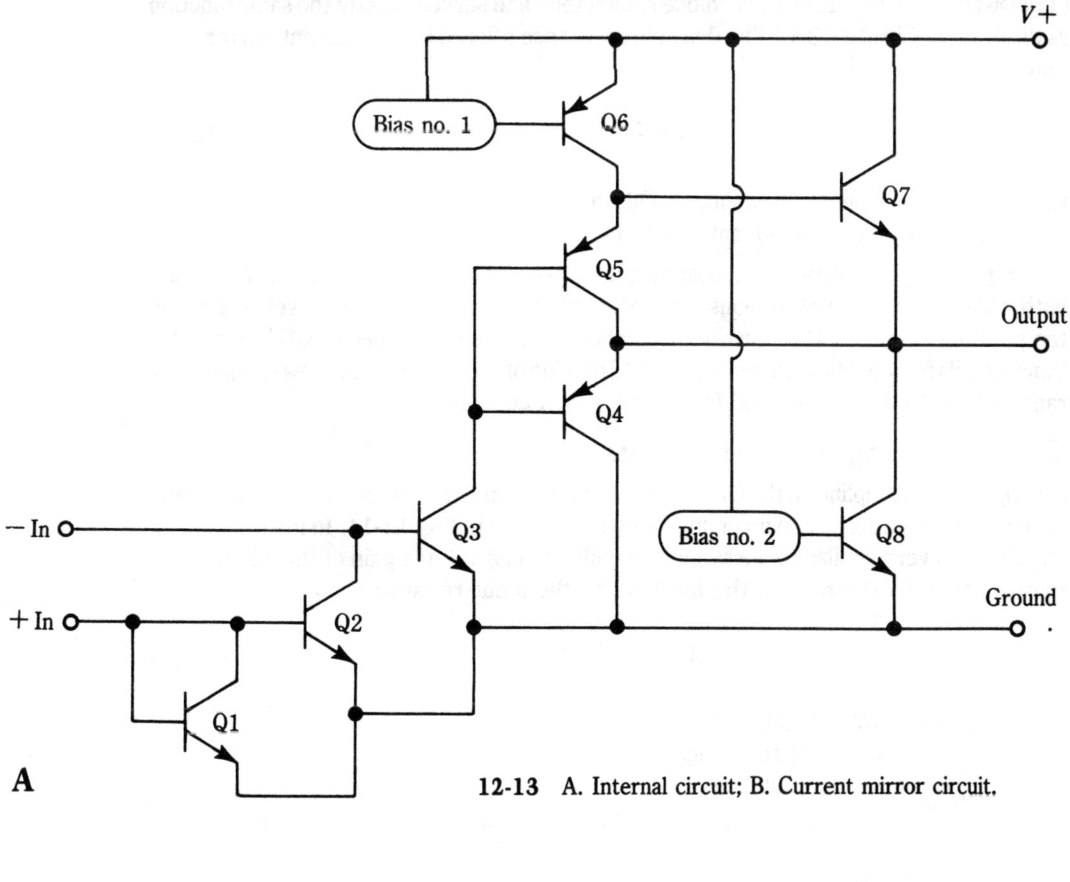

12-13   A. Internal circuit; B. Current mirror circuit.

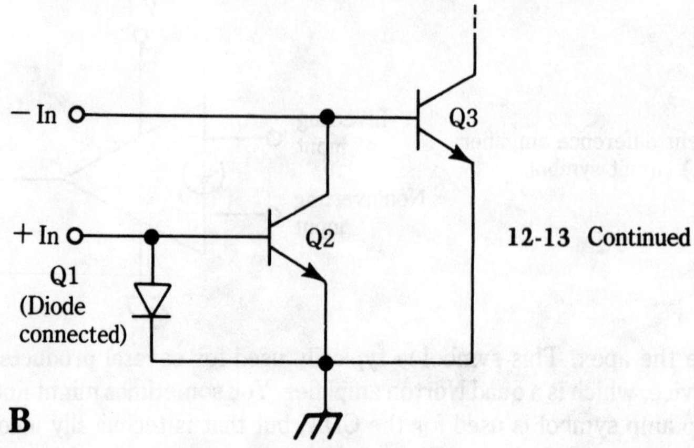

Transistor Q7 in Fig. 12-13A forms the output transistor, while Q5 is the driver. Both the npn output transistor and the pnp driver transistor operate in the emitter follower configuration. Transistors Q4, Q5, Q6, and Q8 are connected to serve as current sources. The input transistor is Q3, and it operates in the common emitter configuration. The base of Q3 forms the inverting (−IN) input for the CDA.

The noninverting input of the CDA is formed with a "current mirror" transistor, Q1. (Transistor Q1 in Fig. 12-13A is "diode connected" and serves exactly the same function as diode D1 in Fig. 12-13B.) The dynamic resistance offered by the current mirror transistor (Q2) is given by:

$$r = \frac{26}{I_b} \tag{12-24}$$

Where: r is the dynamic resistance of Q2, in ohms
I_b is the base bias current of Q3, in milliamperes

Equation 12-24 is used only at or near normal room temperature because I_b will vary with wide temperature excursions. For most common applications, however, the room-temperature version of the equation will suffice. Data sheets for specific CDAs give additional details for amplifiers that must operate outside of the relatively narrow temperature range (25°C to 30°C) specified for the simplified equation.

CDA inverting follower circuits

Like operational amplifiers, the CDA can be configured in either inverting or noninverting follower configurations. The inverting follower is shown in Fig. 12-14. In many respects, this circuit is very similar to operational amplifiers. The voltage gain of the circuit is set approximately by the ratio of the feedback to the input resistor:

$$A_v = \frac{-R_2}{R_1} \tag{12-25}$$

Where: A_v is the voltage gain
R_2 is the feedback resistance
R_1 is the input resistance

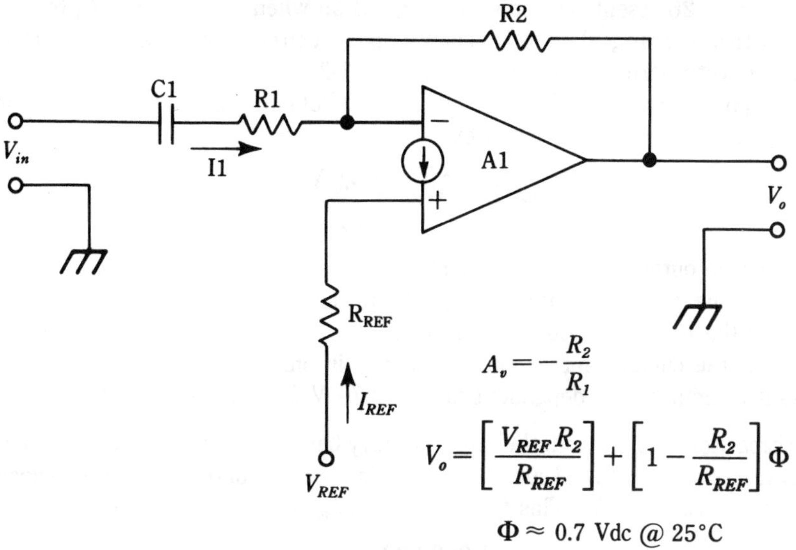

$$A_v = -\frac{R_2}{R_1}$$

$$V_o = \left[\frac{V_{REF}\, R_2}{R_{REF}}\right] + \left[1 - \frac{R_2}{R_{REF}}\right]\Phi$$

$$\Phi \approx 0.7 \text{ Vdc @ } 25°C$$

12-14   Inverting follower circuit.

(*Note: R_1 and R_2 are in the same units. The minus sign indicates that a 180-degree phase reversal occurs between input and output signals.*)

We must provide a bias to the current mirror transistor (i.e., Q2 in Fig. 12-13A), so resistor R_{ref} is connected in series with the noninverting input of the CDA and a reference voltage source, V_{ref}. In many practical circuits, the reference voltage source is merely the V+ supply used for the CDA. In other cases, however, some other potential might be required, or alternatively the reference current is required to be regulated more tightly (or with less noise) than the supply voltage. Ordinarily the reference current is set to some convenient value between 5 μA and 100 μA. For V+ power supply values of +12 Vdc, for example, it is common to find a 1 MΩ resistor used for R_{ref}. In that case, the input reference current is:

$$I_{ref} = 12/1,000,000$$
$$= 12 \ \mu A.$$

A constraint placed on CDAs is that the input resistor (R1) used to set gain must be high compared with the value of the current mirror dynamic resistance. The CDA becomes nonlinear (i.e., distorts the input signal) if the input resistor value approaches the current mirror resistance, *r*. In that case, the voltage gain is not $-R_2/R_1$, but rather:

$$A_v = \frac{R_2}{R_1 + r} \tag{12-26}$$

Where: A_v is the voltage gain
      R_2 is the feedback resistance
      R_1 is the input resistance
      r is the current mirror resistance

Equation 12-26 essentially reduces to Eq. 12-25 when we can force R_1 to be very much larger than r. This goal is easily achieved in most circuits because r is tiny. (Work a few examples with normal bias currents in Eq. 12-25.)

The output voltage of the CDA will exhibit an offset potential even when the ac input signal is zero. This potential is given by:

$$V_o = \left(\frac{V_{ref} R_2}{R_{ref}} + 1\right) - \left(\frac{R_2}{R_{ref}}\right)\phi \qquad (12\text{-}27)$$

Where: V_o is the output potential, in volts
$\quad$ V_{ref} is the reference potential, in volts (usually V+)
$\quad$ R_2 is the feedback resistance, in ohms
$\quad$ R_{ref} is the current mirror bias resistance, in ohms
$\quad$ ϕ is a temperature dependent factor (0.70 V for room temperature)

The capacitor in series with the input circuitry has the effect of limiting the low-end frequency response. The -3 dB cutoff frequency is a function of the value of this capacitor and the input resistance, R_1. This frequency, F, is given by:

$$F = \frac{1,000,000}{2\pi R_1 C_1} \qquad (12\text{-}28)$$

Where: F is the lower end -3 dB frequency, in hertz (Hz)
$\quad$ R_1 is in ohms (Ω)
$\quad$ C_1 is in microfarads (μF)

In some CDA circuits, there is also a capacitor in series with the output terminal. The purpose of that capacitor is to prevent the dc offset that is inherent in this type of circuit from affecting following circuits. The output capacitor also will limit the low-end frequency response. The same form of equation (i.e., Eq. 12-28) is used to determine this frequency, but using the input resistance of the load as the "R" term.

As is often the case with equations presented in electronics books, Eq. 12-28 is not necessarily in the most useful form. In most cases, you will know the input resistance (R_1) from the application. It is typically not less than ten times the source impedance and forms part of the gain equation. Consideration of driving source impedance and voltage gain tends to determine the value of R_1.

The required low-end frequency response is usually determined from the application. You generally know (or can find out) the frequency spectrum of input signals. From the lower limit of the frequency spectrum, you can determine the value of F. Thus, you will determine F and R_1 from other considerations than the circuit. You therefore need a version of Eq. 12-28 that calculates C_1 from knowledge of R_1 and F. Using algebra:

$$F = \frac{1,000,000}{2\pi R_1 C_1}$$

Design example

Design a gain of 100 ac amplifier based on a CDA in which the input impedance is at least 10 kΩ and the -3 dB frequency response is 3 Hz or lower. Assume dc power supplies of ± 15 Vdc. (See Fig. 12-14 for the final circuit.)

1. Set the reference current to the noninverting input to a value between 5 and 100 μA. Select 15 μA.

$$R_3 = V/I_{ref}$$
$$= \frac{15\ \text{Vdc}}{(0.000015)}$$
$$= 1\ \text{M}\Omega$$

2. Set the gain resistors. R_1 can be 10,000 Ω in order to meet the input impedance requirement.

    (From $A_v = R_2/R_1$, we know that $R_2 = A_vR_1$)

$$R_2 = A_{vR_1}$$
$$= (100)(10,000)$$
$$= 1\ \text{m}\Omega$$

3. Find the value of input capacitor C1 when $F_{-3\ \text{dB}}$ is 3 Hz.

$$C1 = \frac{1,000,000}{2\pi R_1 F}$$

$$= \frac{1,000,000}{(2)\,(3.14)\,(10,000)\,(3)}$$

$$= 5.3\ \mu\text{F}$$

Because 5.3 μF is a nonstandard value, select the next higher standard value (which is 6.8 μF).

The value of output capacitor C2 can be arbitrarily set to 6.8 μF if the load impedance is 10,000 Ω. If the load impedance is higher than that value, then use 4.7 μF or 6.8 μF. If the load impedance is very much higher than 10,000 Ω, or is lower than 10,000 Ω, then calculate the value using the same equation as for C1, but with the load resistance substituted for the input impedance.

Noninverting amplifier circuits

The noninverting amplifier CDA configuration is shown in Fig. 12-15. This circuit retains the reference current bias applied to the noninverting input but rearranges some of the other components. As in the inverting amplifier configuration, the noninverting amplifier uses R2 to provide negative feedback between the output terminal and the inverting input. Unlike the inverting CDA circuit, however, input resistor R1 is connected in series with the noninverting input. The gain of the noninverting CDA amplifier is given by:

$$A_v = \frac{R_2}{\left(\dfrac{26\ R_1}{I_{ref}\,\text{mA}}\right)} \tag{12-29}$$

Where: A_v is the voltage gain
    R_1 and R_2 are in ohms (Ω)
    I_{ref} is the bias current, in milliamperes (mA)

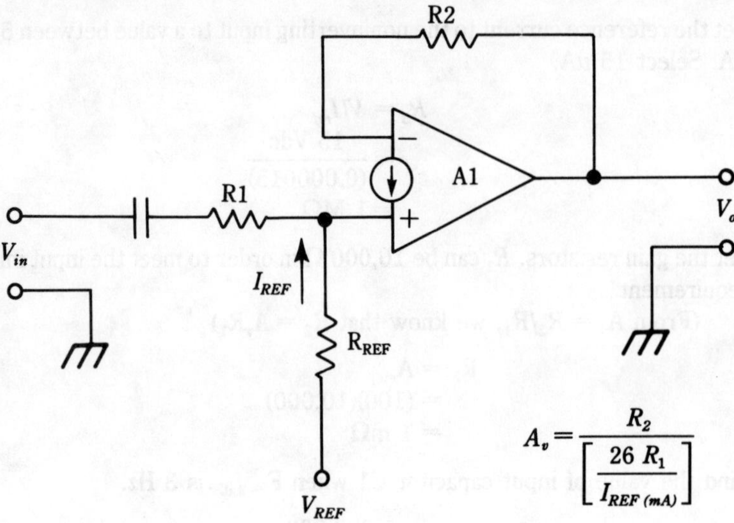

$$A_v = \dfrac{R_2}{\left[\dfrac{26\,R_1}{I_{REF\,(mA)}}\right]}$$

12-15   Noninverting follower circuit.

The reference current, I_{ref}, is set to a value between 5 and 100 μA. Unlike the situation in the inverting amplifier, the value of this current is partially responsible for setting the gain of the circuit. Some clever designers have even used this current as a limited-gain control for some CDA stages.

The value of the resistor that provides the reference current (R_{ref}) is set by Ohm's law, considering the required value of reference current and the reference voltage, V_{ref}. In most common applications, the reference voltage is merely one of the supply voltages. The value of R_{ref} is determined from:

$$R_{ref} = \frac{V_{ref}}{I_{ref}} \tag{12-30}$$

Where: R_{ref} is the reference resistor, in ohms
V_{ref} is the reference potential, in volts
I_{ref} is the reference current, in amperes
   (*Note:* 1 μA = 0.000001 ampere)

The value of input impedance is approximately equal to R_1, provided that R_1 is much higher than the dynamic resistance of the current mirror inside the CDA (which is typically the case). As was true in the inverting follower case, the input capacitor (C1) sets the low-end frequency response of the amplifier. The -3 dB frequency is given by exactly the same equation as for the inverting case (see Eqs. 12-27 and 12-28).

Figure 12-16 shows a modification of the noninverting follower circuit that allows for a noisy reference source. This type of amplifier circuit might be used where the dc power supplies that are used for the reference voltage are electrically noisy. Such noise could come from other stages in the circuit, or from outside sources.

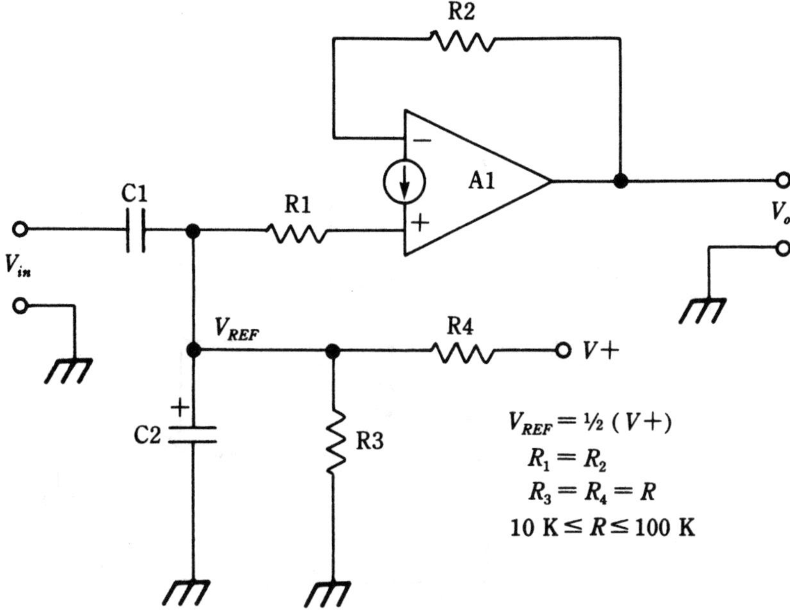

12-16   Operation from a single supply.

The purpose in Fig. 12-16 is to form a reference voltage from a resistor voltage divider circuit consisting of R3 and R4. The value of V_{ref} will be:

$$V_{ref} = \frac{(V+) R_3}{R_3 + R_4} \tag{12-31}$$

Where: V_{ref} is the reference potential, in volts (V)
$V+$ is the supply potential, in volts (V)
R_3 and R_4 are in ohms (Ω)

Inspection of Eq. 12-24 reveals that $V_{ref} = (V+)/2$ when $R_3 = R_4$, which is the usual case in practical circuits. The reference current is:

$$I_{ref} = \frac{V_{ref}}{R_1} \tag{12-32}$$

Project 12-7   Super-gain amplifier

There is a practical limit to voltage gain using standard resistor values and standard circuit configurations. (A similar problem also exists for operational amplifiers.) Figure 12-17 shows a means for overcoming the limitations. This supergain amplifier circuit forms a noninverting follower in a manner similar to the earlier circuit, except that feedback resistor R2 is driven from an output voltage divider network, rather than directly from the output terminal of the CDA. The voltage gain of the circuit of Fig. 12-17 is given by:

$$A_v = \left(\frac{R_2}{R_1}\right) \left(\frac{R_3 + R_4}{R_3}\right) \tag{12-33}$$

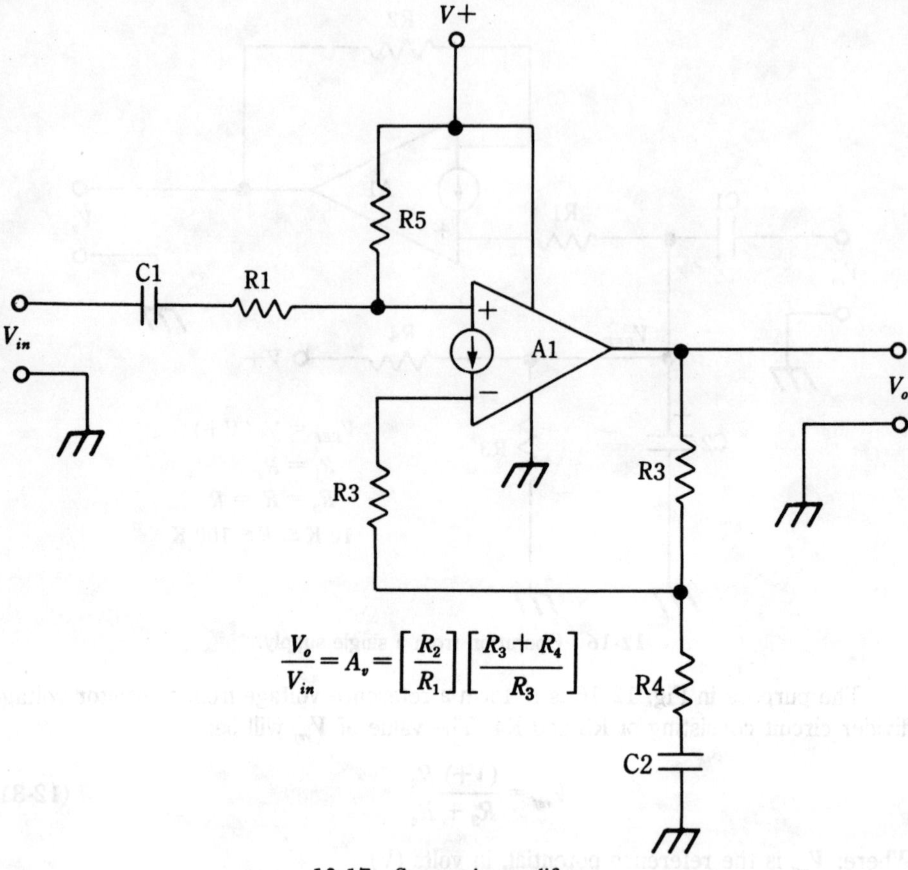

$$\frac{V_o}{V_{in}} = A_v = \left[\frac{R_2}{R_1}\right]\left[\frac{R_3 + R_4}{R_3}\right]$$

12-17 Supergain amplifier.

C_1 is set using the same Eq. 12-27 that was used previously, while C_2 is set to have a capacitive reactance of $R_4/10$ at the lowest frequency of operation (in other words, the low-end -3 dB point).

CDA differential amplifiers

A *differential amplifier* is one that will produce an output that is proportional to the gain and the difference between potentials applied to the inverting and noninverting inputs. Figure 12-18 shows the circuit of a CDA differential amplifier. It is similar to the operational amplifier version of this simple circuit in several respects. For example, the two input resistors are equal, and the differential voltage gain is the ratio of the negative feedback resistor (R3) and the input resistor:

$$Av = \frac{R_3}{R_2} \tag{12-34}$$

if $R_1 = R_2$, and $R_3 = R_4 = R_5$.

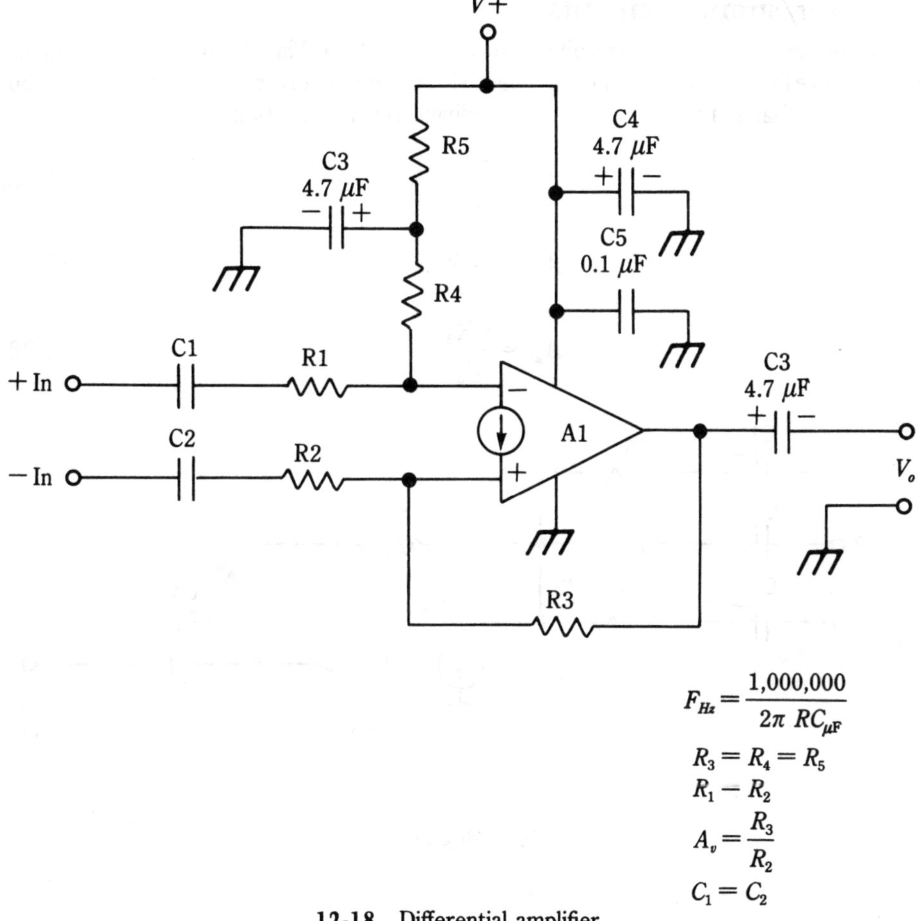

$$F_{Hz} = \frac{1,000,000}{2\pi\ RC_{\mu F}}$$

$$R_3 = R_4 = R_5$$
$$R_1 = R_2$$
$$A_v = \frac{R_3}{R_2}$$
$$C_1 = C_2$$

12-18 Differential amplifier.

The input impedance (differential) of this circuit is twice the value of the input resistances: $R_{in} = R_1 + R_2$. The bias current is provided through the two series resistors, R4 and R5.

Assuming that $R_1 = R_2 = R$, and $C_1 = C_2 = C$, we can calculate the low-end -3 dB frequency from the equation:

$$F = \frac{1,000,000}{2\pi RC} \tag{12-35}$$

Where: F is the -3 dB frequency, in hertz
     R is in ohms
     C is in microfarads

When this circuit is used as a 600 Ω line receiver, we can make the two input resistors 330 Ω each (or 270 Ω) if a small mismatch can be tolerated. Ideally, the input resistors will be 300 Ω each (which might require two resistors for each input resistor).

ac mixer/summer circuits

The CDA mixer, or summer, circuit is shown in Fig. 12-19. This circuit is used to combine two or more inputs into one channel. The basic circuit is an inverting follower. Each input sees a gain that is the quotient of the feedback resistor to its input resistor:

$$A_{v1} = \frac{-R_2}{R_3} \tag{12-36}$$

$$A_{v2} = \frac{-R_2}{R_4} \tag{12-37}$$

$$A_{v3} = \frac{-R_2}{R_5} \tag{12-38}$$

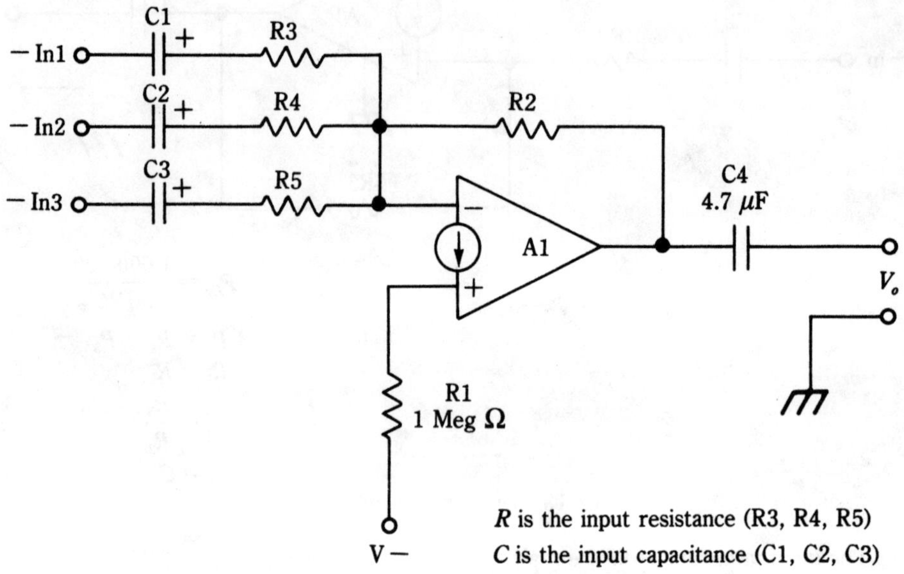

12-19   Audio mixer circuit.

From Eqs. 12-36 through 12-38 we can deduce that the output voltage is found from:

$$V_o = R_2\left(\frac{V_1}{R_3} + \frac{V_2}{R_4} + \frac{V_3}{R_5}\right) \tag{12-39}$$

The frequency response of each channel is found from the usual equation for -3 dB frequency:

$$F_{-3\ dB} = \frac{1,000,000}{2\pi\ RC} \tag{12-40}$$

Where: F is the -3 dB frequency, in hertz (Hz)

R is the input resistance (R_3, R_4, or R_5), in ohms (Ω)

C is the input capacitance (C_1, C_2, or C_3), in microfarads (μF)

Project 12-8   Differential output 600 Ω line driver amplifier

The 600 Ω line used in broadcasting electronics and professional audio recording requires either a center-tapped output transformer or a linear amplifier with a push-pull output to drive the line. Figure 12-20 shows the circuit of a CDA 600 Ω line driver amplifier. This circuit basically consists of two separate amplifiers: one an inverting follower and the other a noninverting follower.

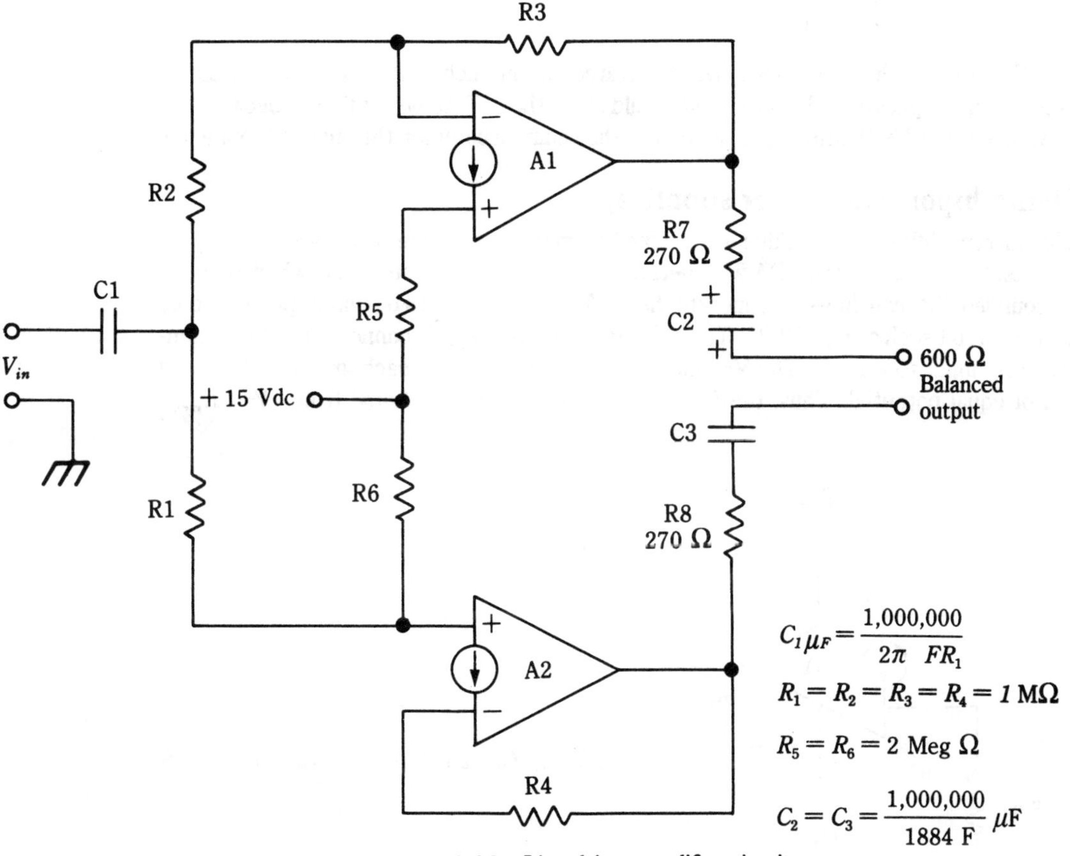

12-20   Line-driver amplifier circuit.

The bias resistors (R5 and R6) are set to provide a small bias current in the 5 to 100 μA range. This current is found from Ohm's law, $I_{ref} = (V+)/R$. In the example shown in Fig. 12-20, the resistors are set to 2 MΩ for a supply voltage of +15 Vdc.

The capacitors in the circuit set the low-end −3 dB point in the frequency-response curve. These capacitor values are set from the following:

Assuming that $R_1 = R_2 = R_3 = R_4$, and $R_5 = R_6$:

$$C_{1\mu F} = \frac{1,000,000}{2\pi FR_1} \qquad (12\text{-}41)$$

Assuming $C_2 = C_3$, and accounting for the constants:

$$C_{1\mu F} = \frac{1,000,000}{600\pi F}$$

$$= \frac{531}{F} \tag{12-42}$$

Where: C_1 and C_2 are in microfarads (μF)
      F is in hertz (Hz)
      R_1 is in ohms (Ω)

For best results, a multiple CDA integrated circuit such as the LM-3900 should be used for this application. Such a circuit would allow the drift to be controlled because the two halves would both drift at the same rate (they share a common thermal environment).

Using bipolar dc power supplies

The current difference amplifier is designed primarily for single-polarity power supply circuits. In most cases, the CDA will operate with a V+ dc power supply, in which one side is grounded. You can, however, operate the CDA in a circuit with a bipolar dc power supply using a circuit such as Fig. 12-21. The reference resistor, R_{ref}, is connected from the non-inverting input to ground. The V− and V+ power supplies are each ground-referenced and of equal potential. Thus, the 5 to 100 μA bias current is found from $(V+)/R_{ref}$.

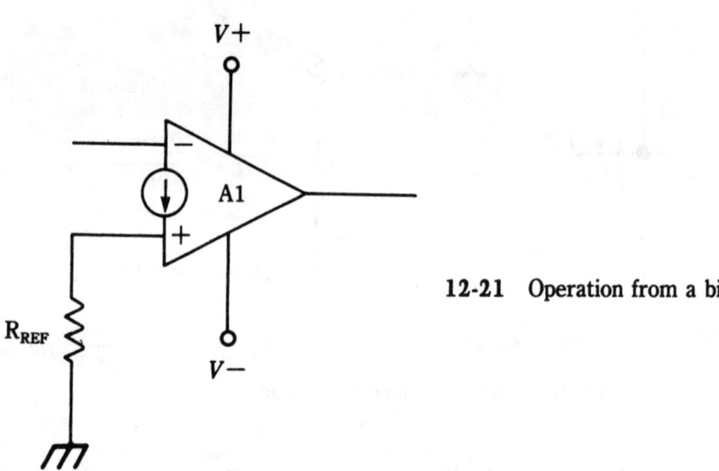

12-21   Operation from a bipolar supply.

Project 12-9   20 dB gain CDA ac amplifier

An inverting amplifier is needed that has an ac gain of 100 (i.e., 20 dB) with an input impedance of at least 10 kΩ. The LM-3900 CDA is available, so it is selected for the application. The circuit for this project is shown in Fig. 12-22. The pinouts can be selected from the four internal devices shown in Fig. 12-23.

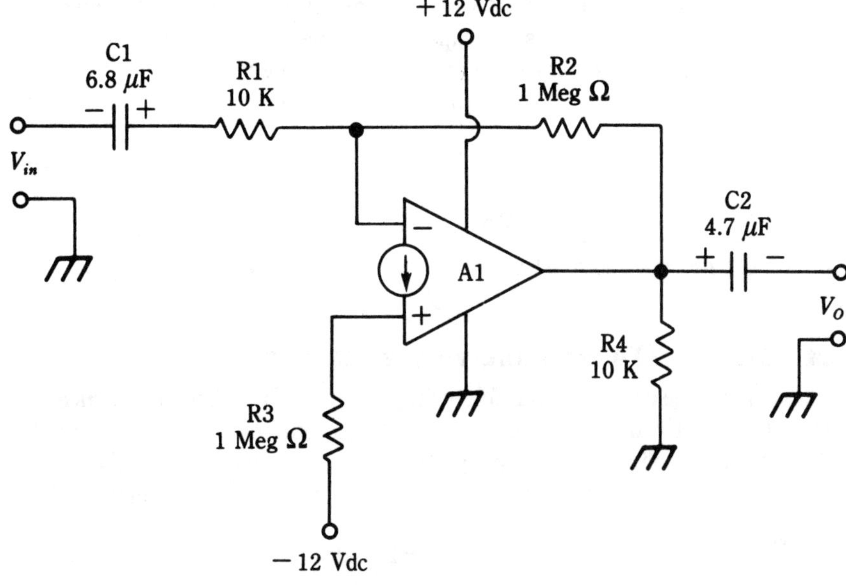

12-22 Gain-of-100 amplifier.

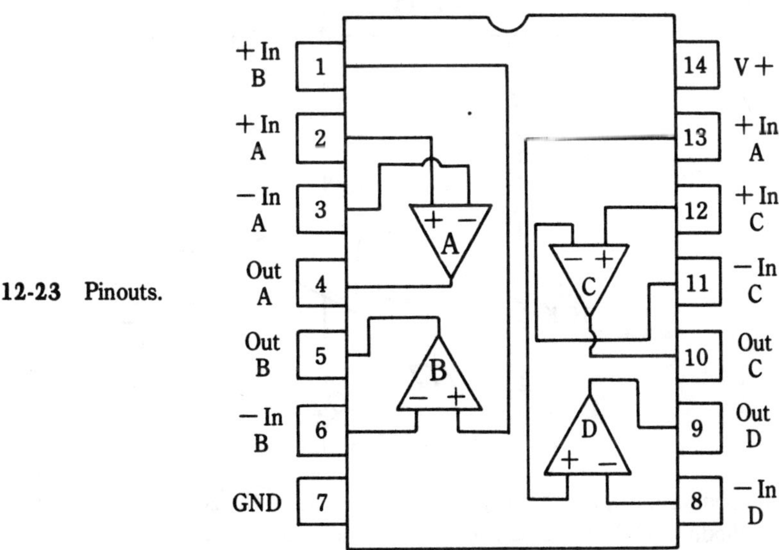

12-23 Pinouts.

The reference resistor is set to 1 MΩ. The gain (A_v) is set by the ratio of the feedback and input resistances: $-R_2/R_1$. Because the input impedance needs to be at least 10 kΩ, we set $R_1 = 10$ kΩ. The gain is 100, so the feedback resistor must be:

$$R_2 = A_v R_1 \qquad (12\text{-}44)$$
$$= (100)\ (10,000\ \text{ohms})$$
$$= 1,000,000\ \text{ohms}$$

The dc power supplies are set to ±12 volts, a common value that is obtained easily.

The input and output terminals are capacitor-coupled. These capacitors, and their associated resistors, form high-pass filters that set the lower end −6 dB points in the frequency-response characteristic. In each case, the frequency is:

$$F = \frac{1}{2\pi RC} \tag{12-45}$$

The last capacitor in the circuit (C3) is used to decouple the dc power supply line. This capacitor is a 0.1 μF unit and must be mounted as close as possible to the body of the LM-3900.

Project 12-10   CDA variable-voltage source

Figure 12-24 shows the circuit of a variable-voltage source (VVS) that will produce output potentials of 0.6 to 20 V, depending on the setting of potentiometer R1. This circuit also can be used as a fixed-voltage source by replacing the potentiometer with a fixed resistor of appropriate value. Like other circuits in this chapter, the VVS is based on the LM-3900 device.

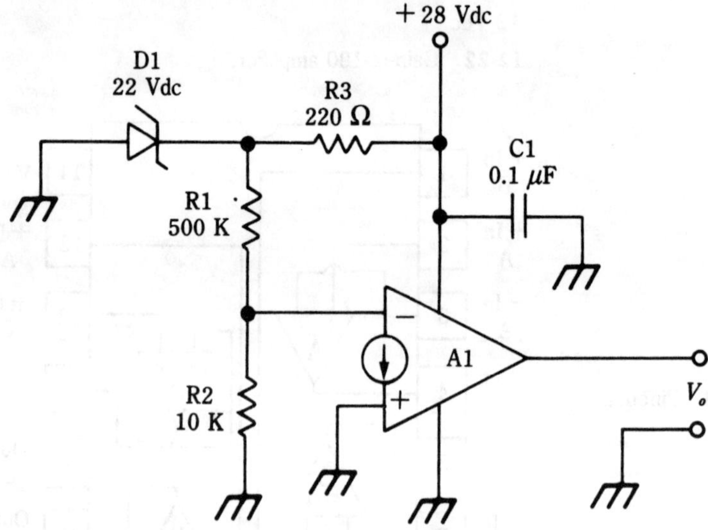

12-24   CDA variable voltage source.

The maximum voltage that is produced depends on the zener potential of the zener reference diode (D1). In the case shown here, the zener is a 22 V unit, so the maximum output voltage will be something between 20 and 22 V. Lower values of maximum output voltage can be accommodated by using a diode with a lower zener potential for D1. The advantage of this change is that the resolution is improved; i.e., the voltage change per turn of potentiometer R1.

Project 12-11   CDA over/under temperature alarms

The CDA can be used as a voltage comparator, so it is also useful for certain alarm circuits that use a voltage comparison at the heart. Figures 12-25A and 12-25B show two such

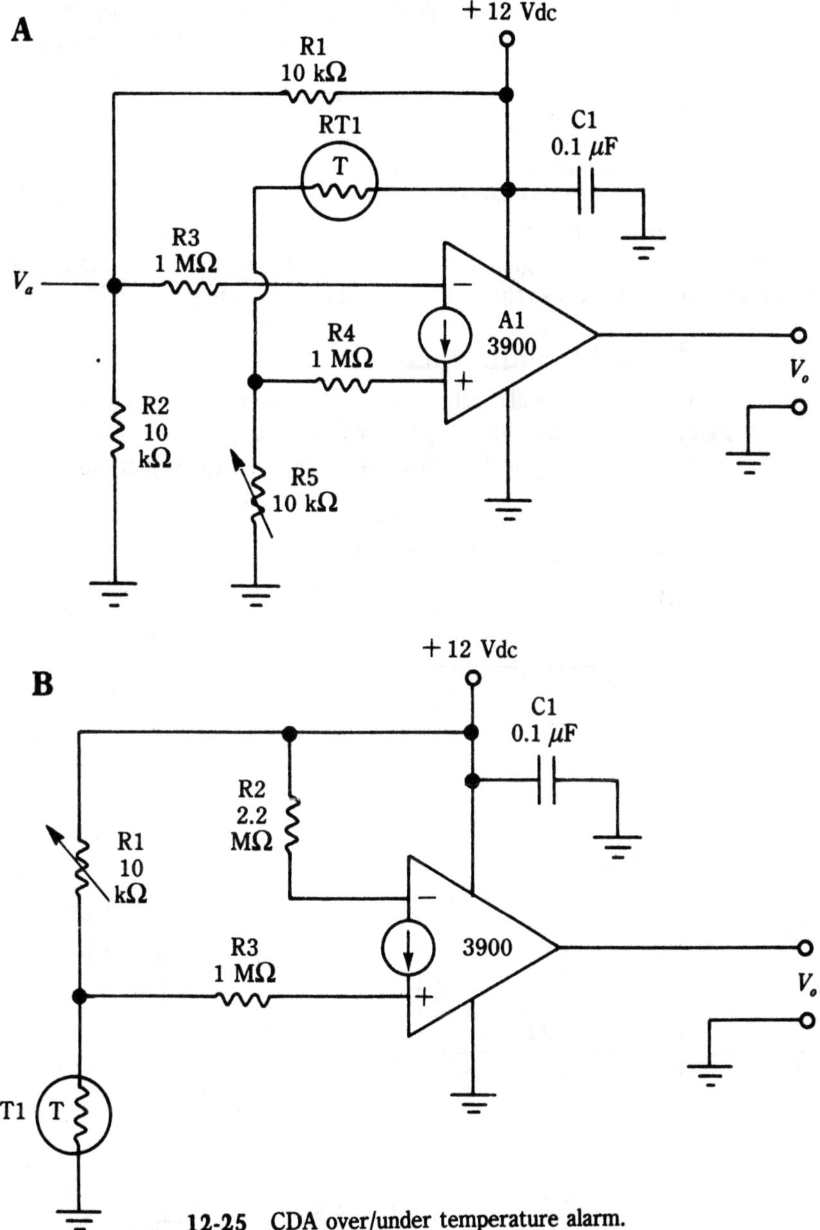

12-25 CDA over/under temperature alarm.

circuits; Fig. 12-25A is an overtemperature alarm, and Fig. 12-25B is an undertempera-
ture alarm.

In Fig. 12-25A, two reference circuits are provided. First is a 2:1 voltage divider
consisting of R1 and R2, which produces a voltage, V_a, of ½ (V+), or about 6 V. This potential produces an input current through resistor R3 (which is set to 1,000,000 Ω, partly
to avoid loading the voltage divider).

The other reference circuit is a voltage divider consisting of a thermistor, RT1, and a potentiometer, R5. The potentiometer is used to set the trip point at which the output changes to indicate the alarm condition.

The thermistor should be a negative temperature coefficient type, in which the resistance will be approximately 10 kΩ in the temperature range of interest.

Calibration of the circuit is done using a liquid such as water, into which both the thermistor and a mercury thermometer are immersed. The water is either heated or chilled to the desired temperature (which takes a little time), or is left to equilibrate at room temperature, if that is the goal. Let both the thermistor and the thermometer come to equilibrium, then adjust R5 so that the output of the CDA snaps to HIGH at that point.

Project 12-12   CDA audio mixer

An audio mixer is a circuit that will combine two or more audio signal sources into one signal channel. Figure 12-26 shows an audio mixer that is based on the current difference amplifier. The LM-3900 can be used for this project, and you can select the pinouts from the diagram in Fig. 12-23.

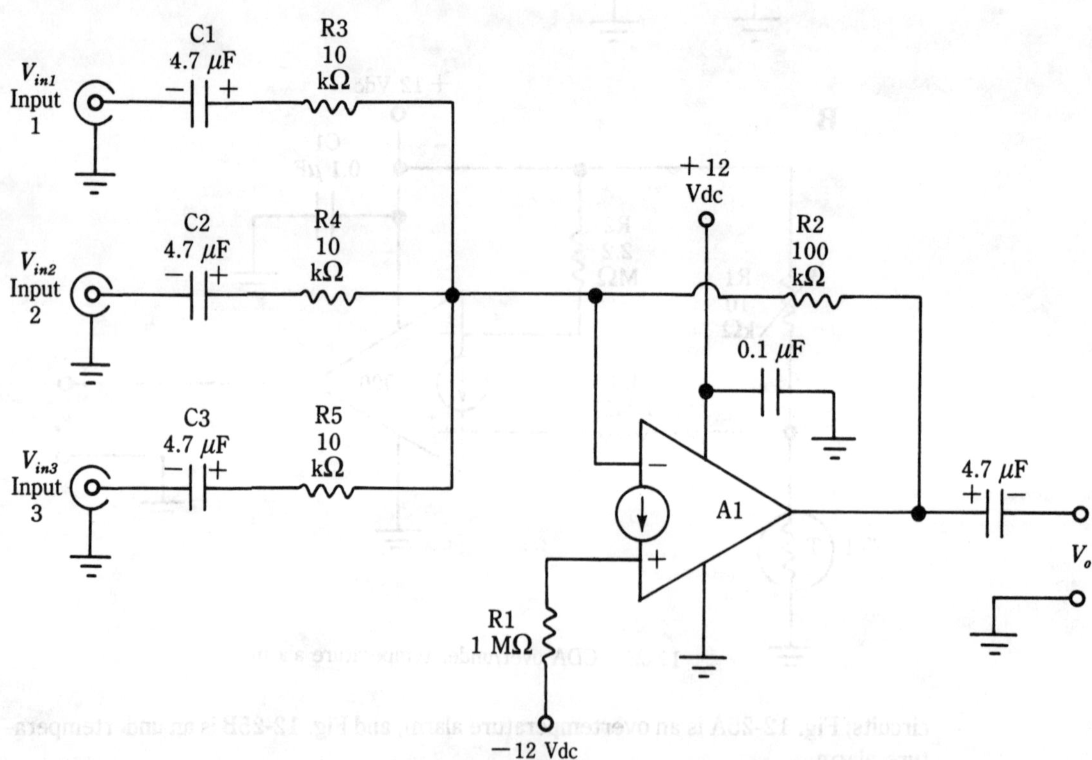

12-26   Audio mixer.

The crux of this circuit is the three input networks, principally R3, R4, and R5. These resistors are connected to different input sources (labeled Input-1, Input-2, and Input-3). These three resistors are joined into one point at the inverting input of the CDA. The gain is approximately:

$$A_v = \frac{-R_2}{R_x} \tag{12-46}$$

Where R_x is the value of any one input resistor.

The output voltage is:

$$V_o = -R_2 \left(\frac{V_{in1}}{R_3} + \frac{V_{in2}}{R_4} + \frac{V_{in3}}{R_5} \right) \tag{12-47}$$

If R2 is made variable, then the potentiometer used for R2 will serve as a master gain control for the audio mixer.

13

Audio amplifiers

AUDIO AMPLIFIERS OPERATE IN THE RANGE OF HUMAN HEARING, PLUS A little beyond. These amplifiers are intended to convey sound to the human ear from an electrical signal. They are found in high-fidelity reproduction equipment, communications equipment, and radio receivers. In this chapter, you will take a look at some of the more popular audio amplifier circuits based on solid-state devices.

Frequency response

It is impossible to use one frequency-response characteristic as universal among all audio electronic equipment. The typical frequency response might be quite varied among the various types of equipment.

In a simple AM radio, there is little need for a frequency response greater than about 5 kHz since many AM broadcast stations are limited in audio response. Because there might be no recoverable program material above 5 kHz, but there will be other forms of undesirable interference in this range, it is actually a disadvantage to have an AM radio with a wideband audio amplifier. However, FM broadcasters enjoy a wider latitude of the audio spectrum. These stations are permitted to transmit audio frequencies up to 15 kHz. In fact, not only are the higher frequencies transmitted, they are actually given a degree of pre-emphasis in order to improve the signal-to-noise ratio at the receiver. An FM radio, therefore, might easily have a frequency response wide enough to include frequencies up to 15 kHz.

Since most modern FM sets also include an AM band, it is desirable to be able to reduce the audio bandwidth on AM and increase it on FM. In some sets this is accomplished by the regular tone-control circuitry. The user is expected to adjust the control to give the most pleasing effect. In other sets, the audio amplifiers are designed to offer the 15 kHz FM response characteristic. The AM response is limited by incorporating an RC low-pass filter between the AM detector and the audio input. In certain other designs, the requirement is handled either by a compromise response of around 8 kHz or by ignoring the problem altogether.

Power

For the fan who loves super-power audio systems, it might come as quite a surprise to learn that most radio and tape players seldom produce more than a few watts (continuous) per channel. In fact, some of the lower priced models offer something less than 1 watt of power. Even the most expensive sets will not offer much more than 10 watts per channel maximum, and most American radios are rated closer to 5 watts.

Feedback

Negative feedback plays an important part in automotive electronic designs. Figure 13-1 shows two very common feedback circuits. In Fig. 13-1A, we see what is called the *second collector to first emitter* circuit, while in Fig. 13-1B we see the *second emitter to first base* circuit. In some cases you will find that Fig. 13-1A is used between two stages in the radio, while Fig. 13-1B is being used simultaneously — but somewhere else in the radio. Combination feedback methods are very common in most brands of car radios.

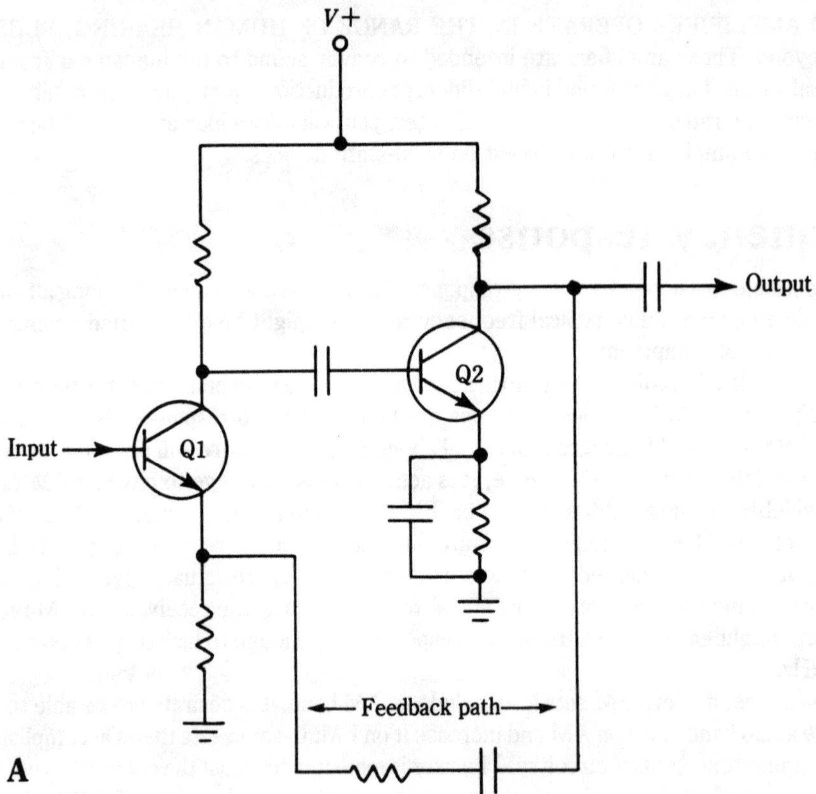

13-1   Feedback methods: A. Second collector to first emitter; B. Second emitter to first base.

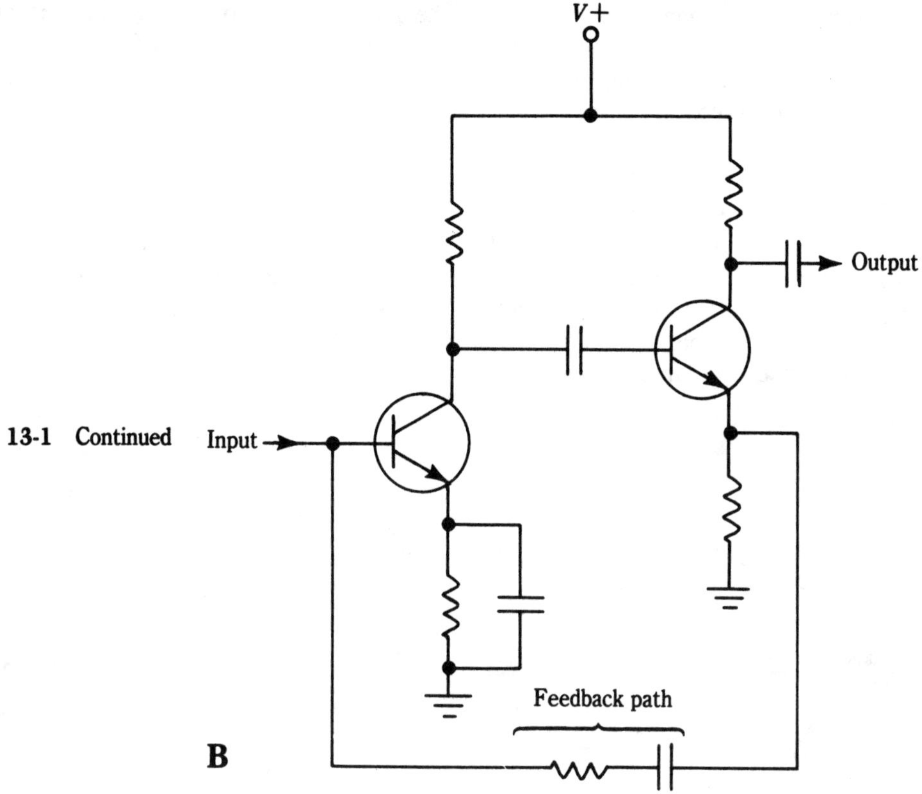

Audio stages

Figure 13-2 shows the block diagram of an audio amplifier section to a radio. The signal from the i-f amplifier is fed to either a simple AM (diode) detector or one of the more complex FM detectors. The detected audio signal is fed to the preamplifier stage via a volume control and a variable tone-shaping circuit. The preamplifier strengthens the signal and passes it on to the driver stage.

The driver amplifier is frequently another voltage amplifier stage. In some designs,

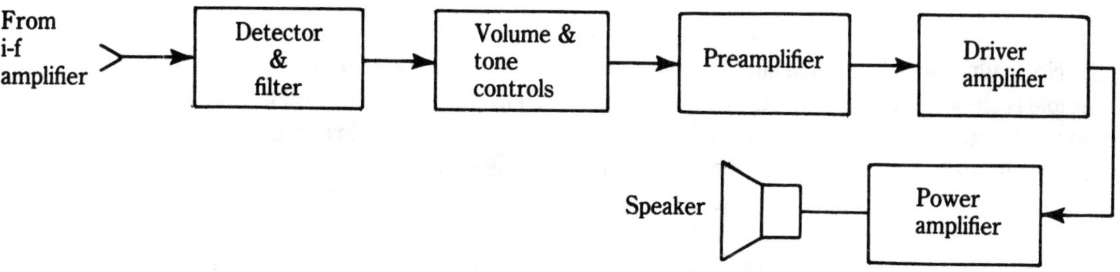

13-2 Audio amplifier circuit in block diagram form.

however, it is also a minor power amplifier. The purpose of the driver stage is to further develop the audio signal so that it is strong enough to drive the power-amplifier transistor.

The power amplifier supplies the audio power required by the speaker. These stages almost universally contain larger and more powerful transistors than the preamplifier and driver stages.

Volume and tone controls

The input side of the audio section usually contains the volume control. Most American radios use a circuit similar to the circuit shown in Fig. 13-3. Audio is taken from the control wiper terminal to the audio preamplifier input via a coupling capacitor. Its function is to isolate the control from the preamplifier dc bias network. Feedback from the collector of the output transistor frequently is applied to this circuit.

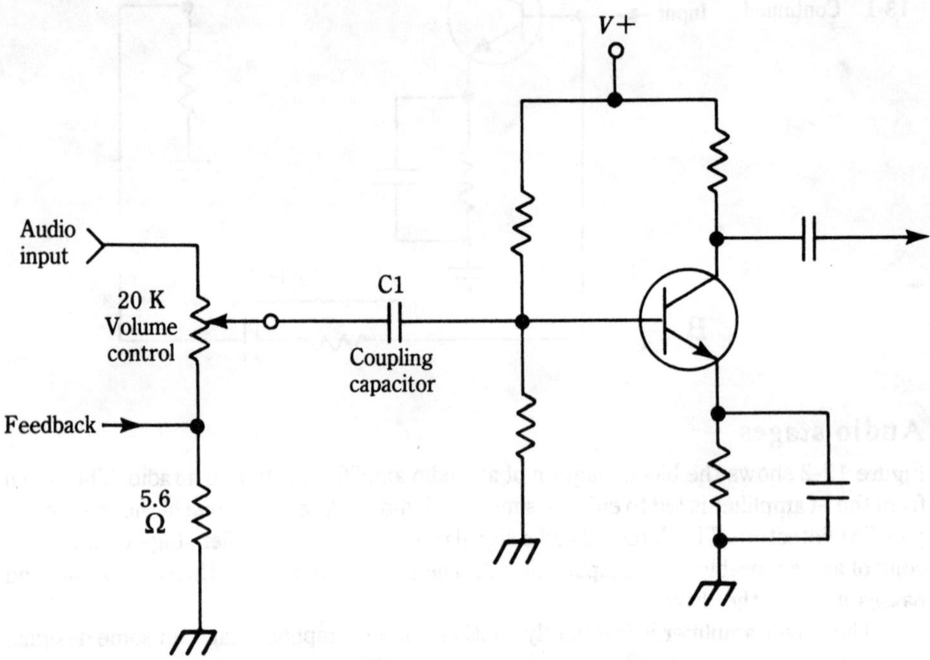

13-3   Volume and tone controls.

Normally, tone controls in simple, low-cost radios are of the *treble rolloff* variety; an example is shown in Fig. 13-4. The setting of the tone potentiometer determines the degree and frequency characteristics of the rolloff curve. Only a very few low-cost radios or tape players use the negative feedback "Baxandall" tone-control circuit.

Figure 13-5 shows the complete control section for one channel of a stereo audio section. The relative position of the balance control wiper determines the amplitude of the signals at the respective inputs. When the balance control is set all the way toward one channel, the signal line to the other channel will be shorted to ground, while the signal in the live channel will be only slightly attenuated.

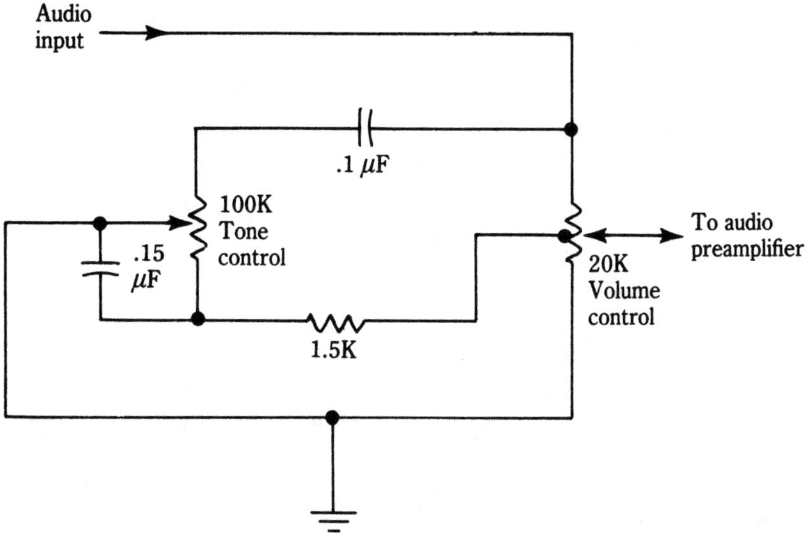

13-4 Volume and tone controls with loudness tap.

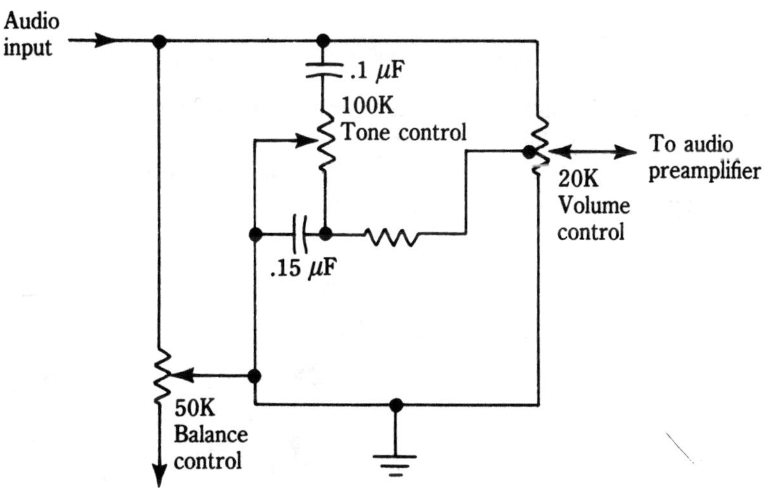

13-5 Volume, tone, and stereo balance controls.

Preamplifiers

Figure 13-6 illustrates one of the earliest types of solid-state preamplifier stages. This circuit functions as both driver and preamplifier. The input signal from the volume control is coupled by a capacitor to the base terminal of Q4. The output signal is coupled to the base of the power-amplifier transistor by an interstage transformer, in much the same manner as previous vacuum-tube designs. The bias for these stages is usually class-A. De-

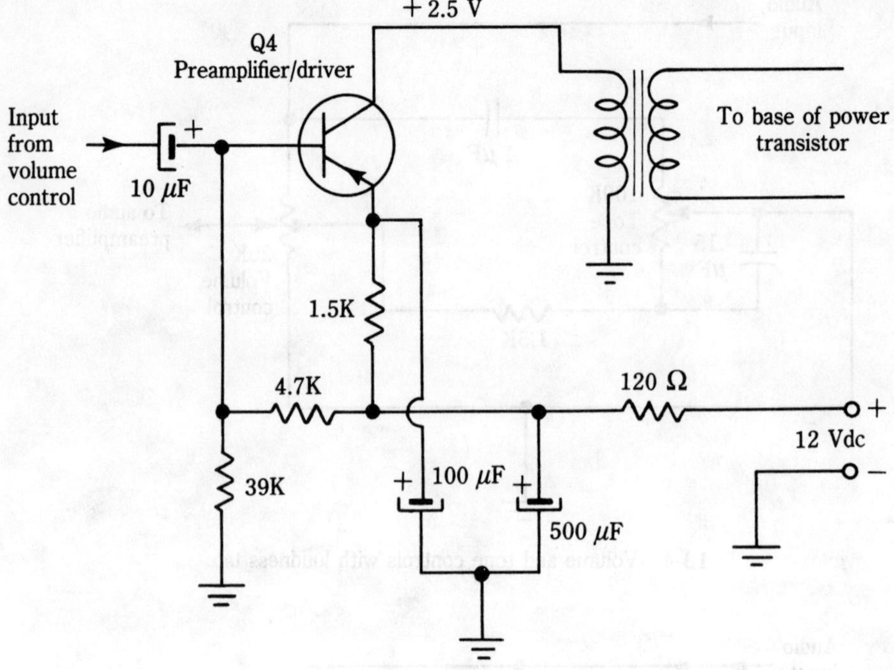

13-6 Preamplifier circuit.

coupling is accomplished by high-value electrolytic capacitors. These capacitors are generally part of a multisection filter capacitor.

A significant improvement in gain is offered by the preamplifier circuit shown in Fig. 13-7, which uses two direct-coupled transistors. Transistor Q4 acts as the preampli-

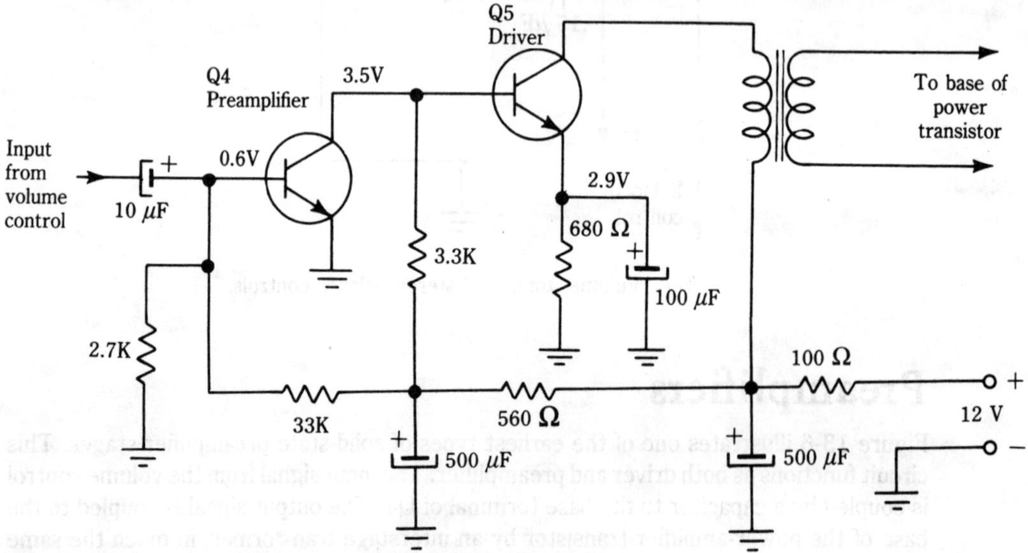

13-7 Preamplifier/driver circuit.

fier, while Q5 functions as the driver stage. Note that the collector voltage at Q4 is the same as the base voltage of Q5, allowing the elimination of coupling devices such as capacitors and transformers. The input signal applied to the base of Q4 causes the collector voltage to vary. These variations are read by the driver transistor as an ac input signal, then Q5 further amplifies this signal and couples it to the power amplifier through an interstage transformer.

Audio circuits are among the most popular weekend and one-evening projects for electronic hobbyists. Part of the reason for this popularity is that audio circuits are so useful to so many. Another reason is that audio circuits are generally well behaved, so can be built with low-cost components. In this section we will take a look at some audio projects and circuits that you can build as experiments.

Audio preamplifiers

Several companies make IC preamplifiers especially for audio applications. Typical of them are the LM-381 and LM-382 devices (Fig. 13-8).

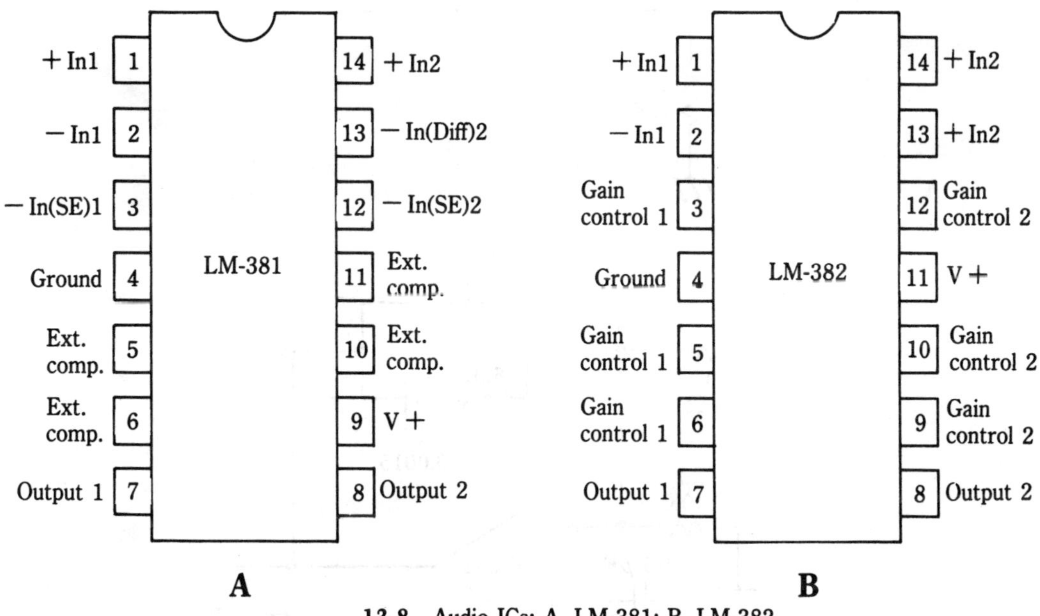

13-8   Audio ICs: A. LM-381; B. LM-382.

Figure 13-9 shows two circuits based on the LM-381 device. This device is similar to a dual operational amplifier, except for the single-polarity dc power supply used. The circuit in Fig. 13-9A is a wideband, low-distortion audio preamplifier. With the values shown it will yield a voltage gain of 10 at a total harmonic distortion (THD) of less than 0.05 percent. The frequency response of this circuit is essentially flat throughout the audio spectrum (up to the limit of the amplifier).

The circuit shown in Fig. 13-9B is a tape preamplifier. A large number of cassette tape players use the LM-381 device as the preamplifier, and I suspect that was the intended market. The tape head is coupled to the noninverting input of the LM-381 device

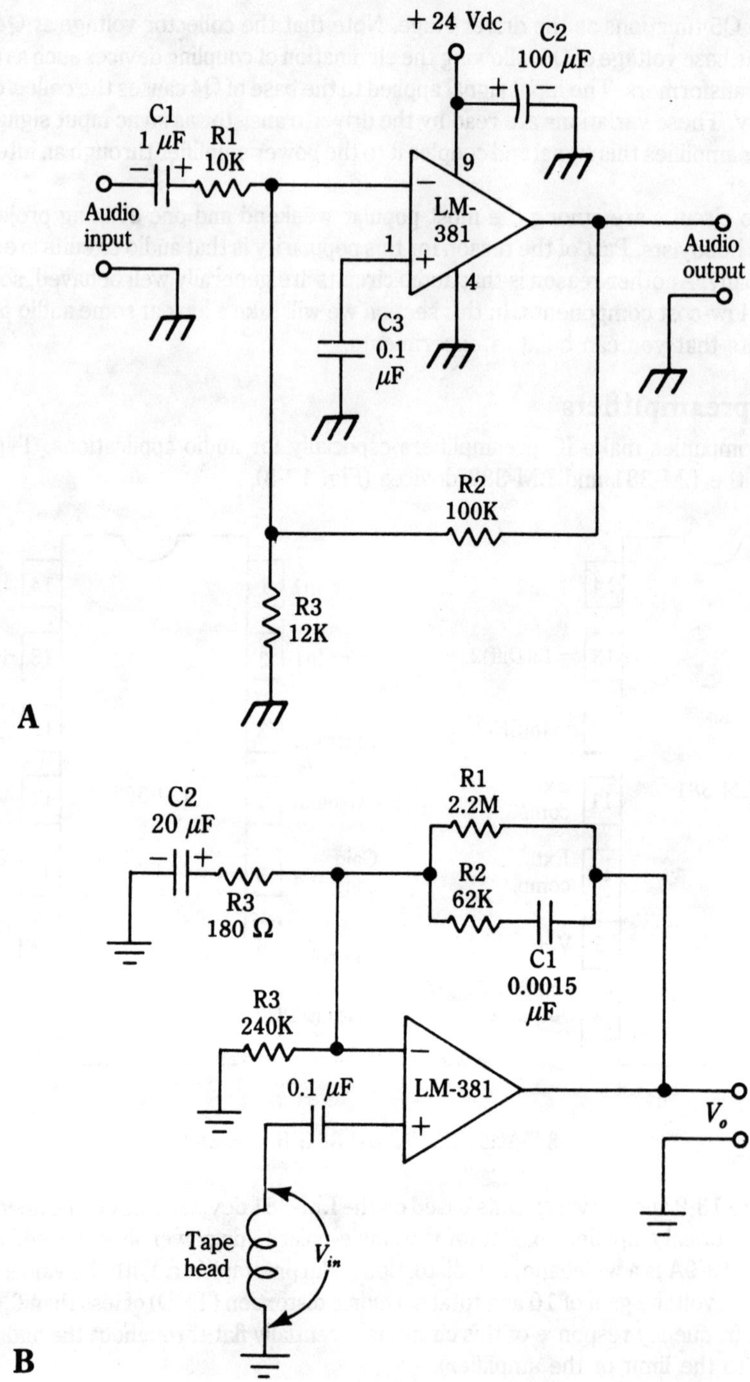

13-9   A. LM-381 preamplifier circuit; B. LM-381 tape preamplifier.

through a 0.1 μF capacitor. Because of the low input bias of the preamplifier, no resistor is needed from the noninverting input to ground. On many versions of this circuit, a resistor is needed in order to prevent charging of the capacitor by input bias currents.

The low frequency response of the circuit is set by the combined action of resistor R3 and capacitor C2 according to the expression:

$$F = \frac{1,000,000}{2\pi R_3 C_2} \tag{13-1}$$

Where: F is the low end -3 dB frequency, in hertz (Hz)
$\quad\quad$ C_2 is expressed, in microfarads (μF)
$\quad\quad$ R_3 is expressed, in ohms (Ω)

The values shown for R_3 and C_2 in Fig. 13-9B yield a low-end -3dB frequency of about 45 Hz. The shape of the frequency response of this preamplifier is set to correspond to that required for the cassette tape equalization curve.

Operational amplifier preamplifiers

The classical operational amplifier is one of the most useful ICs in the electronics catalog. It is easy to apply, is usually well behaved, and is low in cost. Designing circuits with the operational amplifier is generally so easy that one wag was tempted to correctly claim that it makes ". . . the contriving of contrivances a game for all." Circuits that only very advanced hobbyists could tackle are now open to even the newcomer.

Figure 13-10 shows the basic operational amplifier configurations used in audio circuits. Figure 13-10A is the inverting follower and Fig. 13-10B is the noninverting follower. In the inverting circuit, the output signal will be 180 degrees out of phase with the input signal (i.e., reversed). In the noninverting signal, on the other hand, the output signal is in phase with the input signal. In both cases, the gain is set by the combined action of the input resistor, R1, and a feedback network. The expressions for the special cases where

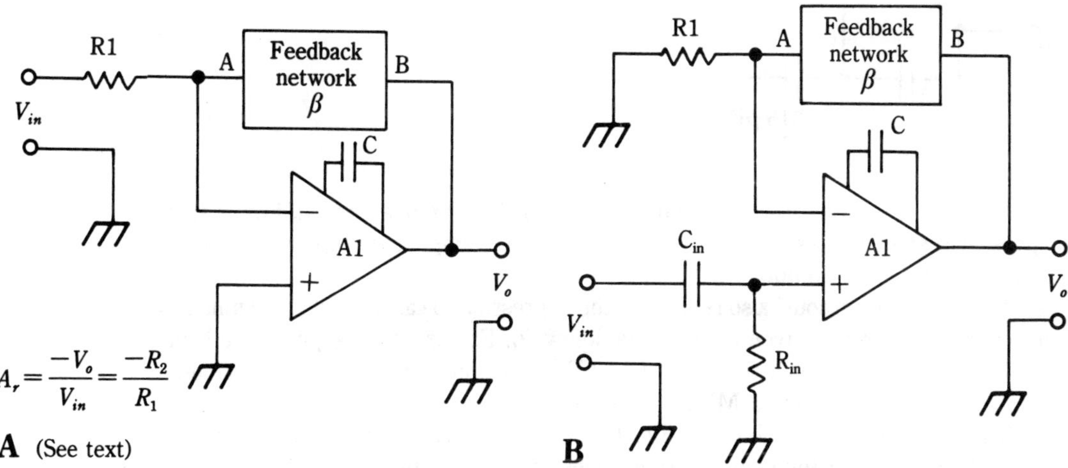

13-10   A. Op amp inverting follower used for audio; B. Op amp noninverting follower used for audio.

the feedback network is a single resistor (see Fig. 13-11A) are shown with each circuit.

Several popular variations of the feedback network that find application in audio preamplifiers are shown in Fig. 13-11. The version in Fig. 13-11A is for a wideband amplifier that has a frequency response with no tailoring except by the natural bandwidth of the amplifier. You can calculate the approximate available bandwidth from the gain-bandwidth product (F_t) of the device, which is the frequency at which gain drops to unity (1). The expression is:

$$F_t = Gain \times Bandwidth \qquad (13\text{-}2)$$

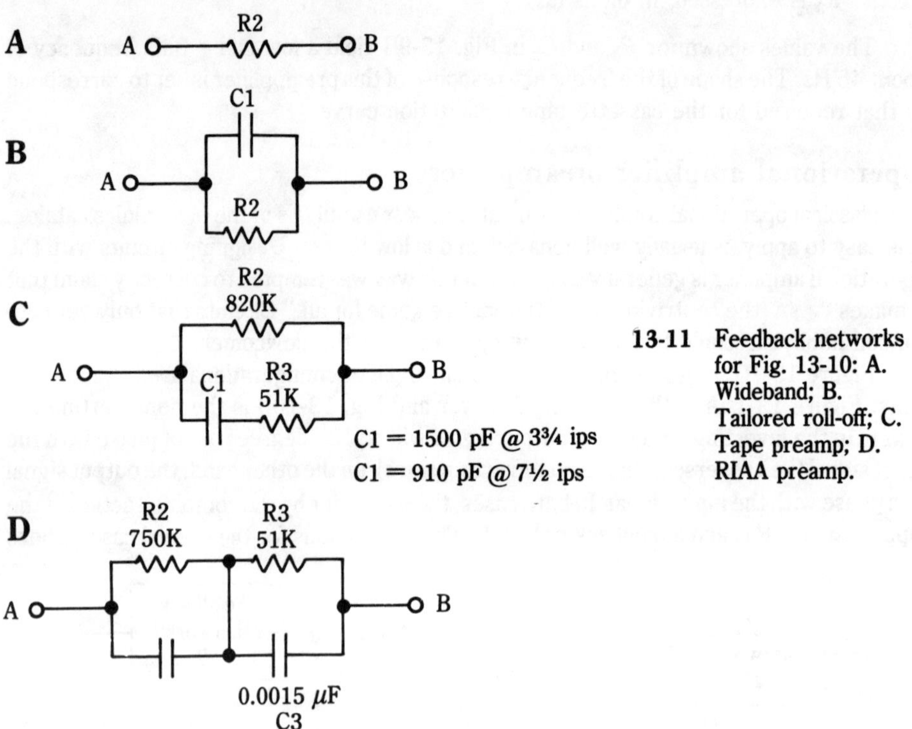

13-11 Feedback networks for Fig. 13-10: A. Wideband; B. Tailored roll-off; C. Tape preamp; D. RIAA preamp.

One use of this expression is to determine which op amp is needed. For example, suppose we need a frequency response of 20,000 Hz in an amplifier with a gain of 150. The required F_t is $150 \times 20,000 = 3$ MHz.

Alternatively, we could also rearrange the expression to calculate the maximum frequency response of any given circuit if we know F_t. For example, suppose we have a 1 MHz op amp in a circuit with a gain of 100. The maximum frequency response of the amplifier will be $F_t/Gain = 3$ MHz/100 = 30 KHz.

The feedback network shown in Fig. 13-11B produces a flat gain at low frequencies equal to that of the resistor alone. At frequencies above a certain point, however, the gain of the amplifier will roll off at −6 dB per octave. The "breakpoint" between the low-frequency gain and the rolled-off segment (i.e., the high-end -3 dB point) is found from Eq. 13-1.

The remaining feedback networks are used in special preamplifiers. That shown in Fig. 13-11C is used for NAB-compensated tape preamplifiers, while that of Fig. 13-11D is used for RIAA-compensated phonograph preamplifiers.

Figure 13-12 shows an example of a stereo audio preamplifier based on the op amp. It was once called the MC-1303 when Motorola was the source of the chip, but other manufacturers now make it, so it is now available from many sources under the type number LM-1303. The chip contains two operational amplifiers that are completely independent except for the V− and V+ dc power supply connections.

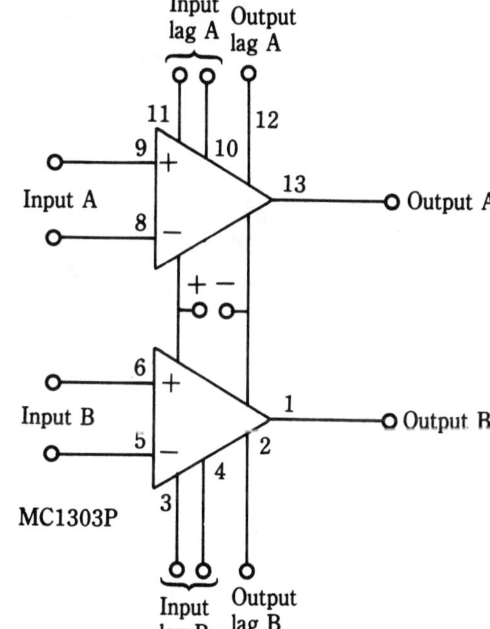

13-12 LM-1303/MC-1303P stereo audio device.

Some additional projects

Most readers of this book are probably activists in that they prefer to go to the workbench and try some of the things that they learned in the text. In this section several practical circuits are presented that can be used in audio and wideband amplifier applications.

Audio mixer projects

An *audio mixer* is a circuit that combines audio frequency signals from two or more input sources into a single channel. Application examples include multiple microphone public address systems, multiple electric guitar systems (music?), or a radio station console service where inputs from tape players, record players, or two (or more) microphones are combined into a single line that goes to the transmitter modulator input connection.

Audio mixer 1   The first audio mixer project is an operational amplifier version shown in Fig. 13-13. This circuit is basically nothing more than a unity-gain inverting follower circuit with multiple inputs. Three audio lines are identified here: AF1, AF2, and AF3. Each of these sources is applied to the input of the operational amplifier, and gains of R_4/R_1, R_4/R_2, and R_4/R_3, are realized. The overall transfer equation for this circuit is:

$$V_o = -V_{in}R_4 \left(\frac{1}{R_1} + \frac{1}{R_2} + \frac{1}{R_3} \right) \tag{13-3}$$

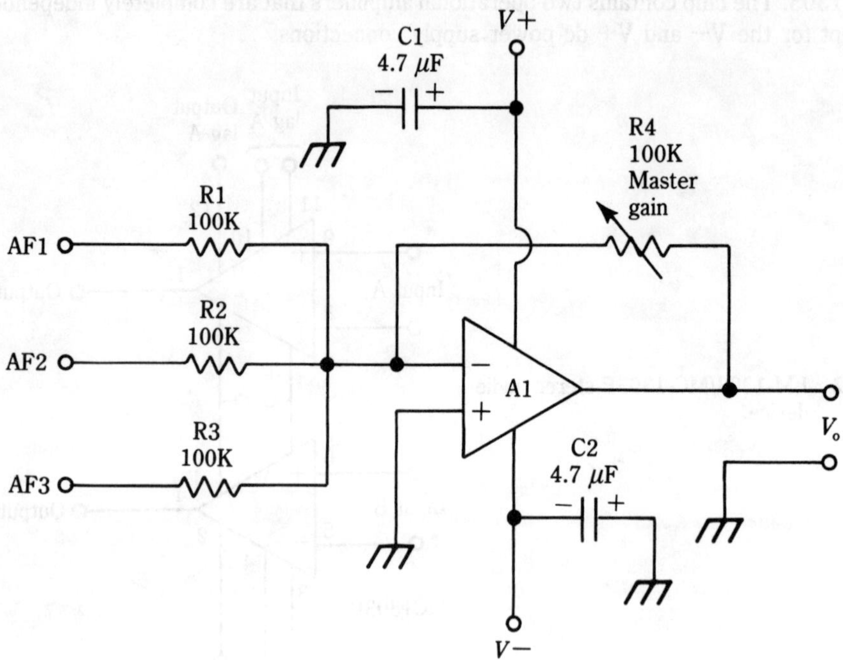

13-13   Op amp mixer circuit.

Because all of the resistors in Fig. 13-13 are 100 Ω, the overall gain is unity.

Gain can be customized on a channel-by-channel basis by varying the input resistance value. The gain of any given channel will be $100 \ k\Omega/R$, where R is the input resistance (e.g., R_1, R_2, or R_3) in kΩ. Be careful not to reduce the input resistance so far that the source is loaded down. If the source is another operational amplifier preamplifier (or voltage amplifier), then the input resistance can be reduced to several kilohms without causing a problem. If the source is a high-impedance phonograph cartridge or microphone element, however, then 50 kΩ (or even higher) is more preferable.

In some cases, it might be beneficial to increase the value of the feedback resistance to 1 MΩ or so, in order to make the corresponding input resistances higher for any given gain value. Remember, the input impedance seen by any single channel is the value of the input resistor.

A master gain control is provided by making the feedback resistor variable from 0 to 100 kΩ. If no gain control is needed, then make the resistor fixed. An audio taper potentiometer is used for most applications. If the application calls for a one-time set-and-forget gain control adjustment (as might be true in a radio station application), then make R4 a trimmer type potentiometer; otherwise, it should be a panel-mounted type, with a shaft appropriate for a knob (e.g., quarter-inch half or full round).

The operational amplifier selected for this circuit can be almost any good operational amplifier with a gain bandwidth (GBW) product that is sufficient for audio applications. Because the gain is unity, any GBW over 20 KHz will suffice (which almost all devices except the 741 family, which can be used in communications applications only, deliver).

Audio mixer 2   The circuit in Fig. 13-14 is an improved audio mixer that is based on the GE/RCA CA-3048 amplifier array. This circuit provides appproximately 20 dB of

Pinouts

Amplifier	−In	+In	Output
A	3	4	1
B	7	8	6
C	10	9	11
D	14	13	16

13-14   CA-3080 amplifier circuit.

gain at each channel. The CA-3048 device is a 16-pin DIP integrated circuit that contains four independent ac amplifiers. Offering a gain of 53 dB with a GBW of 300 KHz (typical), the CA-3048 has 90 kΩ input impedance and an output impedance of less than 1 kΩ. It will produce a maximum low distortion output signal of 2 volts RMS, and can accept input signals up to 0.5 volt RMS.

Each dc power supply terminal can accept up to $+16$ Vdc. There are actually two V+ and two ground connections. These multiple power-supply connections are used to reduce the internal coupling between amplifiers. The two V+ terminals are tied together externally, and the two ground terminals are also tied together. The V+ terminals are bypassed with a dual capacitor. C5 in Fig. 13-14 is a 0.01 μF unit, and is used for decoupling the high frequencies, while C6 is a 4.7 μF unit that decouples low frequencies. Both capacitors must be mounted as close as physically possible to the body of the CA-3048, with C5 taking precedence for closeness over C6 (high frequencies are more critical).

An RC network from the amplifier output to ground (R3/C2) is used to stabilize the amplifier, thus preventing oscillations. Like the power-supply bypass and decoupling capacitors, these components need to be mounted as close as possible to the body of the amplifier.

Only one channel is shown in detail in Fig. 13-14 because of space limitations. Each of the other three channels are identical to the circuit shown, and are joined together with each other at the output capacitor (C4). Each of the four channels has its own level control (R1), which also provides a high input impedance for the amplifier.

600 Ω audio circuits

Professional audio and broadcast amplifiers generally use a 600 Ω balanced line between devices in the system. For example, a remote preamplifier will have a 600 Ω balanced output and will connect to the next stage through a three-wire line (Fig. 13-15). Such a

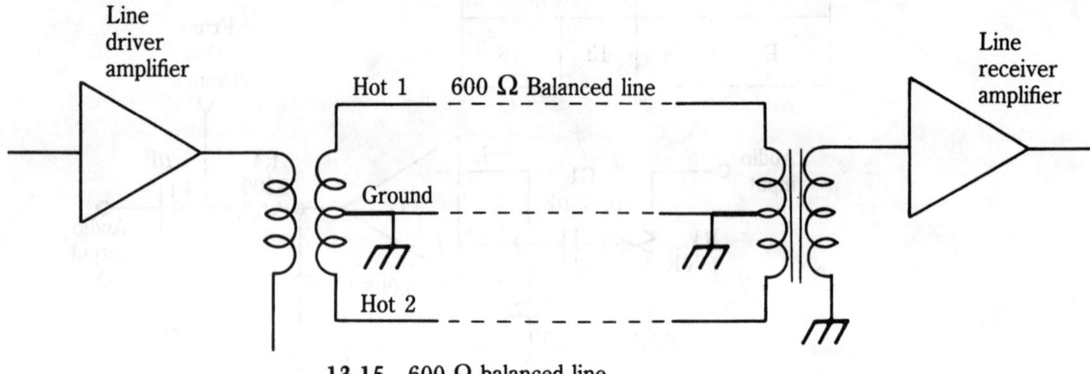

13-15   600 Ω balanced line.

system has two "hot" wires and a ground line as the stage-to-stage interconnection. An amplifier with a 600 Ω balanced output is called a *line driver*, while an amplifier with a 600 Ω balanced input is called a *line receiver*. Some amplifiers are both line drivers and line receivers.

Figure 13-16 shows a line receiver amplifier that is based on the LM-301 operational amplifier IC device connected in a unity gain, noninverting circuit configuration. The input circuit is a line transformer with a 600 Ω balanced primary. If the turns ratio of transformer T1 is 1:1, then the overall circuit gain is unity. If, however, the transformer has a turns ratio other than 1:1, the gain is essentially the turns ratio of the transformer. For example, suppose a transformer is selected with a 600 Ω balanced input winding, and a 10,000 Ω secondary winding. Such a transformer has a secondary-to-primary ratio of 10,000/600, or about 17:1, so the turns ratio is on the order of the square root of 17, or

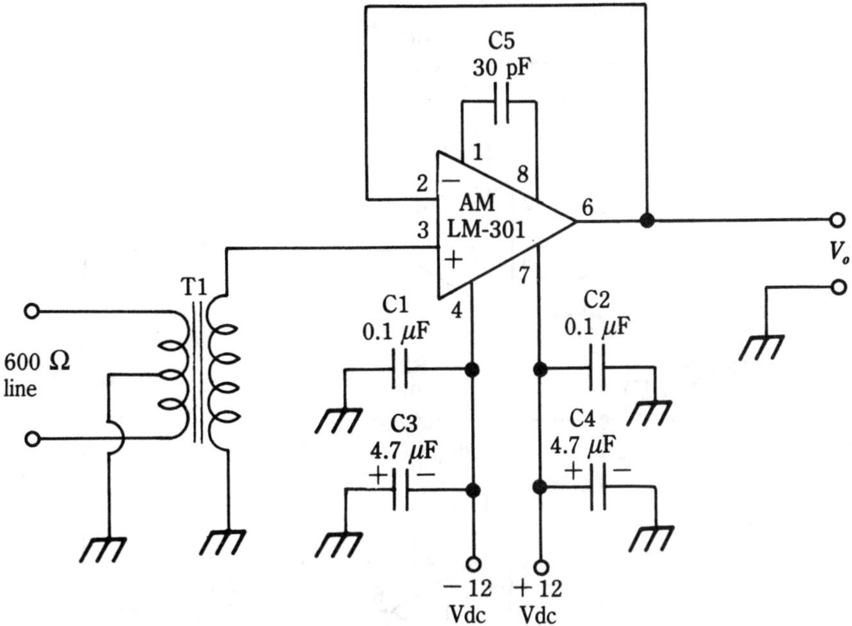

13-16 Line receiver amplifier.

about 4.1. Thus, the voltage actually applied to the noninverting input will be 4.1 times higher than the input signal voltage received from the 600 Ω line.

Like most other operational amplifier circuits, Fig. 13-16 requires two dc power supplies: V− and V+. Typically, these power supplies can be anywhere from ±6 to ±15 volts. As in other circuits, each dc power supply is decoupled with a pair of capacitors: a 0.1 μF unit for higher frequencies and a 4.7 μF unit for lower frequencies.

The output of the circuit is an ordinary single-ended voltage amplifier (as in other op-amp circuits), so will typically have a very low impedance. Some designers use a trans-former-coupled output circuit. It is possible to get away with a 600 Ω 1:1 transformer if the natural output impedance of the op-amp is on the order of 50 Ω or less. The general rule of thumb is that for best voltage transformation, the primary impedance of any trans-former selected should be ten times the natural output impedance of the device.

Another way to make a 600 Ω line input amplifier is to use the simple dc differential amplifier circuit. Use 300 Ω input resistor (note: 330 Ω resistors are standard values).

Figure 13-17 shows a line driver amplifier based on a pair of operational amplifiers. The dc power supply connections are deleted for sake of simplicity, but are the same as in the previous example. The output circuitry is balanced because it is made from two single-ended op amps driven 180 degrees out of phase with respect to each other. The low output impedance of the op amp, plus the 270 Ω resistors in series with each output, makes a balanced output impedance of approximately 600 Ω.

The circuit of Fig. 13-17 is a good example of the clever use of one of the properties of the ideal operational amplifier. Recall from chapter 11 that one of the ideal properties is that "inputs stick together." In other words, applying a voltage to one input causes the same voltage to appear at the other input.

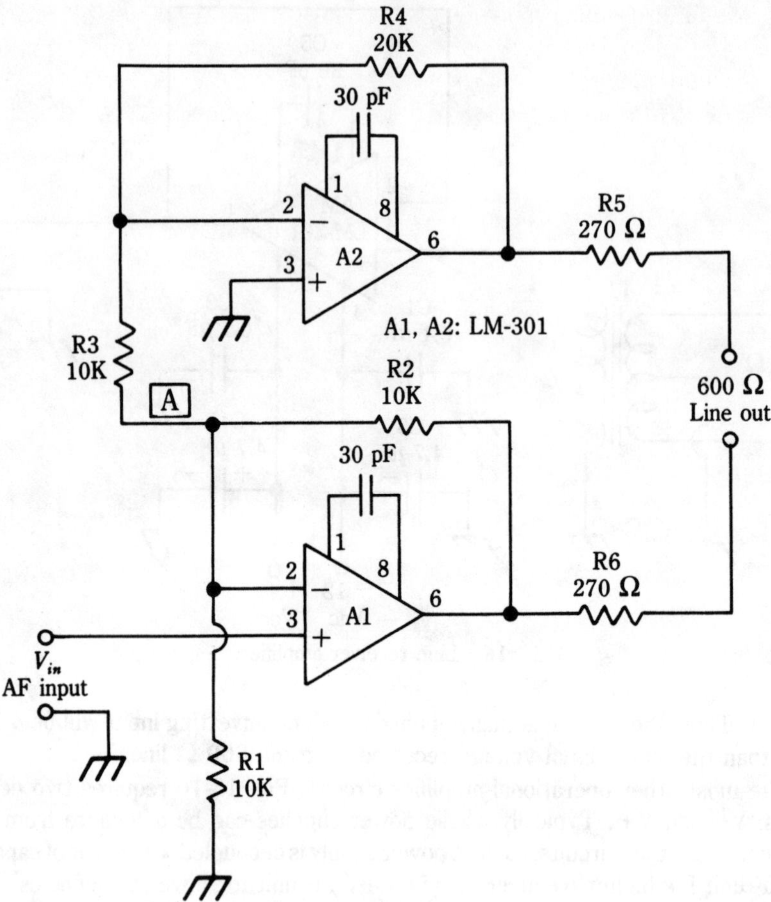

13-17   Line driver amplifier.

In this case, for example, the audio frequency input signal voltage applied to the non-inverting input of amplifier A1 also appears on the inverting input of the same amplifier. Thus, we will see V_{in} both on the noninverting input and at point "A" in Fig. 13-17. We can therefore use point "A" to feed the other half of the balanced circuit, amplifier A2. Because A1 is a noninverting gain-of-two circuit and A2 is a gain-of-two inverting circuit, the two sides are out of phase with each other — which is the condition required of the two balanced output lines.

Audio preamplifier circuits

A *preamplifier circuit* is an audio amplifier that gives some initial amplification to the signal before passing on to another circuit for additional amplification or signal processing. For example, a microphone has a low-level output of several millivolts to 0.2 volts. A preamplifier typically boosts the microphone signal to 100 to 1,000 mV before it is applied to the input of a power amplifier (for the loudspeaker), the input of a transmitter modulator, or a graphic equalizer or other circuit.

Microphone preamplifier Figure 13-18 shows a simple gain-of-100 microphone preamplifier for communications and non-hi-fi public address uses. This circuit uses an LM-301 operational amplifier in the noninverting follower gain configuration. With the values of feedback and input resistors (R1 and R2) shown, the gain is 101 plus or minus tolerance variations in the resistors.

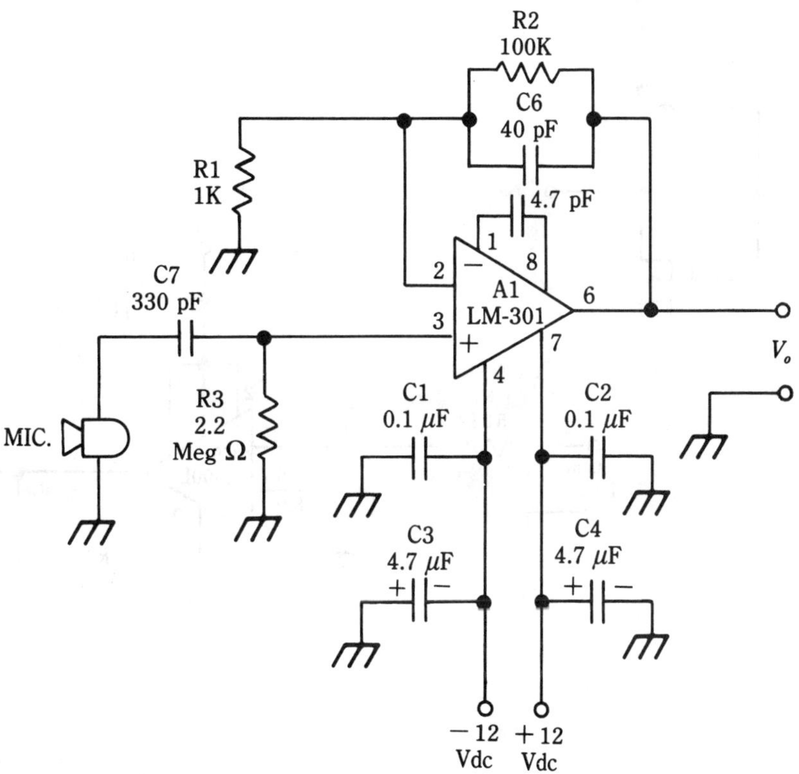

13-18 Microphone preamplifier.

The input circuit of this preamplifier is capacitor-coupled to the microphone. In order to keep the input bias current of the op amp from charging the capacitor, and thereby causing latch-up of the op amp, use a 2.2 MΩ resistor (R3) from the noninverting input to ground.

This circuit is relatively generic, but can be modified toward the less complex if you use a dynamic microphone only. Dynamic microphones use either a high- or low-impedance coil (like a loudspeaker, but reversed in function) that is permanently connected into the circuit. In that case, delete R3 and C7, and connect the microphone between ground and pin no. 3 of the op amp.

If the microphone is to be disconnected from time to time, then you must retain R3 in the circuit in order to prevent the op amp from saturating at or near V+ when the noninverting input goes open.

The frequency response of the circuit is tailored by capacitor C5 and by capacitor C6 shunting the feedback resistor. With the values of capacitance shown, the upper

—3 dB point in the frequency response curve is a little over 3,000 Hz, and falls off at a rate of approximately −6 dB/octave above that frequency.

General preamplifier   Figure 13-19 shows two general-purpose preamplifiers based on the RCA CA-3600E device. This IC is a complimentary COS/MOS transistor

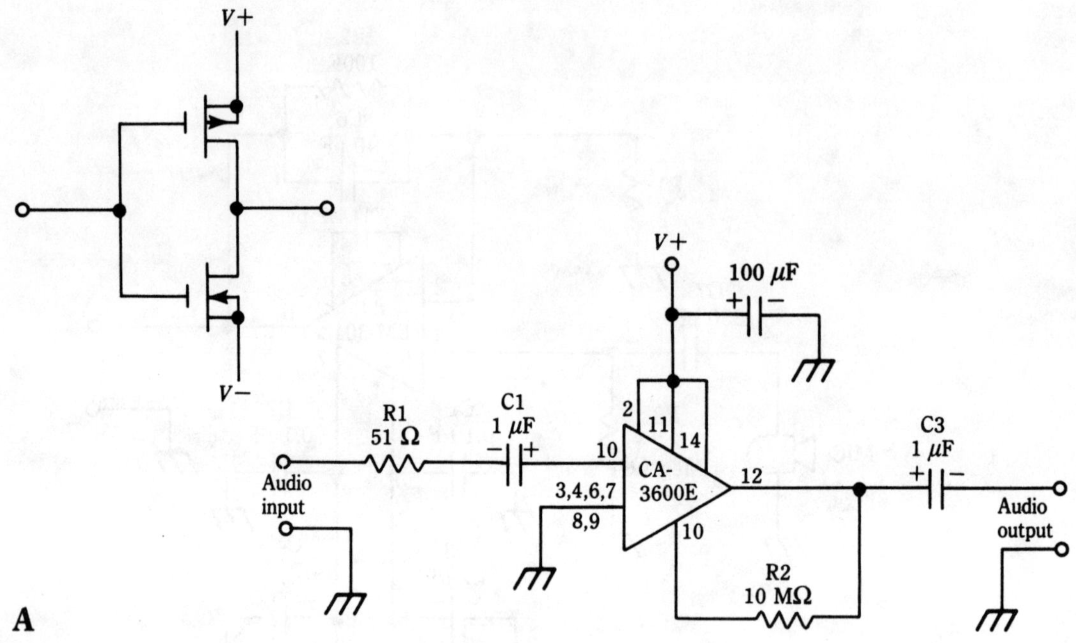

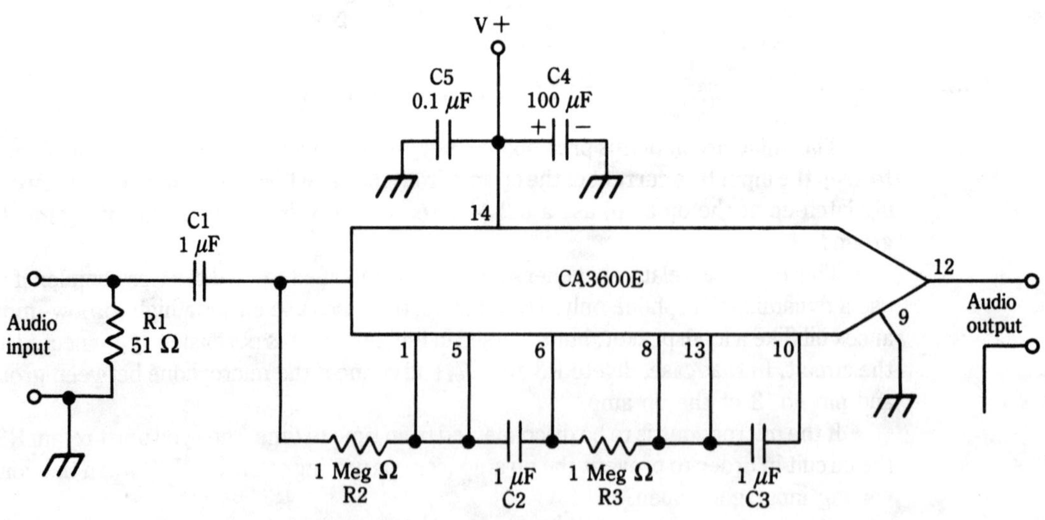

13-19   A. CA-3600E wideband audio amplifier; B. wideband amplifier.

array. A single-stage design is shown in Fig. 13-19A, while a multistage design is shown in Fig. 13-19B. The internal transistor-array equivalent circuit for one transistor pair is shown in the inset to Fig. 13-19A. The single-stage design is capable of up to 30 dB voltage gain at a V+ voltage of 15 Vdc, and slightly more at lower dc power-supply potentials. The higher gain, however, sacrifices some of the 1 MHz -3 dB point.

The multistage design shown in Fig. 13-19B is capable of voltage gains up to 100 dB at frequencies, with dc power-supply potentials up to 10 Vdc. The gain drops to 80 dB for $+15$ Vdc supplies.

This gain and frequency response is quite useful in audio and other applications, but must be approached with caution when actually built. Be sure to keep the input and output sides of the preamplifier separated from each other. Also, make sure that the power-supply decoupling capacitors are mounted as close as possible to the body of the IC.

The 50 Ω input impedance of this amplifier removes it from the audio amplifier category (which usually wants to see higher impedances), unless a preamplifier stage is provided that will have a higher impedance. A good candidate is the circuit of Fig. 13-19A.

Compression amplifiers

A *compression amplifier* is one that reduces its gain on input signal peaks and increases the gain in the input signal valleys. These circuits are used by electronic music fans and by broadcasters to raise the average power in the signal without appreciable distortion (Fig. 13-20).

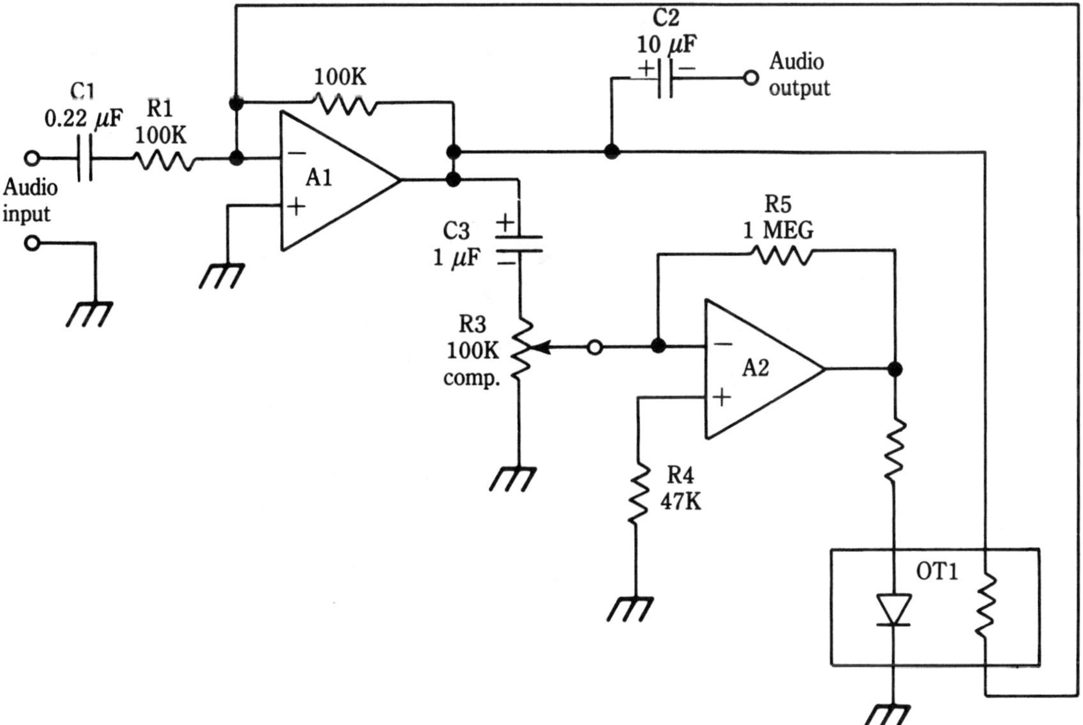

13-20   Compression amplifier.

The amplifier (A1) is any good audio operational amplifier, such as the LM-301. The gain of the circuit is set by input resistor R1 and a feedback resistance that consists of the parallel combination of R2 and the resistance of the optocoupler (OT1) output element. The OT1 resistance is set by the intensity of the light-emitting diode (LED) brightness, which is in turn set by the signal amplitude produced by A2. Because the A2 output signal is proportional to the A1 output signal, overall gain "reduces itself" (i.e., *compresses*). Optoisolator OT1 can be any resistance output device such as the Clairex, or a modern type such as the H11 that uses a JFET for the resistance element.

14
Small-signal
RF amplifiers

THE RADIO FREQUENCY (RF) AMPLIFIER, WHICH IS THE FIRST STAGE IN THE
radio, is fed its input signal by the antenna. Since signals from the antenna are extremely
weak, the RF amplifier must strengthen the signal of interest and attenuate all others. In
all radios, a transistor or integrated circuit — which contains transistors — provides the
amplification, while LC tank circuits consisting of inductors (L) and capacitors (C) provide
the signal selection (Fig. 14-1).

RF amplifiers

The circuit shown in Fig. 14-2 is typical of many solid-state RF amplifier designs in radios.
This circuit and close variations have been in use for over 10 years.

This particular circuit is from a Delco car radio designed and produced by the Delco
Electronic Division of General Motors. Notice that variable tuning is accomplished by adjusting the powdered-iron slugs of the tuning coils, rather than by the use of variable capacitors. The variable capacitors, shown in Fig. 14-2, are used to trim the coils so that
they will all track on the same frequency. This is standard procedure on car radios.

Home radios normally use two or three variable capacitors coupled together, or
ganged, as the main tuning control. In those sets, you must trim the coil and sometimes a
very low value trimmer capacitor to obtain proper tracking. Car radios, however, have
followed a different procedure because of the design requirements of push-button tuners.
The mechanical assembly that gangs all of the coil slugs is called a *permeability tuning
mechanism* (PTM). The PTM is driven by a linkage system connected to either a manual
or a push-button tuning assembly.

Resistor R1 serves to stabilize the emitter circuit of the RF amplifier stage. Bias to
this type of stage is supplied by the agc voltage connected to the base terminal of transistor Q1 via resistor R2 and the secondary winding of coil L2. The dc return is through coil
L3 and resistor R3, connected between ground and the collector terminal of transistor
Q1.

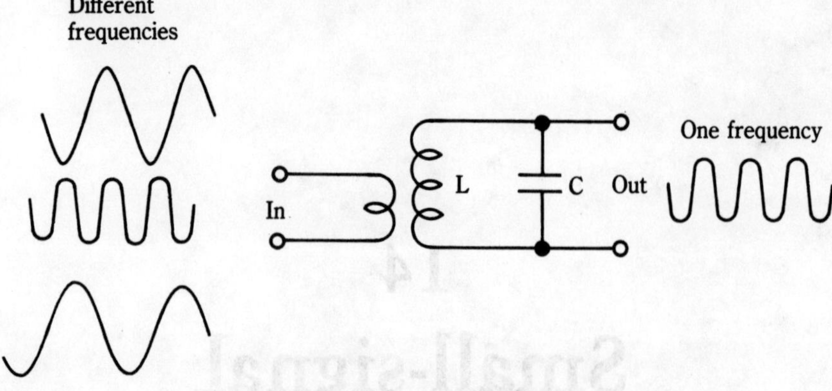

14-1 Operation of a tuned LC RF circuit.

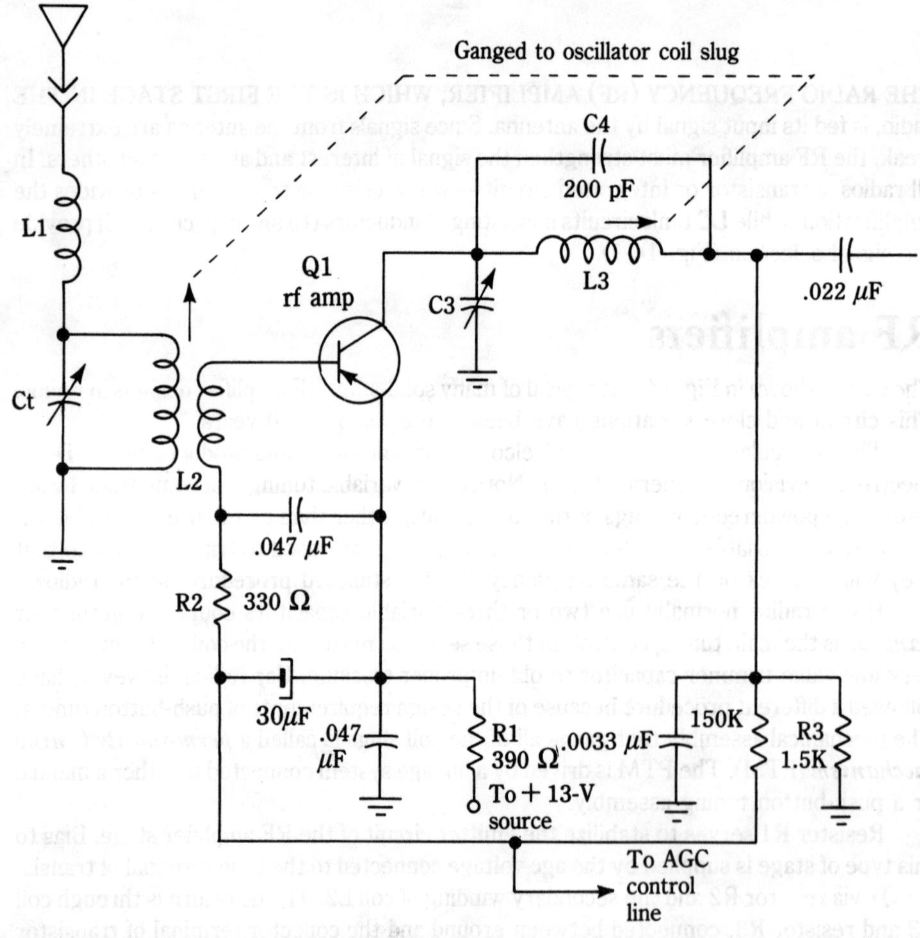

14-2 Tuned pnp transistor RF amplifier.

Fig. 14-3 shows a simplified schematic of the RF amplifier stage. For clarity, all the circuitry, except for that in the dc path, has been removed. The total current drawn by the stage flows down through resistor R1. Inside the transistor, two currents combine to create this total current. The majority of the current, in excess of 95 percent, flows down from the collector circuit. The remaining current flows into the base from the agc line.

From a servicing technician's point of view, there are three significant voltages in this type of circuit. One is the base-to-emitter bias voltage. In radios using germanium transistors, this will run close to 0.2 volt, while in radios with silicon transistors, the value will be closer to 0.6 volt. The last two significant voltages — the voltage drop across the emitter resistor and the collector-to-ground voltage — offer a good indication of how much current the state is drawing. The bias voltage and the two conduction voltages will vary directly with the agc control voltage. The agc control voltage is proportional to the strength of the received signal.

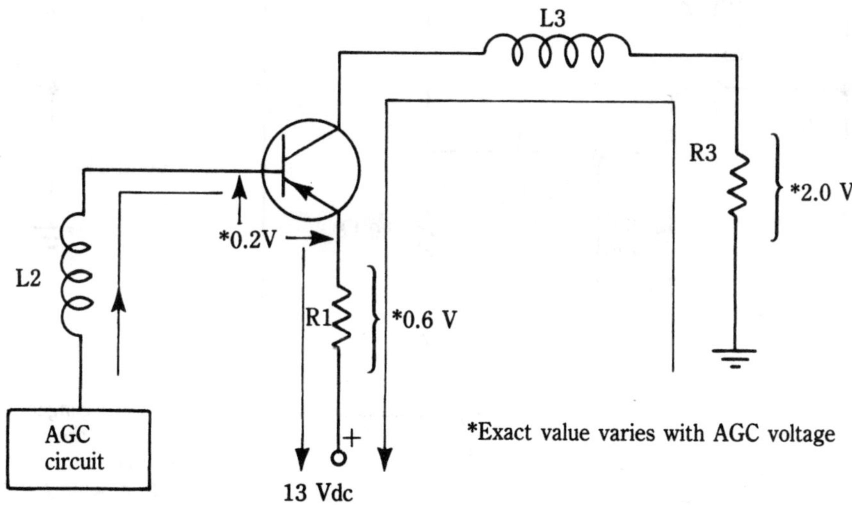

14-3   Conduction in RF amplifier circuit.

Figure 14-4 shows an RF stage using an npn transistor. RF signals are decoupled or bypassed to common ground by capacitor C1. The function of this capacitor is to place the lower end of the input coil secondary at ac ground potential, while keeping it above ground as far as the dc bias voltage is concerned. It does so because a capacitor will pass ac while at the same time it blocks dc.

Notice that in this type of stage the emitter is connected close to dc ground (negative side of the battery on most American automobiles), while the collector is tied to the positive side of the battery. In the pnp stage of Fig. 14-2, the opposite is true due to the polarity differences of the respective types of transistor. Other than that, the two stages are essentially the same.

There are two facets to the ac signal path in the RF amplifier stage: the bypassing or decoupling ac path and the frequency-selective (tuning circuits) signal path. Figure 14-5 illustrates two examples of tuned circuits.

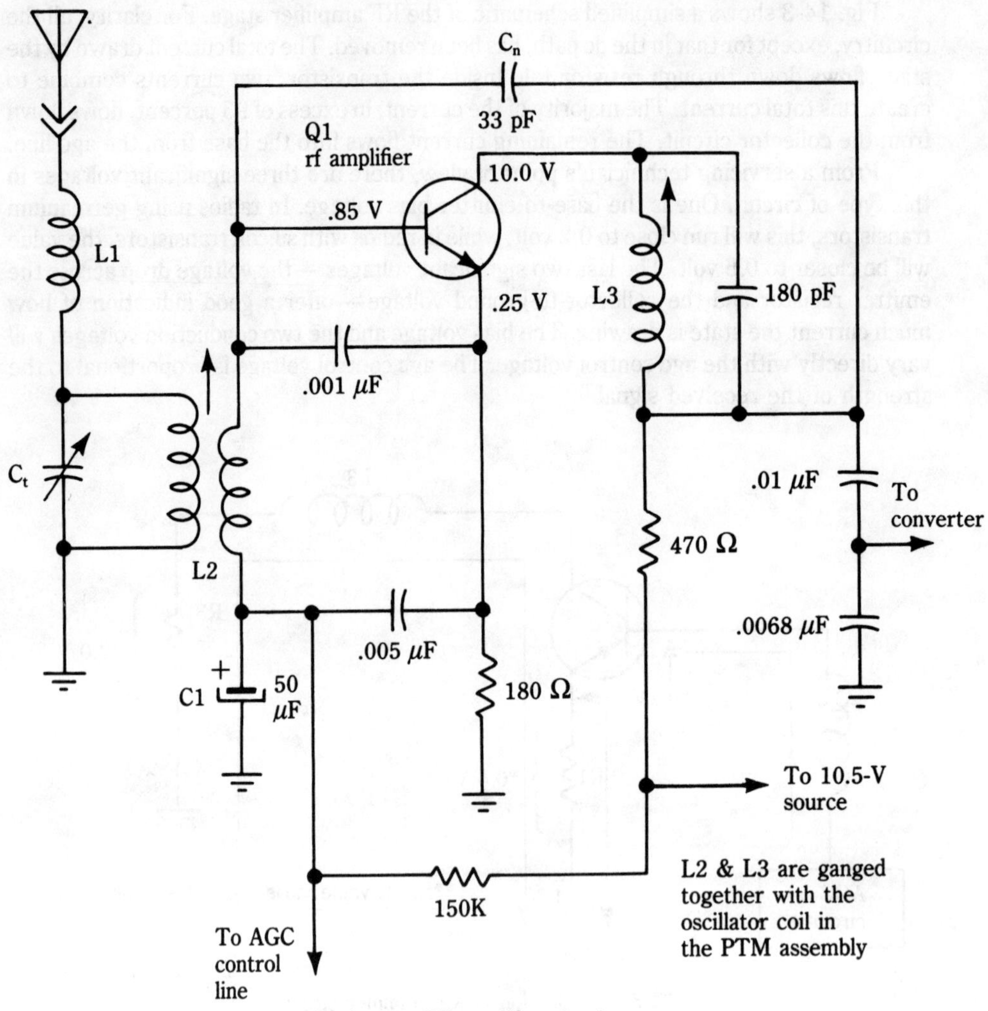

14-4   npn RF amplifier circuit.

Although the preceding RF amplifiers show only two apparent tuned circuits, there are in reality three. In Fig. 14-2, the primary of coil L2 is tuned to the frequency of the station being received. Capacitor C_t is called the *antenna trimmer capacitor*. In addition to tuning the coil, it also acts to compensate for differences in antenna cable capacitance. It is usually adjusted when the radio is originally installed in a vehicle.

The inductor L3, which is in the collector circuit of the RF amplifier transistor, is tuned to two different frequencies. Capacitor C3 tunes L3 to the same frequency to which L2 is tuned. This is the frequency of the radio station being received. Capacitor C4 also tunes L3. This capacitor tunes L3 to the image frequency, a frequency other than the received frequency.

The *image frequency* is the signal frequency (F_s) plus two times the i-f frequency ($F_{i\text{-}f}$) or $F_s + 2F_{i\text{-}f}$. In the standard American-made car radio, the image frequency will be

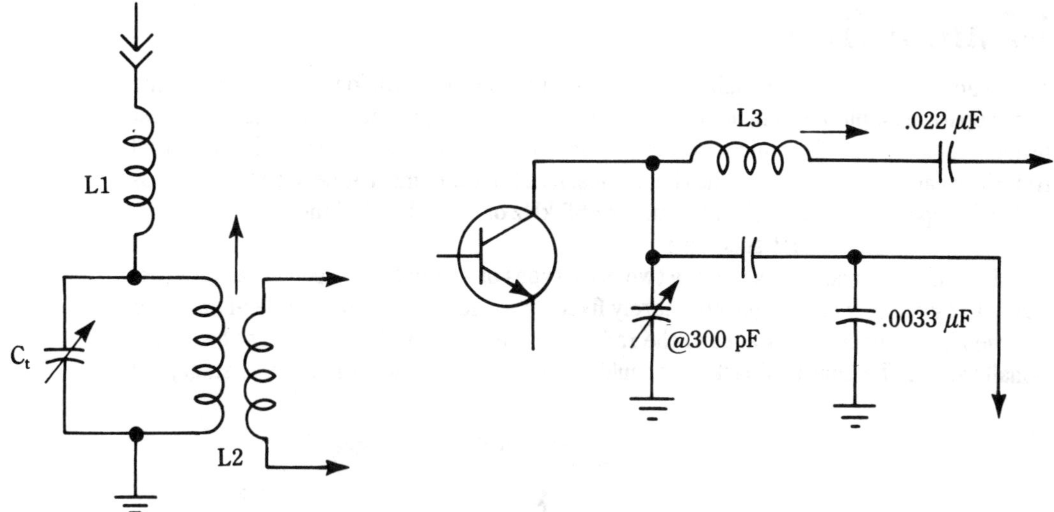

14-5   Tuned circuit and image trap.

equal to $F_s + (2 \times 262.5 \text{ kHz})$. If the radio were tuned to 1,000 kHz, the image frequency would appear at 1,000 kHz + (2 × 262.5 kHz), or 1,000 kHz + 525 kHz = 1,525 kHz. The radio would respond to the 1,525 kHz when it was tuned to 1,000 kHz. On the crowded AM broadcast band, this would produce chaos for the listener.

The tuned circuit (C4 and L3) suppresses the image frequency and greatly improves the so-called image response of the radio. Without this image trap, the heterodyne image easily could interfere with the proper operation of the radio. The trap is parallel resonant at the image frequency. The purpose of a parallel-resonant circuit is to offer a high impedance to its own resonant frequency, while offering a low impedance to all other frequencies. This arrangement allows the trap to pass the RF signal while blocking the image signal.

A capacitive voltage divider couples the output from the RF amplifier to the input of the converter stage. Its function is to match the output impedance of the RF amplifier to the input impedance of the converter transistor. Not all car radios will use this voltage divider, however.

Choke coil L1 (Figs. 14-2 and 14-4) is one of those parts that seems out of place until its value and function are made known. The RF choke has a very low inductance. As a result, L1 will have only a negligible inductive reactance at frequencies occupied by the AM broadcast band. It will, of course, offer a higher reactance at higher frequencies ($X_L = 2\pi f L$).

Noise signals from automobile ignitions and other sparking sources are in the form of pulses with an irregular waveshape. Such signals are extremely rich in high-frequency harmonics. The purpose of L1 is to suppress these high-frequency harmonics, thereby attenuating the noise pulses. It offers opposition to the noise pulse while passing the input signal virtually unaffected. Since car radios are prone to motor noise by the proximity of the auto ignition system, they almost universally incorporate this choke into the antenna circuit. Home radios almost never use such a choke.

i-f amplifiers

The i-f amplifier provides the bulk of amplification required in a radio receiver to bring the signal up to an amplitude sufficient to drive the detector stage. The i-f amplifier also must provide most of the discrimination against unwanted signals (selectivity). In almost all American car radios, 262.5 kHz has been chosen as the i-f. In most American home radios and in European car radios, the i-f is either 455 kHz or 460 kHz. In American car radios, however, it is rare to find such an i-f.

Figures 14-6 and 14-7 show the two basic types of i-f amplifiers used in car radio designs. In essence, these stages are merely fixed-tuned RF-style stages. The i-f amplifier has the highest gain of all stages in the radio. For this reason, the stage must be exceptionally stable. To limit feedback that could cause unwanted oscillation of this stage, the

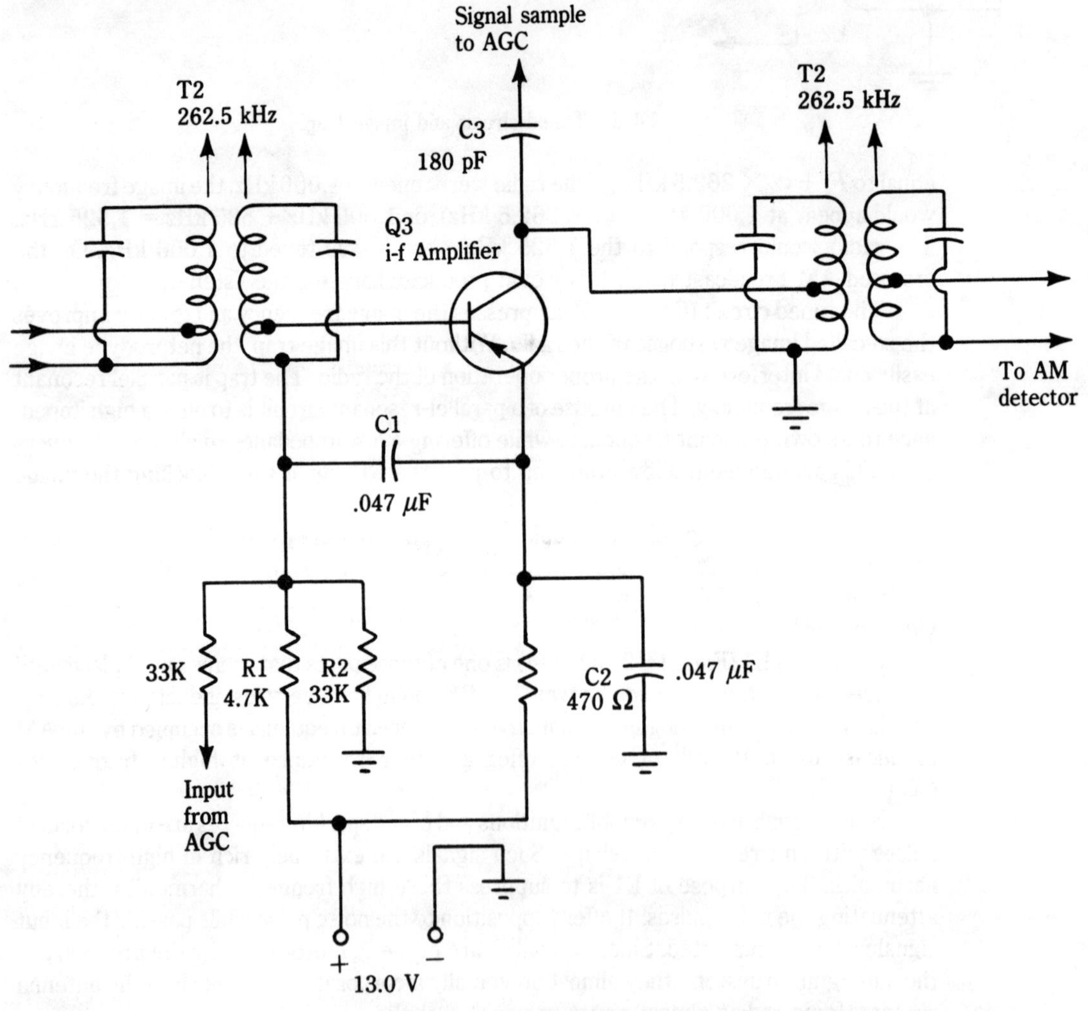

14-6 pnp i-f amplifier circuit.

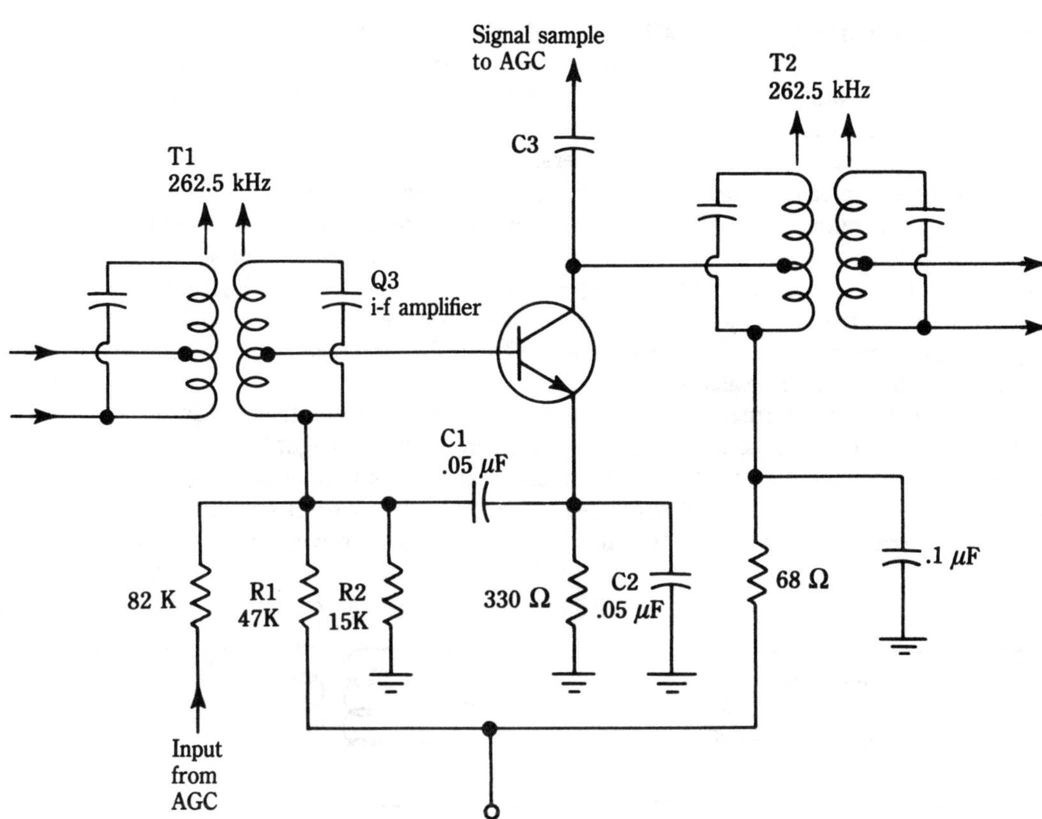

14-7   npn i-f amplifier circuit.

input and output i-f transformers are shielded by a metal housing or *can*. Decoupling, as well as higher gain, is provided by bypass capacitors C1 and C2.

Bias for the i-f amplifier transistor is supplied from two sources: the resistor voltage divider network (consisting of R1 and R2) and the dc control voltage supplied by the agc circuit. The latter bias source is proportional to the strength of the received signal. When the radio is tuned to a strong station, this bias reduces the gain of the i-f and RF amplifier transistors. On weaker stations, the bias increases the gain of the two amplifiers. The effect is to reduce the loudness difference experienced by the listener as he tunes across the bank encountering signals of widely varying strengths.

In the pnp stages, the base of the transistor becomes more negative than the emitter to increase the stage gain. To reduce the stage gain, the agc circuit drives the base to a less negative voltage (as measured in relation to the emitter).

Selectivity in a radio is the property of rejecting all stations except that station to which the radio dial is tuned. Although no one has yet invented a way to receive one station while rejecting all others, the modern superhet approaches that ideal.

The majority of receiver selectivity is a product of the i-f amplifier. The tuned i-f transformers have a high "Q" and are subcritically coupled, producing a narrow selectivity curve, or *bandpass*. Since AM band channels are 10 kHz wide, the i-f amplifier must pass a range of frequencies up to 5 kHz either side of the center frequency. It also must attenuate signals outside of this range. Most car radio i-f amplifiers have passbands up to 10 or 12 kHz wide, which allow them to pass the carrier and sidebands of one station while rejecting stations on adjacent channels. A few very novel AM car radio designs have replaced the i-f transformers with ceramic filters.

The signal level sample used by the agc circuit to generate the dc control voltage generally is taken from the collector of the i-f amplifier stage. This method is preferred because the i-f collector signal has a large enough value to be relatively free of noise and other components that could deteriorate the performance of the agc. It is also high enough to provide an effective agc, which might not be possible in a stage with a lower signal level. The sampling capacitor, C3, will have a value in the 100 to 500 pF range, with most of the capacitors being close to a value of 200 pF.

The i-f output transformer is usually a tapped-impedance resonant tank circuit such as shown in Figs. 14-6 and 14-7. The input transformers, however, might be one of several popular designs. Figure 14-8 shows three common types of i-f impedance circuits. Figure 14-8A is the tapped impedance type, which is identical, except for impedance levels in some cases, to the i-f output transformer. Figure 14-8B, however, shows a some-

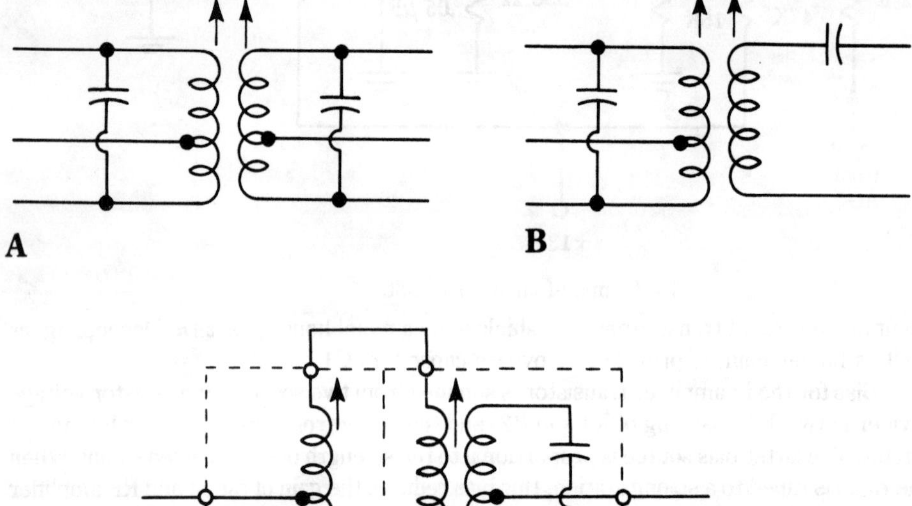

14-8 Tuned rf/i-f tank circuits: A. Tapped impedance type; B. Resonating capacitor in series with coil; C. Three-winding type.

what different type — a resonating capacitor is in series with the secondary winding. It not only resonates with the coil, but it also blocks the dc bias from flowing to ground via the transformer winding.

Figure 14-8C illustrates a circuit used in one American radio and in many European and Japanese radios. Bendix began using this type of i-f transformer in their 1972 radios. In this circuit, there are actually three windings, two of which are in a transformer configuration. The other is used as a collector load for the i-f amplifier transistor or mixer stage.

Some projects

One of the themes of this book, indeed the entire *TAB Mastering* series, is that electronics can be learned through experiments and projects. This section deals with several RF amplifier projects that can be used as preamplifiers or preselectors to radio receivers operating in the medium-wave through low-VHF regions.

The active RF amplifier circuits in this section are based on either of two devices: the MPF-102 junction field effect transistor (JFET), and the 40673 metal oxide semiconductor field effect transistor (MOSFET). Both of these devices are available from both mail-order sources and from local distributor replacement lines (for example, the MPF-102 is the NTE-312 and the 40673 is the NTE-222). These transistors were selected because they are both easily obtained and are well behaved into the VHF region.

RF amplifiers should be built inside shielded metal boxes in order to prevent RF leakage around the device. Select boxes that are either die-cast or are made of sheet metal and have an overlapping lip. Do not use the lower cost tab-fit sheet metal type box.

Project 14-1   JFET RF amplifier circuits

Figure 14-9 shows the most basic form of JFET RF amplifier. This circuit will work into the low-VHF region. This circuit is in the common-source configuration, so the input signal is applied to the gate and the output signal is taken from the drain. Source bias is supplied by the voltage drop across resistor R2, and drain load by a series combination of a resistor (R3) and a radio frequency choke (RFC1). The RFC should be $1,000\,\mu H$ (1 mH) at the AM broadcast band and HF (shortwave), and $100\,\mu H$ in the low-VHF region (> 30 MHz). At VLF frequencies below the broadcast band use 2.5 mH for RFC1 and increase all $0.01\,\mu F$ capacitors to $0.1\,\mu F$. All capacitors are either disk ceramic, or one of the newer "poly" capacitors (if rated for VHF service—not all are!).

The input circuit is tuned to the RF frequency, but the output circuit is untuned. The reason for the lack of output tuning is that tuning both input and output permits the JFET to oscillate at the RF frequency—and that we don't want. Other possible causes of oscillation include layout and a *self-resonance* frequency of the RFC that is too near the RF frequency (select another choke).

The input circuit consists of an RF transformer that has a tuned secondary (L2/C1). The variable capacitor, C1, is the tuning control. Although the value shown is the standard 365 pF "AM broadcast variable," any form of variable can be used if the inductor is tailored to it. These components are related by:

$$f = \frac{1}{2\pi\sqrt{LC}}$$

(14-1)

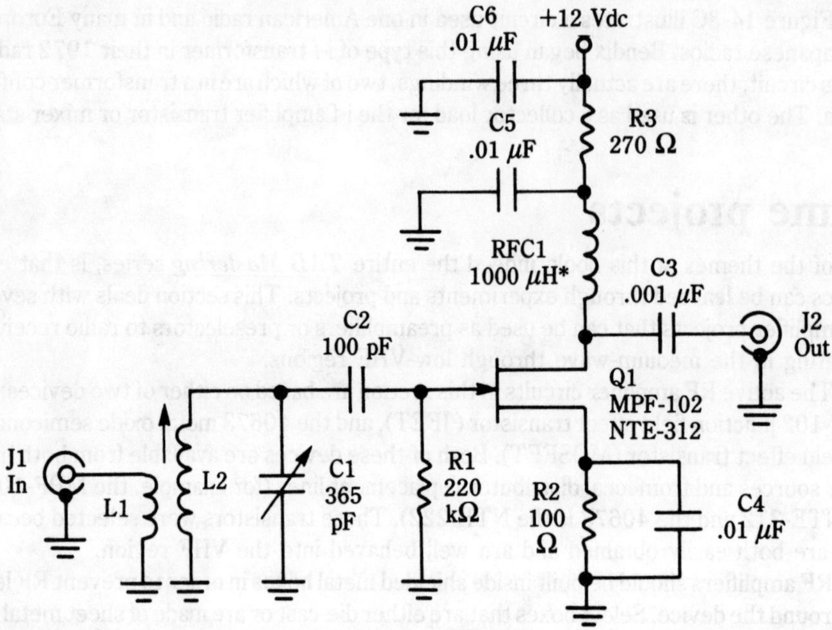

14-9   JFET active preselector.

Where: *f* is the frequency, in hertz (Hz)
      L is the inductance, in henrys (H)
      C is the capacitance, in farads (F)

Be sure to convert inductances from microhenrys to Henrys, and picofarads to Farads. Allow up to 10 pF to account for stray capacitances, although keep in mind that this number is a guess that might have to be adjusted (it is a function of your layout, among other things). You also can solve Eq. 14-1 for either *L* or *C*:

$$L = \frac{1}{39.5\, F^2\, C} \tag{14-2}$$

and,

$$C = \frac{1}{39.5\, F^2\, L} \tag{14-3}$$

Space does not warrant making a sample calculation, but I can report results for you to check for yourself. In a sample calculation, I wanted to know how much inductance is required to resonate 100 pF (90 pF capacitor plus 10 pF stray) to 10 MHz WWV. The solution, when all numbers are converted to hertz and farads, results in 0.00000253 H, or 2.53 μH. Keep in mind that the calculated numbers are close, but are nonetheless approximate—and the circuit might need tweaking on the bench.

The inductor (L1/L2) can be either a variable inductor (as shown) from a distributor such as *Digi-Key** or "homebrewed" on a toroidal core. Most people will want to use the T-50-6 (RED) or T-68-6 (RED) toroids† for shortwave applications. The number of turns required for the toroid is calculated from $N = 100 \times [L_{\mu H}\backslash A_L]^{1/2}$, where $L_{\mu H}$ is in microhenrys, and A_L is 49 for T-50-RED and 57 for T-68-RED. For example, a 2.53 μH coil needed for L2 wound on a T-50-RED core requires 23 turns. Use #26 or #28 enameled wire for the winding. Make L1 approximately 4 to 7 turns over the same form as L2.

Figure 14-10 shows two methods for tuning both the input and output circuits of the JFET transistor. In both cases the JFET is wired in the common gate configuration, so signal is applied to the source and output is taken from the drain. The dotted line indicates that the output and input tuning capacitors are ganged to the same shaft.

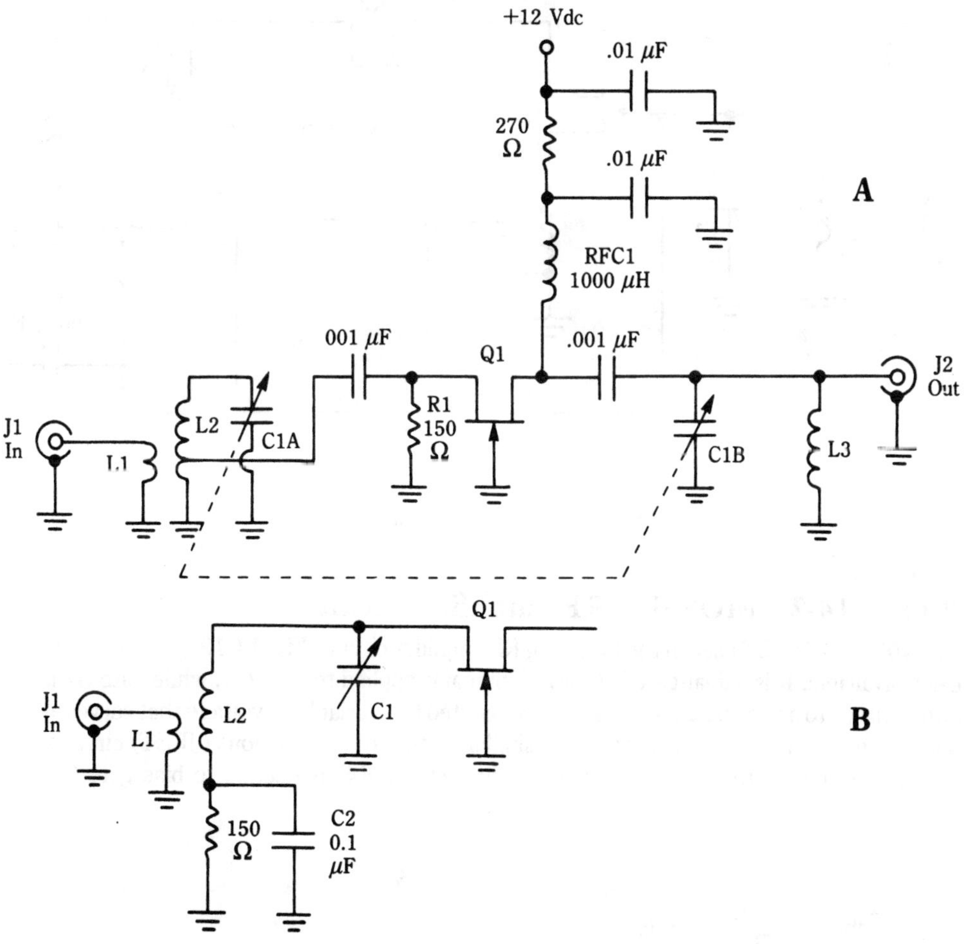

14-10 Grounded gate VHF preselector.

*Digi-Key, P.O. Box 677, Thief River Falls, MN 57601, 1–800–344–4539.
†Amidon Associates, 12033 Otsego St., No. Hollywood, CA 91607.

The source circuit of the JFET is low impedance, so some means must be provided to match the circuit to the tuned circuit. In Fig. 14-10A a tapped inductor is used for L2 (tapped at ⅓ the coil winding), and in Fig. 14-10B a slightly different configuration is used.

The circuit in Fig. 14-11 is a VHF preamplifier that uses two JFET devices connected in *cascode*; i.e., the input device (Q1) is in common source and is direct-coupled to the common gate output device (Q2). In order to prevent self-oscillation of the circuit, a neutralization capacitor (NEUT) is provided. This capacitor is adjusted to keep the circuit from oscillating at any frequency within the band of operation. In general, this circuit is tuned to a single channel by the action of L2/C1 and L3/C2.

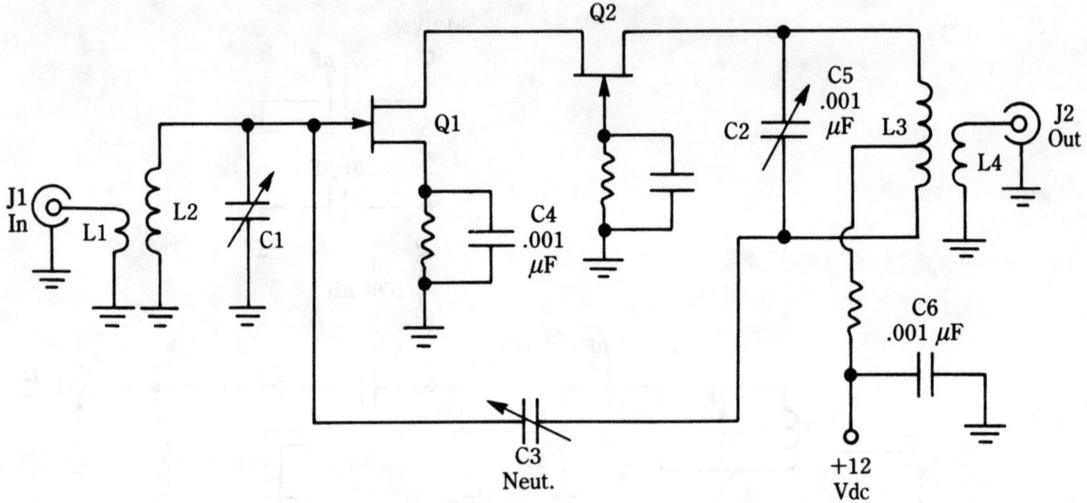

14-11   Cascode RF amplifier.

Project 14-2   MOSFET RF amplifier circuits

The 40673 MOSFET used in the following RF amplifier circuit (Fig. 14-12) is low cost and easily available. It is a dual-gate MOSFET. Signal is applied to gate G1, while gate G2 is either biased to a fixed positive voltage or connected to a variable dc voltage that serves as a gain-control signal. The dc network is similar to that of the previous (JFET) circuits, with the exception that a resistor voltage divider (R3/R4) is needed to bias gate G2.

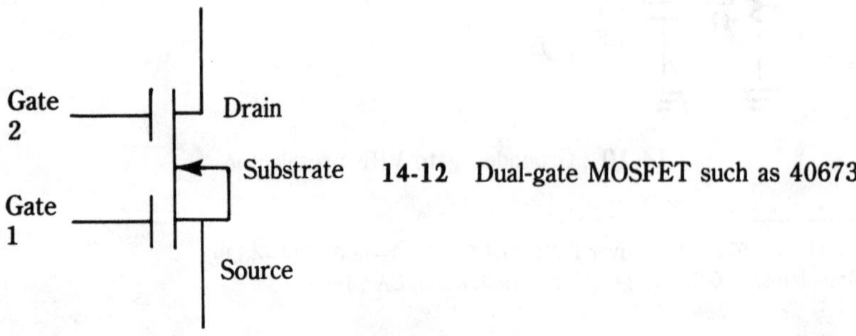

14-12   Dual-gate MOSFET such as 40673.

There are three tuned circuits for this RF amplifier project, so it will produce a large amount of selectivity improvement and image rejection. The gain of the device also will provide additional sensitivity. All three tuning capacitors (C1A, C16, and C1C) are ganged to the same shaft (Fig. 14-13) for "single-knob tuning." The trimmer capacitors (C2, C3, and C4) are used to adjust the tracking of the three tuned circuits—(i.e., to ensure that they are all tuned to the same frequency at any given setting of C1A-C.

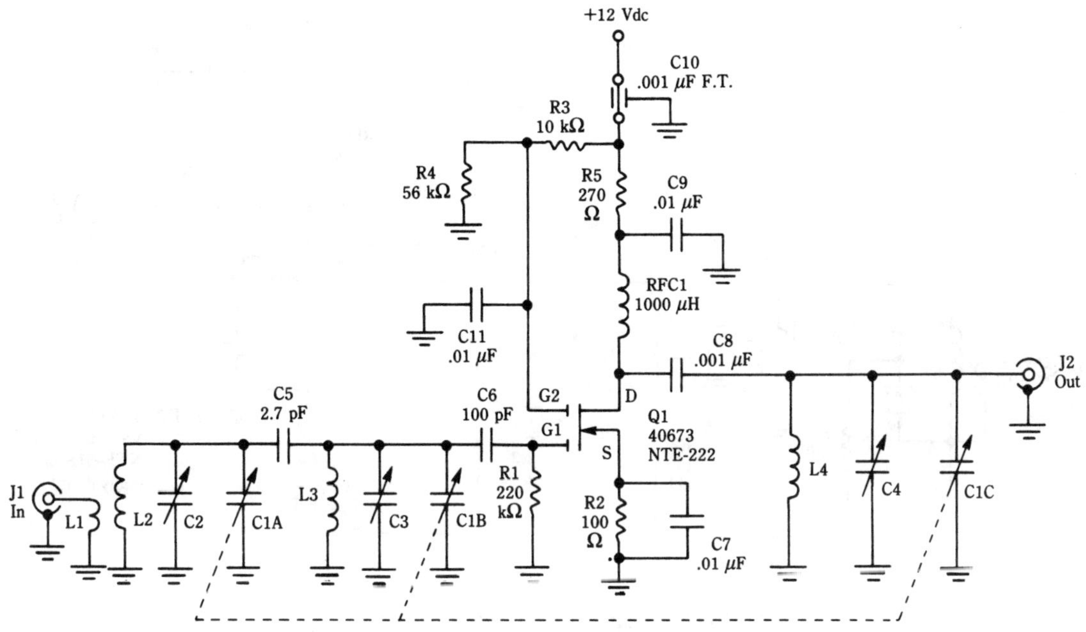

14-13 Dual-gate MOSFET RF amplifier circuit.

The inductors are of the same sort as described previously. It is permissable to put L1/L2 and L3 in close proximity to each other, but they should be separated from L4 in order to prevent unwanted oscillation due to feedback arising from coil coupling.

The circuit in Fig. 14-14 is a little different. In addition to using only input tuning, which lessens the potential for oscillation, it also uses voltage tuning. The hard-to-find variable capacitors are replaced with varactor diodes, also called *voltage variable capacitance diodes.* These pn junction diodes exhibit a capacitance that is a function of the applied reverse bias potential, V_T.

Although the original circuit was built and tested for the AM broadcast band (540 KHz to 1,610 KHz), it can be changed to any band by correct selection of the inductor values. The designated varactor (NTE-618) offers a capacitance range of 440 pF down to 15 pF over the voltage range 0 to +18 Vdc.

The inductors can be either "store-bought" types or wound over toroidal cores. I used a toroid for L1/L2, forming a fixed inductance for L2, and "store-bought" adjustable inductors for L3 and L4. There is no reason, however, why these same inductors cannot be used for all three purposes. Unfortunately, not all values are available in the form that has a low-impedance primary winding to permit antenna coupling.

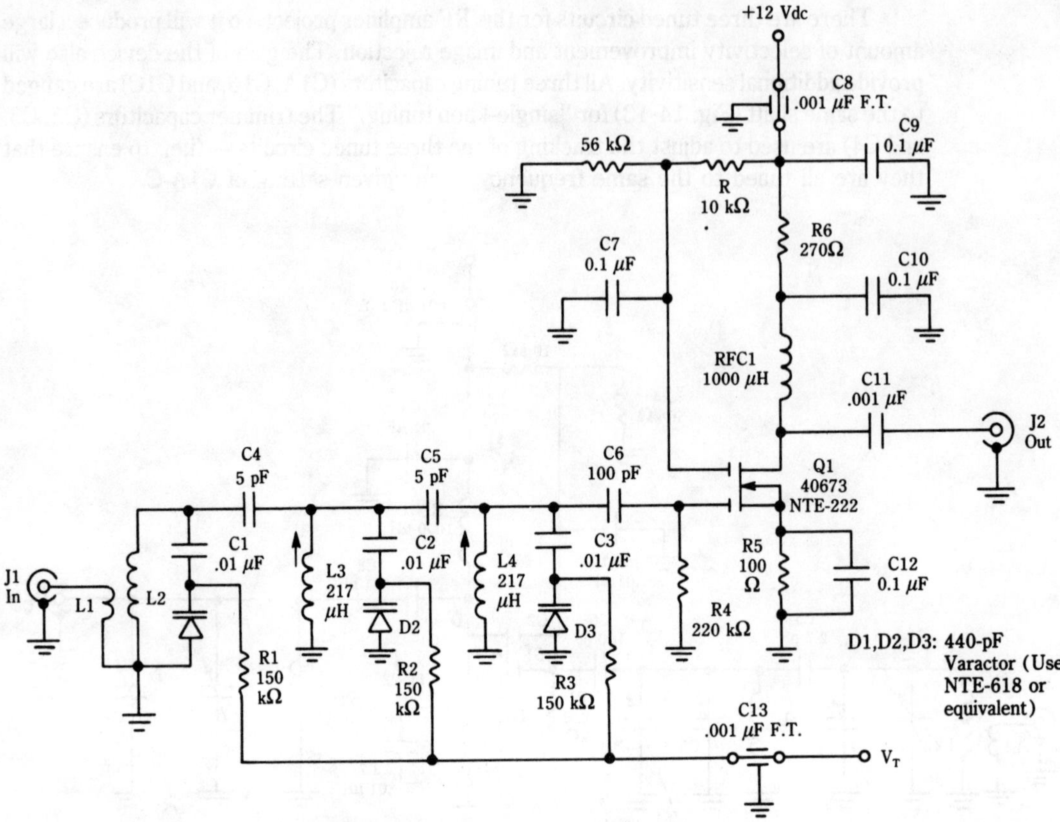

14-14 Voltage tuned preamplifier.

In Fig. 14-14, you see the connection points for the dc power (+ 9 Vdc to + 12 Vdc). These points are 0.001 μF ceramic feedthrough capacitors and are a little hard to find locally, but they can be bought from Newark Electronics or other distributors.

In both of the MOSFET circuits, the fixed-bias network used to place gate G2 at a positive dc potential can be replaced with a variable voltage circuit such as Fig. 14-15. The potentiometer in Fig. 14-15 can be used as an RF GAIN control to reduce gain on strong signals and increase it on weak signals. This feature allows the active RF amplifier to be custom-set to prevent overload from strong signals.

Noise and RF amplifiers

The weakest signal that you can detect is determined mainly by the noise level in the receiver. Some noise arrives from outside sources, while other noise is generated inside the receiver. At the VHF/UHF range, the internal noise is predominant, so it is common to use a low-noise preamplifier ahead of the receiver. This preamplifier will reduce the noise figure for the entire receiver. If you select a commercial ready-built or kit VHF preampli-

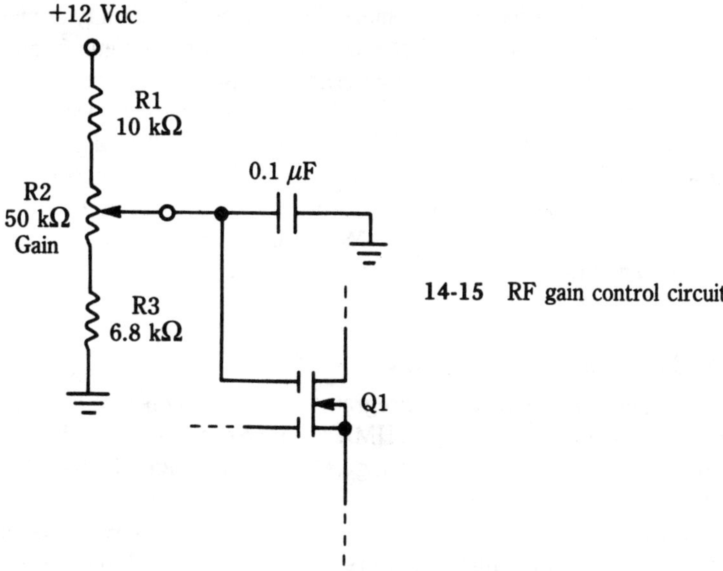

14-15   RF gain control circuit.

fier, such as the units sold by Hamtronics, Inc. (NY)*, then make sure you specify the low-noise variety for the first amplifier in the system.

The low-noise ampifier (LNA) should be mounted on the antenna if it is wideband, and at the receiver if it is tunable (*Note:* the term *RF amplifier* only applies to tuned versions, while *preamplifier* can denote either tuned or wideband models). Of course, if your receiver is used only for one frequency, then it also may be mounted at the antenna. The reason for mounting the antenna right at the antenna is to build up the signal and improve the signal-to-noise ratio (SNR) prior to feeding the signal into the transmission line, where losses cause it to weaken somewhat.

An RF amplifier can improve the performance of your receiver, no matter whether you listen to VLF, AM broadcast band, shortwaves, or VHF/UHF bands. The circuits presented here allow you to "roll your own" and be successful at it.

UHF/microwave RF integrated circuits

Very wideband amplifiers (i.e., those operating from near-dc to UHF or into the microwave regions) have traditionally been very difficult to design and build with consistent performance across the entire passband. Many such amplifiers have either gain irregularities, such as "suck-outs," or peaks. Others suffer large variations of input and output impedance over the frequency range. Still others suffer spurious oscillation at certain frequencies within the passband.

---

*Hamtronics, Inc., 65 Moul Rd., Hilton, NY 14468–9535, 716–392–9430.

Barkhausen's criteria for oscillation requires both a loop gain of unity or more and a) 360-degree (in-phase) feedback at the frequency of oscillation. At some frequency, the second of these criteria can be met by adding the normal 180-degree phase shift inherent in the amplifier to phase shift due to stray RLC components. The result will be oscillation at the frequency where the RLC phase shift was an additional 180 degrees.

In the past only a few applications required such amplifiers. Consequently, such amplifiers were either very expensive or didn't work nearly as well as claimed. Hybrid Microwave Integrated Circuit (HMIC) and Monolithic Microwave Integrated Circuit (MMIC) devices were low-cost solutions to the problem.

What are HMICs and MMICs?

MMICs are tiny "gain block" monolithic integrated circuits that operate from dc or near-dc to a frequency in the microwave region. HMICs, on the other hand, are hybrid devices that combine discrete and monolithic technology. One product (Signetics NE-5205) offers up to $+20$ dB of gain from dc to 0.6 GHz, while another low-cost device (Minicircuits Laboratories, Inc. MAR-x) offers $+20$ dB of gain over the range dc to 2 GHz, depending upon model. Other devices from other manufacturers also are offered, and some produce gains to $+30$ dB and frequencies to 18 GHz. Such devices are unique in that they present input and output impedances that are a good match to the 50 or 75 Ω normally used as system impedances in RF circuits.

Monolithic integrated circuit devices are formed through photoetching and diffusion processes on a substrate of silicon or some other semiconductor material. Both active devices, such as transistors and diodes, and some passive devices can be formed in this manner. Passive components such as on-chip capacitors and resistors can be formed using various thin and thick film technologies. In the MMIC device, interconnections are made on the chip via built-in planar transmission lines.

Hybrids are a level closer to regular discrete circuit construction than ICs. Passive components and planar transmission lines are laid down on a glass, ceramic, or other insulating substrate by vacuum deposition or other methods. Transistors and unpackaged monolithic "chip dies" are cemented to the substrate and then connected to the substrate circuitry with mil-sized gold or aluminum bonding wires. Because the material in this series could sometimes apply to either HMIC or MMIC devices, the convention herein shall be to refer to all devices in either subfamily as microwave integrated circuits (MIC), unless otherwise specified.

Three things specifically characterize the MIC device; first is simplicity. As you will see in the following circuits, the MIC device usually has only input, output, ground, and power supply connections. Other wideband IC devices often have up to 16 pins, most of which must be either biased or capacitor-bypassed. The second feature of the MIC is the very wide frequency range (dc-GHz) of the devices, while the third is the constant input and output impedance over several octaves of frequency.

Although not universally the case, MICs tend to be unconditionally stable because of a combination of series and shunt negative feedback internal to the device. The input and output impedances of the typical MIC device are a close match to either 50 or 75 Ω, so it is possible to make a MIC amplifier without any impedance-matching schemes—a factor that makes it easier to broadband than if tuning was used. A typical MIC device generally produces a standing wave ratio (SWR) of less than 2:1 at all frequencies within the passband, provided that it is connected to the design system impedance (e.g., 50 Ω).

The MIC is not usually regarded as a low-noise amplifier (LNA), but can produce noise figures (NF) in the 3 to 8 dB range. Some MICs are LNAs, however, and the number available should increase in the near future.

Narrow-band and passband amplifiers can be built using wideband MICs. A narrow-band amplifier is a special case of a passband amplifier and is typically tuned to a single frequency. An example is the 70 MHz i-f amplifier used in microwave receivers. Because of input and/or output tuning, such an amplifier will respond only to signals in the 70 MHz frequency band.

Very wideband amplifiers

Engineering wideband amplifiers, such as those used in MIC devices, seems simple but has traditionally caused a lot of difficulty for designers. Figure 14-16A shows the most fundamental form of MIC amplifier; it is a common emitter npn bipolar transistor amplifier. Because of the high-frequency operation of these devices, the amplifier in MICs are usually made of a material such as gallium arsenide (GaAs).

In Fig. 14-16A, the emitter resistor (R_e) is unbypassed, so introduces a small amount of negative feedback into the circuit. Resistor R_e forms series feedback for transistor Q1. The parallel feedback in this circuit is provided by collector-base bias resistor R_f. Typical values for R_f are in the 500 Ω range, and for R_e in the 4 to 6 Ω range. In general, the designer tries to keep the ratio R_f/R_e high in order to obtain higher gain, higher output

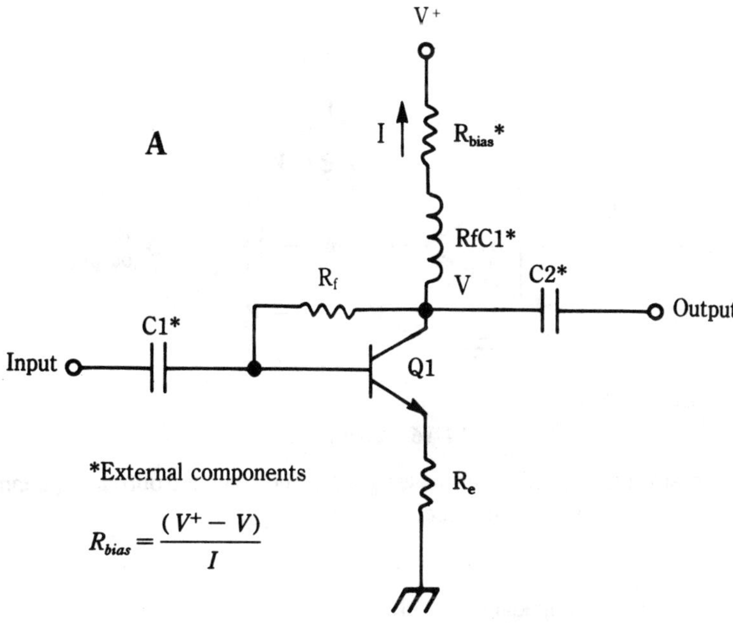

14-16   MMIC internal circuits.

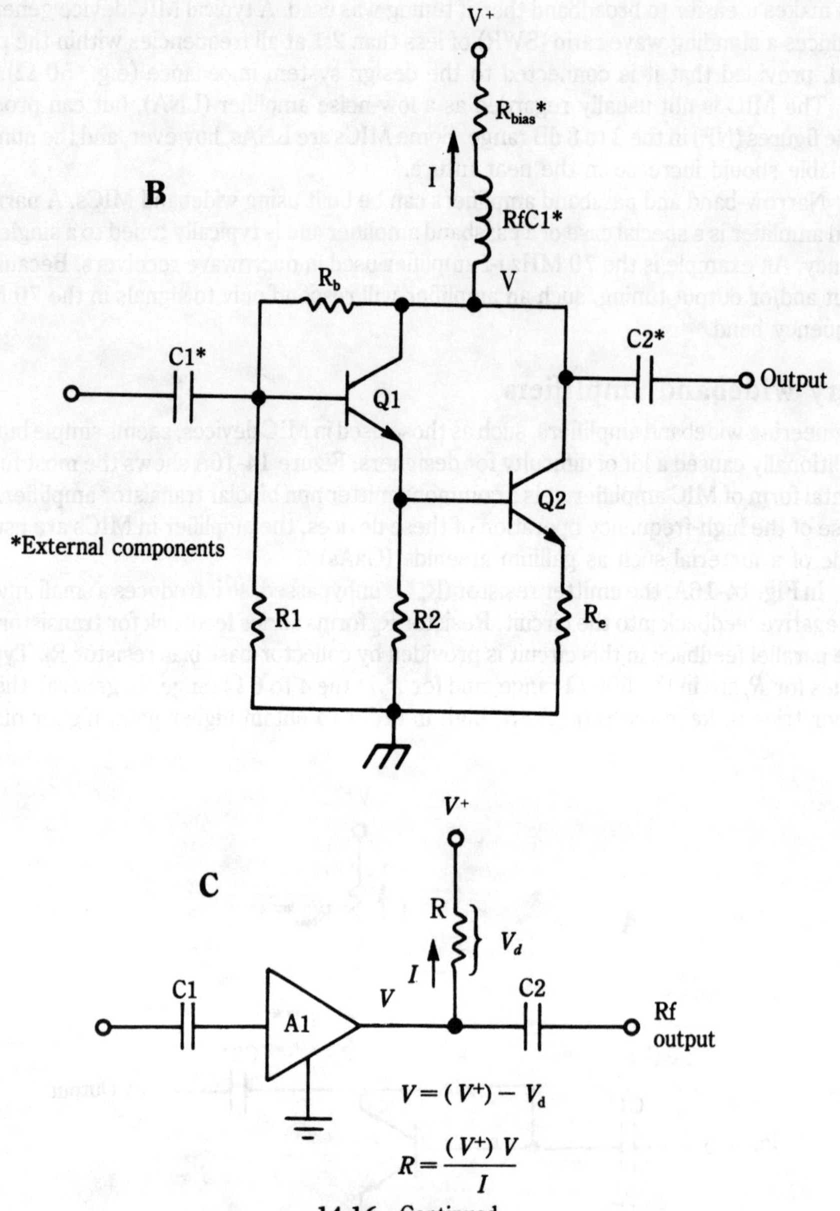

14-16 Continued

power compression points, and lower noise figures. The input and output impedances (R_o) are equal, and defined by the patented equation:

$$R_f + \sqrt{R_f R_e} \tag{14-4}$$

Where: R_o is the output impedance, in ohms
      R_f is the shunt feedback resistance, in ohms
      R_e is the series feedback resistance, in ohms

Figure 14-16B shows a more common form of MIC amplifier circuit. Although based on the Darlington amplifier circuit, this amplifier has the same sort of series and shunt feedback resistors (R_e and R_f) as Fig. 14-16A. All resistors except R_{bias} are internal to the MIC device.

A Darlington amplifier, also called a *Darlington pair* or *superbeta transistor*, consists of a pair of bipolar transistors (Q1 and Q2) connected such that Q1 is an emitter follower driving the base of Q2, and with both collectors connected in parallel with each other. The Darlington connection permits both transistors to be treated as if they were a single transistor with higher than normal input impedance and a beta gain (β) equal to the product of the individual beta gains. For the Darlington amplifier, therefore, β, or H_{fe}, is:

$$\beta_o = (\beta_{Q1})(\beta_{Q2}) \tag{14-5}$$

Where: β_o is the beta gain of the Q1/Q2 pair
$\quad\quad\;\; \beta_{Q1}$ is the beta gain of Q1
$\quad\quad\;\; \beta_{Q2}$ is the beta gain of Q2

You should be able to see two facts. First, the beta gain is very high for a Darlington amplifier that is made with relatively modest transistors. Second, the beta of a Darlington amplifier made with identical transistors is the square of the common beta rating.

External components

Figures 14-16A and 14-16B show several components that are usually external to the MIC device. The bias resistor, R_{bias}, is sometimes internal, although on most MIC devices it is external. RF choke RFC1 is in series with the bias resistor and is used to enhance operation at the higher frequencies. RFC1 is considered optional by some MIC manufacturers.

The reactance of the RF choke is in series with the bias resistance and increases with frequency according to the $2\pi FL$ rule. Thus, the transistor sees a higher impedance load at the upper end of the passband than at the lower end. Use of RFC1 as a "peaking coil" thus helps overcome the adverse effect of stray circuit capacitance that ordinarily causes a similar decreasing frequency-dependent characteristic. A general rule of thumb is to make the combination of R_{bias} and X_{rfc1} form an impedance of at least 500 Ω at the lowest frequency of operation.

The gain of the amplifier may drop about 1 dB if RFC1 is deleted. This effect is caused by the bias resistance shunting the output impedance of the amplifier.

The capacitors are used to block dc potentials in the circuit. They prevent intracircuit potentials from affecting other circuits, as well as preventing potentials in other circuits from affecting MIC operation.

More will be said about these capacitors later, but for now understand that practical capacitors are not ideal; real capacitors are complex RLC circuits. Although the L and R components are neligible at low frequencies, they are substantial in the microwave region. In addition, the LC characteristic forms a self-resonance that can either "suck-out" or enhance gain at specific frequencies. The result is an uneven frequency response characteristic at best, and spurious oscillations at worst.

General MMIC amplifier

Figure 14-16C shows a "generic" circuit representing MIC amplifiers in general. As you will see when we look at an actual product, this circuit is nearly complete. The MIC device usually has only input, output, ground, and power connections; some models don't have separate dc power input. There is no dc biasing, no bypassing (except at the dc power line), and no seemingly "useless" pins on the package.

MICs tend to use either microstrip packages like UHF/microwave small-signal transistor packages or small versions of the miniDIP or metallic IC packages. Some HMICs are packaged in larger transistorlike cases, while others are packaged in special hybrid packages.

The dc bias resistor (R_{bias}) connected to either the power-supply terminal (if any) or the output terminal must be set to a value that limits the current to the device and drops the supply voltage to a safe value. MIC devices typically require a low dc voltage (4 to 7 Vdc) and a maximum current of about 15 to 25 milliamperes (mA), depending upon type. There also might be an optimum current of operation for a specific device. For example, one device advertises that it will operate over a range of 2 to 22 mA, but that the optimum design current is 15 mA. The value of resistor needed for R_{bias} is found from Ohm's law:

$$R_{bias} = \frac{(V_f) - V}{I_{bias}} \tag{14-6}$$

Where: R_{bias} is in ohms

$V+$ is the dc power supply potential, in volts

V is the rated MIC device operating potential, in volts

I_{bias} is the operating current, in amperes

The construction of amplifiers based on MIC devices must follow microwave practices. This requirement means short, wide, low-inductance leads made of printed circuit foil, and stripline construction. Interconnection conductors tend to behave like transmission lines at microwave frequencies, so they must be treated as such. In addition, capacitors should be capable of passing the frequencies involved, yet have as little inductance as possible. In some cases, the series inductance of common capacitors forms a resonance at some frequency within the passband of the MMIC device. These "resonant" circuits sometimes can be detuned by placing a small ferrite bead on the capacitor lead. Microwave "chip" capacitors are used for ordinary bypassing.

MIC technology currently is able to provide very low cost microwave amplifiers with moderate gain and noise figure specifications, and better performance is available at higher cost. The future holds promise of even greater advances. Manufacturers have extended MIC operation to 18 GHz and dropped noise figures substantially. In addition, it is possible to build the entire front end of a microwave receiver into a single HMIC or MMIC, including the RF amplifier, mixer, and local oscillator stages.

Although MIC devices are available in a variety of package styles, those shown in Fig. 14-17 are typical of many. Because of the VHF operation of these devices, MICs are packaged in stripline transistorlike cases. The low-inductance leads for these packages are essential in UHF and microwave applications.

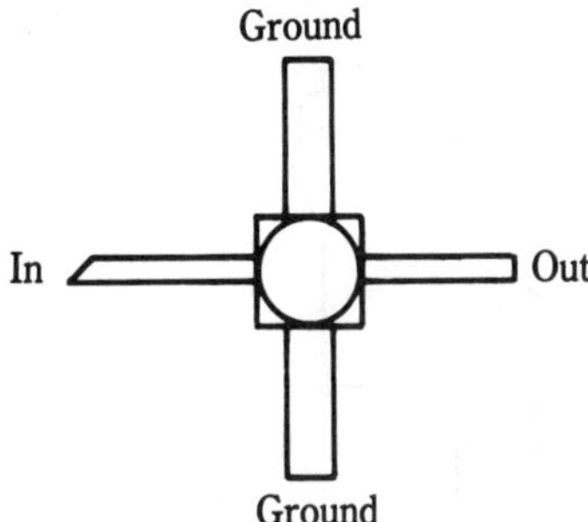

14-17   MMIC packages.

Cascade MIC amplifiers

MIC devices can be connected in cascade (Fig. 14-18) to provide greater gain than is available from only a single device, although a few precautions must be observed. It must be recognized, for example, that MICs possess a substantial amount of gain from frequencies near dc to well into the microwave region.

In all cascade amplifiers, attention must be paid to preventing feedback from stage to stage. Two factors must be addressed. First, as always, is component layout. The output and input circuitry external to the MIC must be separated physically in order to prevent coupling feedback. Second, it is necessary to decouple the dc power supply lines that feed two or more stages. Signals carried on the dc power line can easily couple into one or more stages, resulting in unwanted feedback.

Figure 14-18 shows a method for decoupling the dc power line of a two-stage MIC amplifier. In lower frequency cascade amplifiers, the V+ ends of resistors R1 and R2 normally would be joined and connected to the dc power supply. Only a single capacitor would be needed at that junction to ensure adequate decoupling between stages. As the operating frequency increases, however, the situation becomes more complex, in part because of the nonideal nature of practical components. For example, in an audio amplifier the electrolytic capacitor used in the power-supply ripple filter might provide sufficient decou-

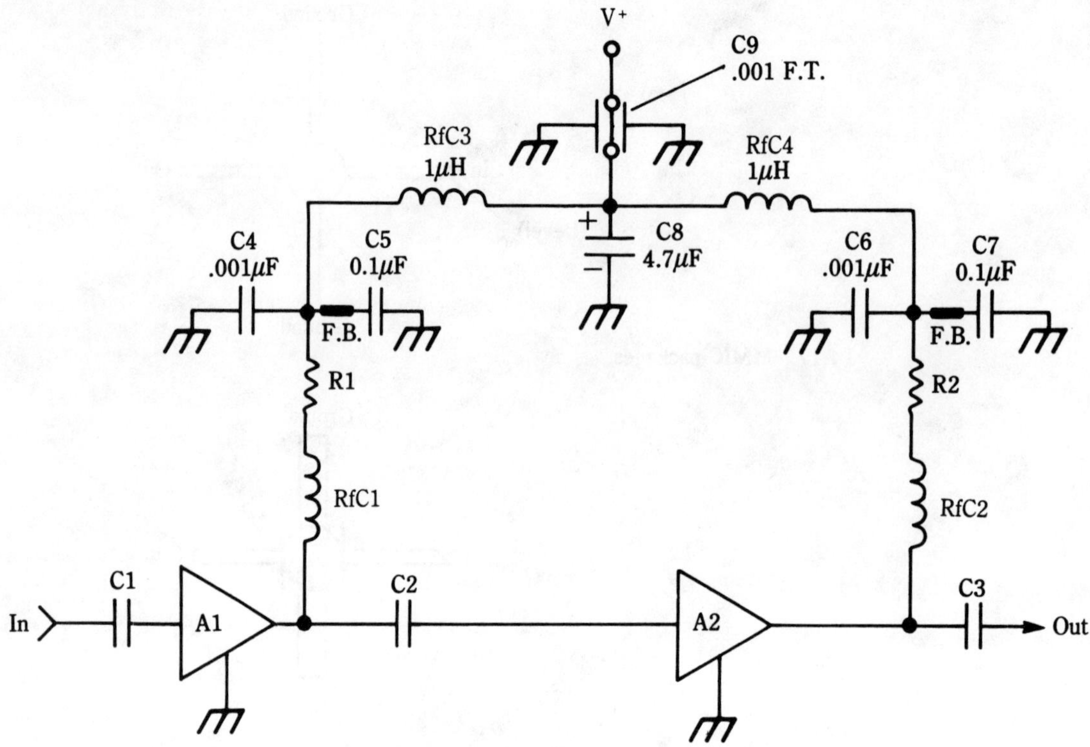

14-18   Cascade MMIC circuit.

pling. At RF frequencies, however, electrolytic capacitors are essentially useless as capacitors. (They act more like resistors at those frequencies.)

The decoupling system in Fig. 14-18 consists of RF chokes RFC3 and RFC4 and capacitors C4 through C9. The RF chokes help block high-frequency ac signals from traveling along the power line. These chokes are selected for a high reactance at VHF through microwave frequencies, while having a low dc resistance. For example, a one microhenry (1 μH) RF choke might have only a few milliohms of dc resistance but (by $2\pi FL$) a reactance of more than 3,000 Ω at 500 MHz. It is important that RFC3 and RFC4 be mounted so as to minimize mutual inductance due to interaction of their respective magnetic fields.

Capacitors C4 through C9 are used for bypassing signals to ground. Note that a wide range of values and several types of capacitors are used in this circuit. Each has its own purpose. Capacitor C8, for example, is an electrolytic type and is used to decouple VLF signals (i.e., those up to several hundred kilohertz). Because C8 is ineffective at higher frequencies, it is shunted by capacitor C9, shown in Fig. 14-18, as a feedthrough capacitor. Such a capacitor is usually mounted on the shielded enclosure housing the amplifier. Capacitors C5 and C7 are used to bypass signals in the HF region. Because these capacitors are likely to exhibit substantial series inductance, they will form undesirable resonances within the amplifier passband. Ferrite beads are sometimes installed on each capacitor lead in order to detune capacitor self-resonances. Like C9, capacitors C4 and C6 are used to decouple signals in the VHF and up region. These capacitors must be of microwave "chip" construction, or they might prove ineffective above 200 MHz or so.

Gain in cascade RF amplifiers

In low-frequency amplifiers, we might reasonably expect the composite gain of a cascade amplifier to be the product of the individual stage gains:

$$G_{total} = G_1 + G_2 + G_3 + \ldots + G_n \tag{14-7}$$

While that reasoning is valid for low-frequency voltage amplifiers, it fails for RF amplifiers (especially in the microwave region) where input-output standing wave ratio (SWR) becomes significant. In fact, the gain of any RF amplifier cannot be accurately measured if the SWR is greater than about 1.15:1.

There are several ways in which SWR can be greater than 1:1, and all of them involve an impedance mismatch. For example, the amplifier might have an input or output resistance other than the specified value. This situation can arise because of design errors or manufacturing tolerances. Another source of mismatch is the source and load impedances. If these impedances are not exactly the same as the amplifier input or output impedance, respectively, then a mismatch will occur.

An impedance mismatch at either input or output of the single-stage amplifier will result in a gain mismatch loss (M.L.) of:

$$M.L. = -10 \text{ LOG} \left[1 - \left(\frac{SWR - 1}{SWR + 1} \right)^2 \right] \tag{14-7}$$

Where: *M.L.* is the mismatch loss, in decibels (dB)

SWR is the standing wave ratio (dimensionless)

In a cascade amplifier is the distinct possibility of an impedance mismatch, hence an SWR, at more than one point in the circuit. An example (Fig. 14-19) might be where neither the output impedance (R_o) of the driving amplifier (A1) nor the input impedance (R_i) of the driven amplifier (A2) are matched to the 50 Ω (Z_o) microstrip line that interconnects the two stages. Thus, R_o/Z_o or its inverse forms one SWR, while R_i/Z_o or its inverse forms the other. For a two-stage cascade amplifier, the mismatch loss is:

$$M.L. = 20 \text{ LOG} \left[1 \pm \left(\frac{SWR_1 - 1}{SWR_1 + 1} \right) \left(\frac{SWR_2 - 1}{SWR_2 + 1} \right) \right] \tag{14-9}$$

The mismatch loss can vary from a negative loss resulting in less system gain (G_b), to a "positive loss" (which is actually a gain in its own right) resulting in greater system gain (G_a). The reason for this apparent paradox is that it is possible for a mismatched impedance to be connected to its complex conjugate impedance.

Attenuators in amplifier circuits?

It is common practice to place attenuator pads in series with the input and output signal paths of microwave circuits in order to "swamp-out" impedance variations that adversely affect circuits which ordinarily require either impedance matching or a constant impedance. Especially when dealing with devices such as LC filters (low-pass, high-pass, band-

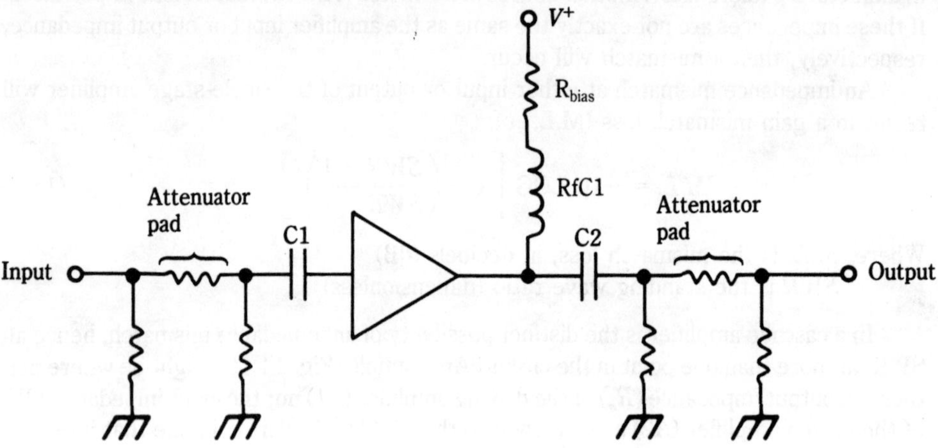

$$SWR_1 = \frac{R_o}{Z_o} \text{ or } \frac{Z_o}{R_o}$$

$$SWR_2 = \frac{Z_o}{R_i} \text{ or } \frac{R_i}{Z_o}$$

14-19 Use of stripline to interconnect MMICs.

pass), VHF/UHF amplifiers, matching networks, and MIC devices, it is useful to insert 1 dB, 2 dB, or 3 dB resistor attenuator pads in the input and output lines.

The characteristics of many RF circuits depend on seeing the design impedance at input and output terminals. With the attenuator pad (see Fig. 14-20) in the line, source and load impedance changes don't affect the circuit nearly as much.

14-20 Input/output attenuators for increased stability.

The attenuator tactic is also sometimes useful when confronted with seemingly unstable very wideband amplifiers. Insert a 1 dB pad in series with both input and output lines of the unstable amplifier. This tactic will cost about 2 dB of voltage gain, but often cure instabilities that arise out of frequency-dependent load or source impedance changes.

Noise figure in cascade amplifiers

It is common practice in microwave systems, especially in communications and radar receivers, to place a low-noise amplifier (LNA) at the input of the system. The amplifiers that follow the LNA need not be of LNA design, so are less costly. The question is sometimes asked: "Why not use an LNA in each stage?" The answer can be deduced from Friis' equation:

$$NF_{total} = NF_1 + \frac{NF_2 - 1}{G_1} + \frac{NF_3 - 1}{G_1 G_2} + \ldots + \frac{NF_n - 1}{G_1 G_2 \ldots G_n} \qquad (14\text{-}10)$$

Where: NF_{total} is the system noise figure
        NF_1 is the noise figure of stage-1
        NF_2 is the noise figure of stage-2
        NF_n is the noise figure of stage-nth
        G_1 is the gain of stage-1
        G_2 is the gain of stage-2
        G_n is the gain of nth stage
        (*Note*: All quantities are dimensionless ratios, rather than decibels.)

A lesson to be learned from this equation is that the noise figure of the first stage in the cascade chain dominates the noise figure of the combination of stages. Note that the overall noise figure increased only a small amount when a second amplifier was used.

Mini-circuits laboratories
MAR-x series devices

The Mini-Circuits Laboratories, Inc. MAR-x series of MIC devices (Fig. 14-21A) offers gains from +13 dB to +20 dB and top-end frequency response of either 1 GHz or 2 GHz, depending upon type. The package used for the MAR-x device (Fig. 14-21B) is similar to the case used for modern UHF and microwave transistors. Pin No. 1 (RF input) is marked by a color dot and a bevel.

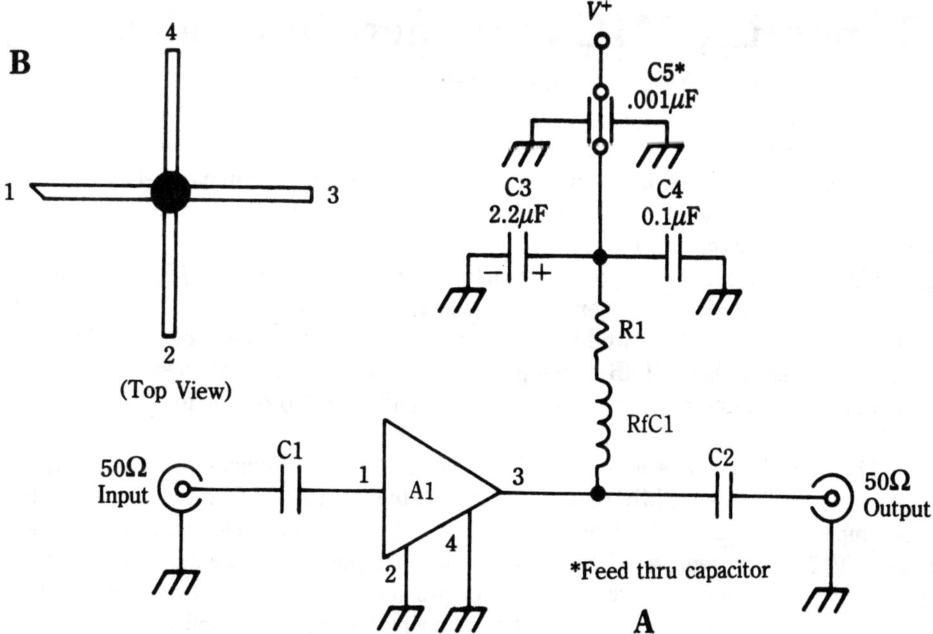

14-21   A. MAR-xx circuitry; B. MAR-xx package.

The usual circuit for the MAR-x series devices is shown in Fig. 14-21A. The MAR-x device requires a voltage of +5 Vdc on the output terminal and must derive this potential from a dc supply of greater than +7 Vdc.

The RF choke (RFC1) is called "optional" in the engineering literature on the MAR-x, but it is recommended for applications where a substantial portion of the total bandpass capability of the device is used. The choke tends to pre-emphasize the higher frequencies, and thereby overcomes the deemphasis normally caused by circuit capacitance. In traditional video amplifier terminology that coil is called a *peaking coil* because of this action; i.e., it peaks up the higher frequencies.

It is necessary to select a resistor for the dc power supply connection. The MAR-x device wants to see +5 Vdc, at a current not to exceed 20 mA. In addition, V+ must be greater than +7 volts. Thus, we need to calculate a dropping resistor (R_d) of:

$$R_d = \frac{(V+) - 5 \text{ Vdc}}{I} \tag{14-11}$$

Where: R_d is in ohms
     I is in amperes

In an amplifier designed for +12 Vdc operation, you might select a trial bias current of 15 mA (i.e., 0.015 amperes). The resistor value calculated is 467 Ω (a 470 Ω, 5 percent, film resistor should prove satisfactory). As recommended previously, a 1 dB attenuator pad is inserted in the input and output lines. The V+ is supplied to this chip through the output terminal.

Combining MIC amplifiers in parallel

Figures 14-22 through 14-23 show several MIC applications involving parallel combinations of MIC devices. Perhaps the simplest is the configuration of Fig. 14-22. Although each input must have its own dc blocking capacitor to protect the MIC device's internal bias network, the outputs of two or more MICs can be connected in parallel and share a common power-supply connection and output coupling capacitor. Several advantages are realized with the circuit of Fig. 14-22.

First, the power output increases even though total system gain (P_o/P_m) remains the same. As a consequence, however, drive power requirements also increase. The output power increases 3 dB when two MICs are connected in parallel, and 6 dB when four are connected (as shown). The 1 dB output power compression point also increases in parallel amplifiers in the same manner: 3 dB for two amplifiers and 6 dB for four amplifiers in parallel.

The input impedance of a parallel combination of MIC devices reduces to R_i/N, where N is the number of MIC devices in parallel. In the circuit shown in Fig. 14-22, the input impedance would be $R_i/4$, or 12.5 Ω if the MICs are designed for 50 Ω service. Because 50/12.5 represents a 4:1 SWR, some form of input impedance matching must be used. Such a matching network can be either broadbanded or frequency-specific, as the need dictates. Figures 14-23 through 14-25 show methods for accomplishing impedance matching.

The method shown in Fig. 14-23 is used at operating frequencies up to about 100 MHz and is based on broadband ferrite toroidal RF transformers. These transformers

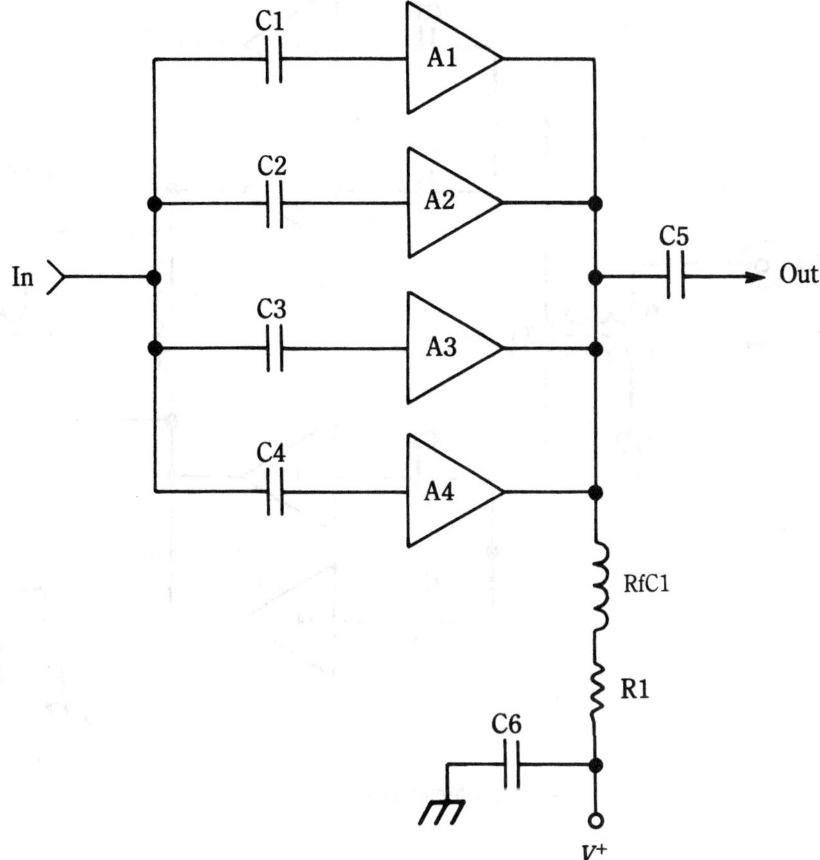

14-22   MMICs in parallel.

dominate the frequency response of the system because they are less broadbanded than the usual MIC device. This type of circuit can be used as a gain block in microwave receiver i-f amplifiers (which are frequently in the 70 MHz region), or in the exciter section of Master Oscillator Power Amplifier (MOPA) transmitters.

Another method is more useful in the UHF and microwave regions. In Fig. 14-24 are several forms of the Wilkinson power divider circuit. An LC network version is shown for comparison, although coaxial or stripline transmission line versions are used more often in microwave applications. The LC version is used to frequencies of 150 MHz.

This circuit is bidirectional, so it can be used as either a power splitter or power divider. RF power applied to port-C is divided equally between port-A and port-B. Alternatively, power applied to ports-A/B are summed together and appear at port-C. The component values are found from the following relationships:

$$R = 2\,Z_o \qquad\qquad (14\text{-}12)$$

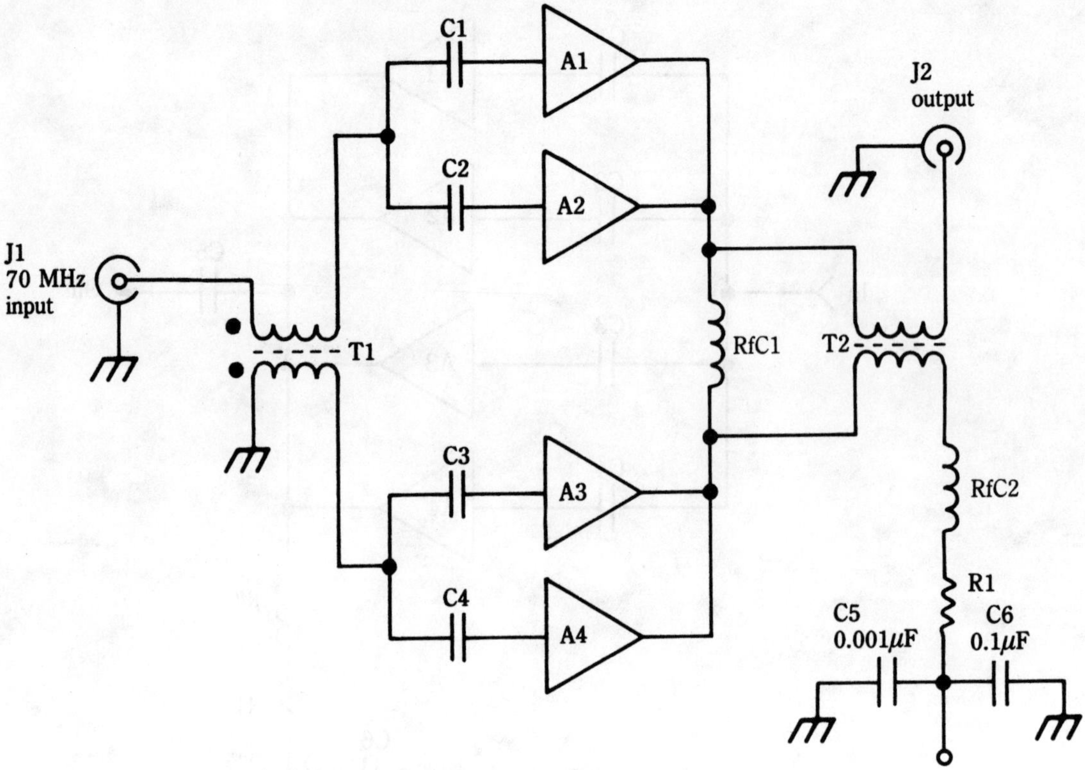

14-23   Push-pull parallel circuit.

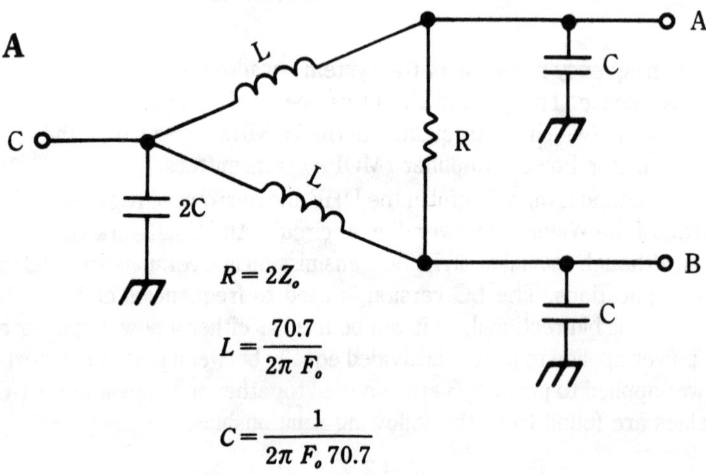

$$R = 2Z_o$$

$$L = \frac{70.7}{2\pi F_o}$$

$$C = \frac{1}{2\pi F_o\, 70.7}$$

14-24   Use of a Wilkinson divider as an MMIC combiner.

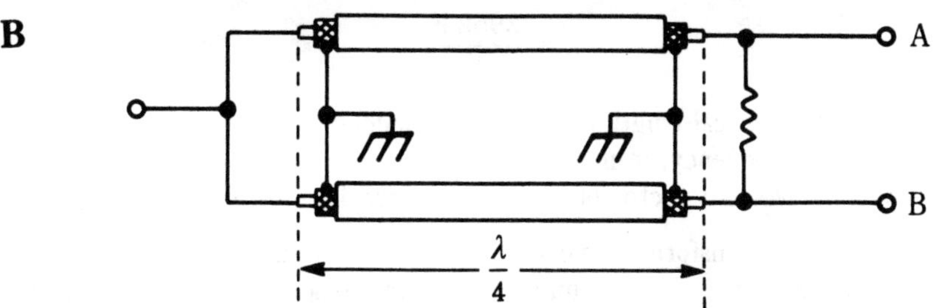

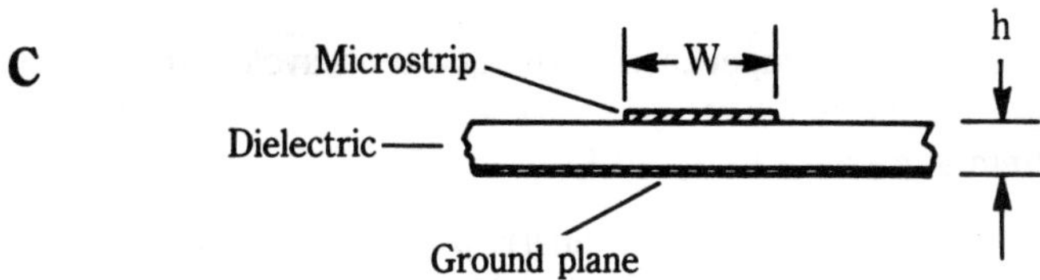

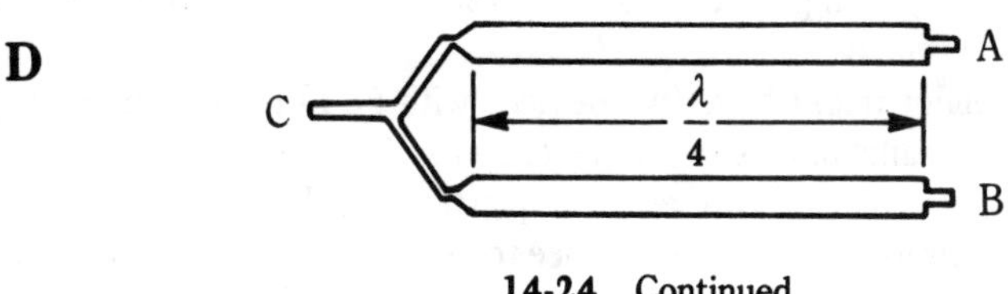

14-24 Continued

$$L = \frac{70.7}{2\pi F_o} \qquad (14\text{-}12)$$

$$C = \frac{1}{141.4\pi F_o} \qquad (14\text{-}13)$$

Where: R is in ohms
L is in henrys
C is in farads
F_o is in hertz

Figure 14-24B shows a coaxial cable version of the Wilkinson divider that can be used at frequencies up to 2 GHz. The lower frequency limit is set by practicality because the transmission line segments become too long to be handled easily. The upper frequency limit is set by the practicality of handling very short lines and by the dielectric losses, which are frequency-dependent. The transmission line segments are each quarter wavelength; their length is found from:

$$L = \frac{2952\,V}{F} \qquad (14\text{-}14)$$

Where: L is the physical length of the line, in inches
  F is the frequency, in megahertz
  V is the velocity factor of the transmission line (0-1)

An impedance transformation can take place across a quarter-wavelength transmission line if the line has a different impedance than the source or load impedances being matched. Such an impedance-matching system is often called a *Q-section*. The required characteristic impedance for the transmission line is found from:

$$(Z_o)' = \sqrt{Z_L Z_o} \qquad (14\text{-}15)$$

Where: $(Z_o)'$ is the characteristic impedance of the quarter wavelength section
  Z_L is the load impedance
  Z_o is the system impedance (e.g., 50 Ω)

In the case of parallel MIC devices, the nominal impedance at port-C of the Wilkinson divider is one-half the reflected impedance of the two transmission lines. For example, if the two lines are each 50 Ω transmission lines, then the impedance at port-C is 50/2 Ω, or 25 Ω. Similarly, if the impedance of the load — i.e., the reflected impedance — is transformed to some other value, then port-C sees the parallel combination of the two transformed impedances.

A parallel MIC amplifier might have two devices with 50 Ω input impedance each. Placing these devices in parallel halves the impedance to 25 Ω, which forms a 2:1 SWR with a 50 Ω system impedance. If, however, the quarter-wavelength transmission line transforms the 50 Ω input impedance of each device to 100 Ω, then the port-C impedance is 100/2, or 50 Ω — which is correct.

At the upper end of the UHF spectrum, and in the microwave spectrum, it might be better to use a stripline transmission line instead of coaxial cable. A stripline (see Fig. 14-24C) is formed on a printed circuit board. The board must be double-sided so that one side can be used as a ground plane, while the stripline is etched into the other side. The length of the stripline is dependent upon the frequency of operation; either halfwave or quarterwave lines are usually used. The impedance of the stripline is a function of three factors: 1) stripline width (w), 2) height of the stripline above the groundplane (h), and 3) dielectric constant (ϵ) of the printed circuit material:

$$Z_o = 377\,\frac{h}{w\,\sqrt{\epsilon}} \qquad (14\text{-}16)$$

Where: h is the height of the stripline above the groundplane
  w is the width of the stripline (h and w in same units)
  Z_o is the characteristic impedance, in ohms

The stripline transmission line is etched into the printed circuit board as in Fig. 14-24D. Stripline methods use the printed wiring board to form conductors, tuned circuits, and the like. In general, for microwave operation the conductors must be very wide (relative to their simple dc and RF power-carrying size requirements), and very short, in order to reduce lead inductance. Certain elements, the actual striplines, are transmission-line segments and follow transmission-line rules.

The printed circuit material must have a large permittivity and a low loss tangent. For frequencies up to about 3 GHz, it is permissable to use ordinary glass-epoxy double-sided board ($\epsilon = 5$), but for higher frequencies a low-loss material such as Rogers Duroid ($\epsilon = 2.17$) must be used.

When soldering connections in these amplifiers, it is important to use as little solder as possible and keep the soldered surface as smooth and flat as possible. Otherwise, the surface wave on the stripline will be interrupted, and operation suffers.

Figure 14-25 shows a general circuit for a multi-MIC amplifier based on a Wilkinson power divider. Because the divider can be used as either splitter or combiner, the same type can be used as input (WD1) and output (WD2) terminations. In the case of the input circuit, port-C is connected to the amplifier main input, while ports-A/B are connected to

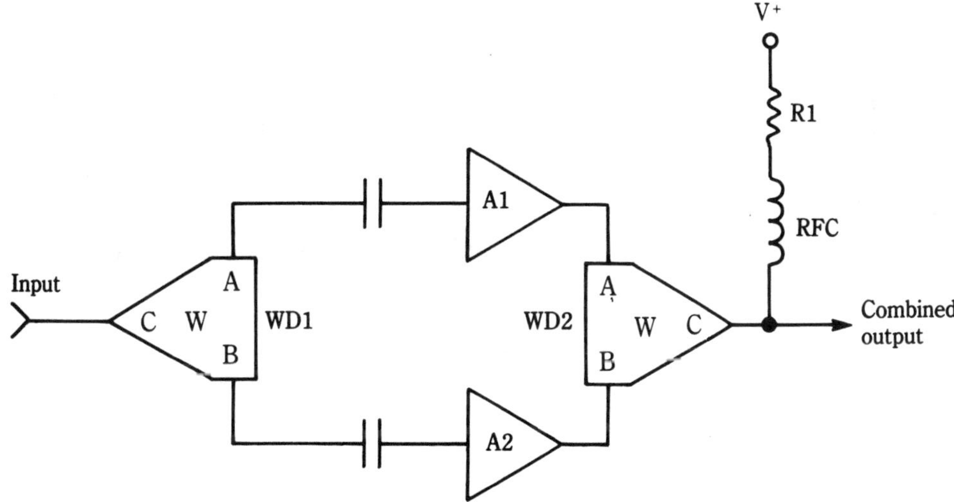

14-25   Block diagram of an MMIC circuit.

the individual inputs of the MIC devices. At the output circuit, another divider (WD2) is used to combine power output from the two MIC amplifiers and direct it to a common output connection. Bias is supplied to the MIC amplifiers through a common dc path at port-C of the Wilkinson dividers.

15

Troubleshooting solid-state circuits

THIS CHAPTER WILL LOOK AT A METHOD FOR TROUBLESHOOTING SOLID-state circuits using a dc voltmeter to isolate the bad stage. The method is not foolproof, but is well-suited to many applications, especially when combined with other methods.

First let's look at Fig. 15-1. Here we have npn and pnp transistors. In most amplifier circuits, the base-emitter voltage will be 0.2 to 0.3 volt for normal germanium (Ge) transistors (used in older equipment) and 0.6 to 0.7 volt for silicon (Si) transistors. The following voltage relationships also exist: In a pnp transistor the base is more negative than the emitter, and in an npn transistor the base is more positive than the emitter. Furthermore, the collector will be more positive than either base or emitter in npn transistors, and more negative in pnp transistors. Keep in mind that the term *more negative* can be interpreted as "less positive" in some cases.

Consider Fig. 15-2. Here we have a cascade chain of three stages in a radio receiver. Each stage consists of a pnp transistor that is powered from a positive dc power supply. The collectors of the transistors are near ground potential, while the emitters and bases are closer to the +10.5-volt "B+" line. If you measure the collector voltage with respect to ground, you will find it very slightly positive, while the emitters are at a much higher potential. Thus, the collector being "less positive" acts exactly like it was "more negative."

Voltmeter "A" will measure the potential between the points being examined (an emitter in Fig. 15-2) and ground. If the voltage is near normal at each stage, then you can assume that there are no massive short circuits — but you cannot get a hint of whether the stage is working properly.

Voltmeter "B" is connected with its positive electrode on the B+ distribution line, and the other used to probe the emitters of the stages. The B+ line often can be located from the service manual, but in some cases you will have to find it. Look for the electrolytic filter capacitor used to decouple the B+ line (C1 in Fig. 15-2). This filter capacitor will denote the proper line (unless you accidentally selected C2!) and usually has enough of a solder tab to allow connection of the voltmeter probe.

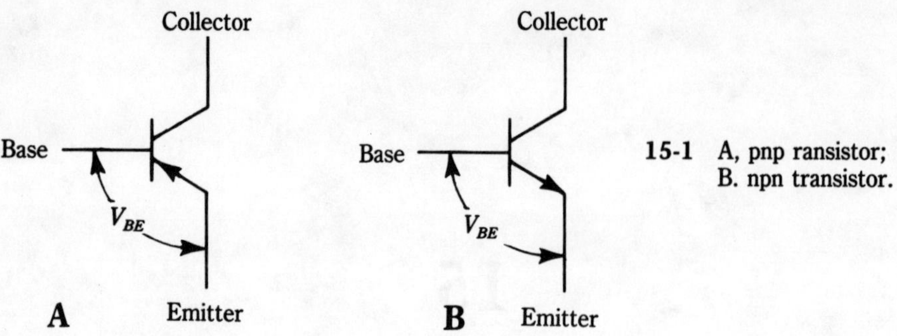

15-1 A, pnp ransistor;
B. npn transistor.

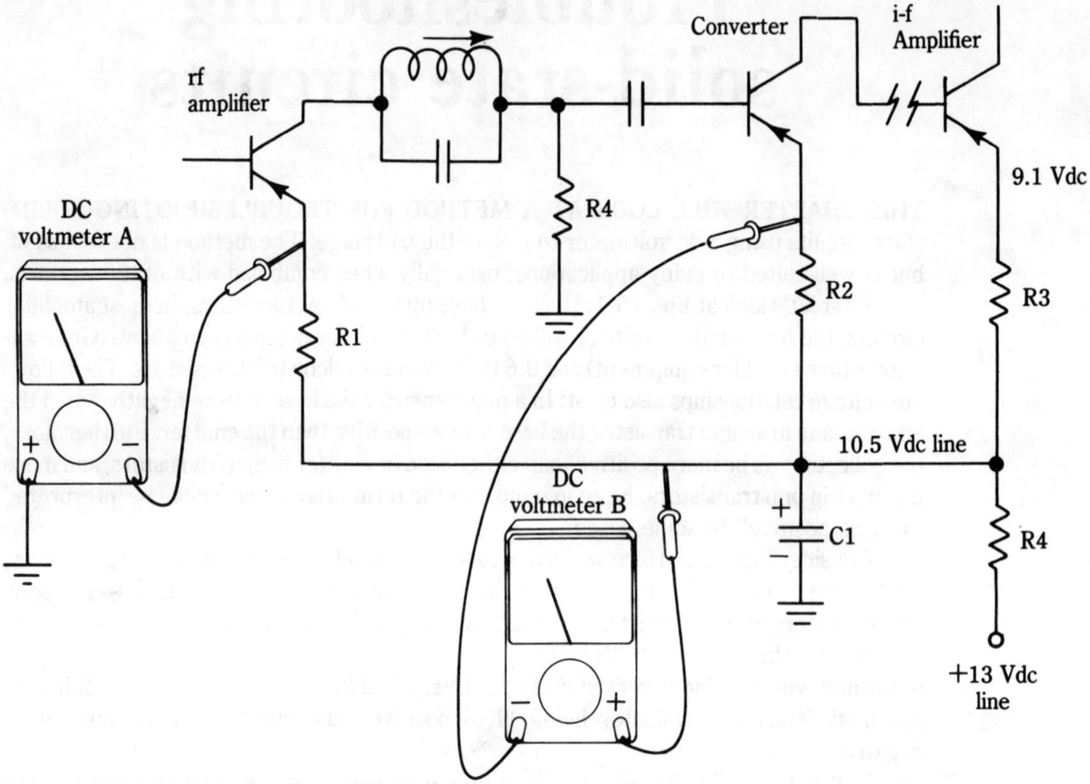

15-2 dc conduction troubleshooting with voltmeter in a pnp circuit.

The voltage drop across the emitter resistor of each stage indicates the *current conduction level* of the stage. If the service manual does not give the normal voltage drop, then calculate it as the difference between the emitter potential printed on the schematic and the B+ voltage. In the case of the i-f amplifier in Fig. 15-2, for example, the emitter voltage is 9.1 volts, while the B+ voltage is 10.5 volts. The normal conduction of this stage will be 10.5 − 9.1, or 1.4 volts. Any radical departure from this value indicates a problem. For example, a shorted transistor would cause that conduction voltage to increase to nearly 10 volts, while a leaky transistor would place the voltage somewhat lower but still larger than 1.4 volts. Similarly, an open emitter (or other condition that cuts off the stage) will reduce the voltage across the emitter resistor to either 0 or nearly 0.

You can isolate the defective stage by looking at each emitter voltage in its turn. Most often, the defective stage will show up from a serious anomaly in the emitter conduction voltages.

Radios with npn transistors in the stages are treated similarly. Figure 15-3 shows a radio with npn transistors powered by a positive-to-ground dc power supply. The collectors of these transistors will be close to the B+ potential, while the emitter and base voltages will be a lot lower.

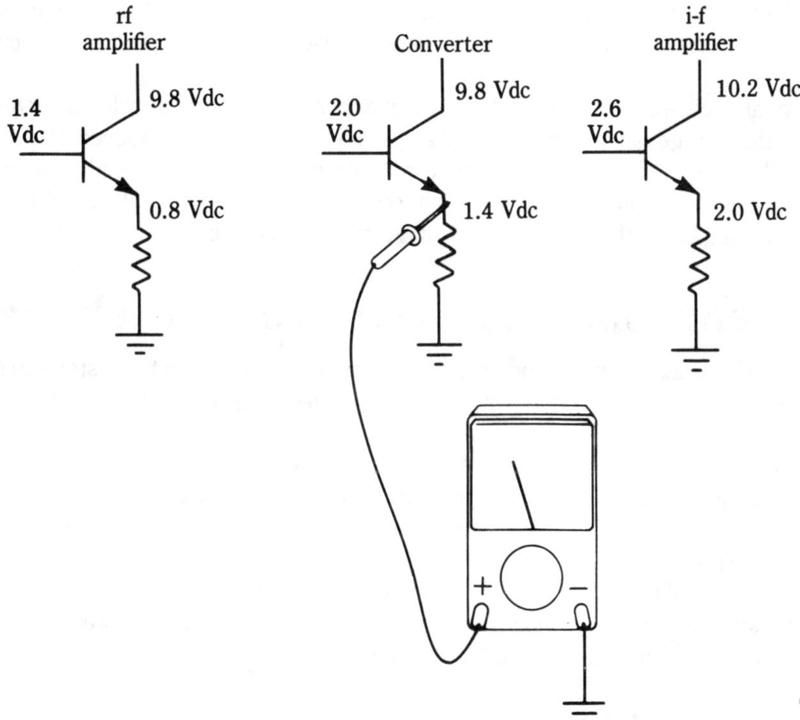

15-3 dc conduction npn transistor troubleshooting.

The values shown in Fig. 15-3 are typical, but they are not to be considered absolute. (There are a lot of design choices that could alter the values, so buy and consult the service manual for the particular equipment.)

As in the case of the pnp transistor stages, the npn emitter conduction voltage denotes the stage activity. In this type of circuit, however, the reference is ground instead of the B+ line. The principle is the same, however. You can check each conduction voltage in its turn and determine if any of them are incorrect. Fortunately, there are fewer calculations to make in this type of circuit. The emitter voltage on the schematic is the conduction voltage.

Both of the methods just discussed assume that there is a dc power supply that is positive with respect to ground. This arrangement is used in all modern American automobiles and most trucks, so it will be found typically in automotive electronic equipment. In home equipment, the dc power supply is often the same positive-with-respect-to-ground as shown here, but that is not necessarily so. The power supply might be negative with respect to ground, or there might be two power supplies — one positive to ground and one negative to ground. The principles in these cases are the same; only the point of voltage reference will change.

Variable frequency oscillators (whether in a radio receiver, transmitter VFO, or piece of test equipment) will behave differently. In most cases, the dc conduction of the oscillator transistor varies with frequency setting. Typically, in LC-tuned oscillators the voltage will be higher at the low end of the band and lower on the high end of the band, with a smooth transition as the dial is tuned. A sudden discontinuity in this transition might indicate a sudden cessation of oscillation or a parasitic oscillation developing at that point on the dial.

Crystal oscillators behave in a similar manner. If the crystal selector switch is changed, the voltage can be expected to change also. If the crystal is removed, the voltage usually will change radically. Don't remove crystals without thinking ahead, however. In some cases, especially in a transmitter when the next stage is a grid-leak biased vacuum-tube power amplifier, there might be damage to the equipment.

The semiconductor tester you already own

With a little knowledge, the analog or digital multimeter becomes a transistor and diode tester. If you have an ohms scale, you can measure the diode's front-to-back resistance, and from that information determine whether or not the diode is leaking. You'll even be able to tell which end of an unmarked diode is the cathode or anode.

Buying "grab-bag" transistors is a popular means of stocking a workshop at low cost. Unfortunately, many of those "bargain" transistors and diodes are less than useless. Some of them are just plain bad — they blow out when installed in a live circuit. Others are unmarked, so you don't know whether the unit is npn or pnp. By correctly tagging these unknowns after evaluation, you can predict with fair reliability which will work in real circuits and which are fit only for the trash heap.

Ohmmeter orientation

To understand our test method, let's first look at the test equipment. The question is, "How do analog and digital ohmmeters work?" In analog ohmmeters, a battery or electronic dc power source is connected in series with the meter movement and some calibration resistors (Fig. 15-4A). Battery current pushes its way through the meter movement, into an external circuit, and back to the battery.

If you make it hard for the current to flow by placing a resistance in series, the meter pointer won't swing as far along the dial as when the probes are simply shorted together. That's why you'll find high-precision resistors in any decent multimeter — they limit the meter pointer travel (and the current) so that the pointer always comes to rest at a specific point on the dial. Short an ohmmeter's leads together, and that place should be 0 Ω. When you connect your multimeter across an external resistance (with the multimeter in the ohms mode), the meter receives even less current than it did before; so there's less swing on the pointer's part.

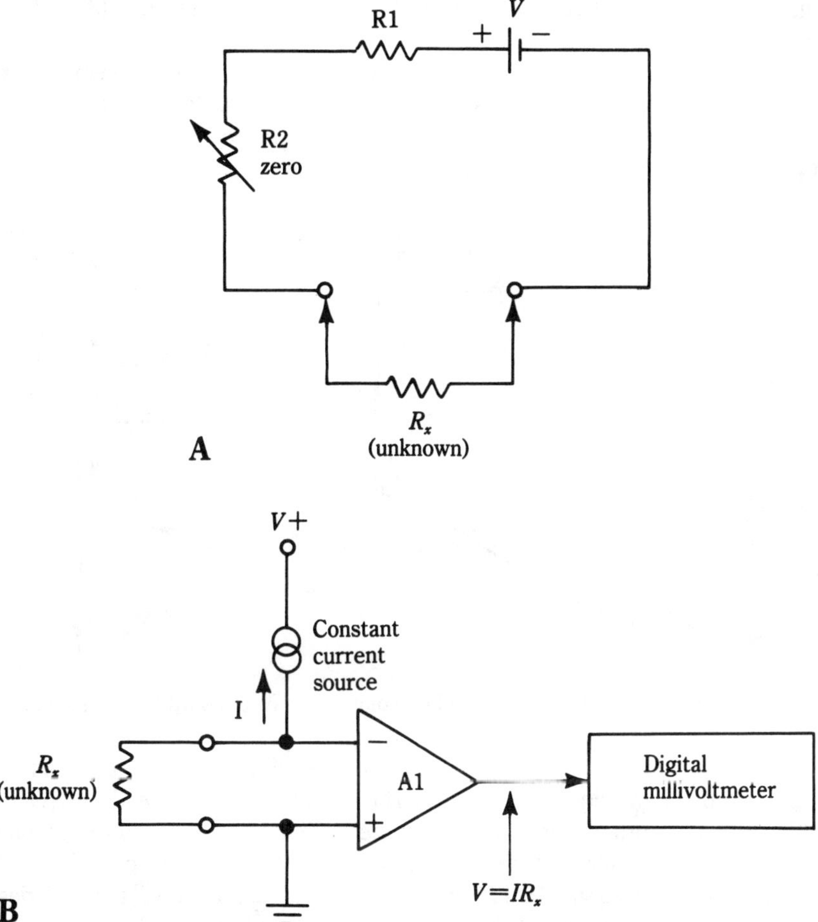

15-4 A. Analog ohmmeter circuit; B. Digital ohmmeter circuit.

Digital multimeters (DMMs) use a different tactic for measuring resistance. In Fig. 15-4B a constant current source is used to provide a precision reference current to the unknown resistance, R_x. The main circuit of the DMM is basically a millivoltmeter with a range of 0 to 1,999 millivolts. If a known constant current is passed through the resistance, you can measure the resistance by measuring the voltage drop. For example, a 1 mA constant current produces a voltage drop of 1 mV/Ω. Typically, the constant current level will be different for each ohmmeter range, but the principle is the same.

Testing diodes

Diodes are the easiest semiconductor devices to test, so we'll examine them first. Besides, this method of transistor testing depends upon the fact that the diode also represents the base-emitter and base-collector junctions of a transistor. We rely on diodes to pass electrical current unidirectionally. This unique ability forms the basis for the test. There are four states of being for any diode: they can be open, shorted, leaky, or o.k. Before you start testing, it would be wise if you had a sheet of paper and a pencil or pen to jot down the readings.

Let's first assume that you are testing a rectifier diode. With the ohmmeter set to the RX1 scale (Fig. 15-5A), place the probes across the diode leads. Record your readings and then swap probes (Fig. 15-5B) and take a second resistance reading. After you take both readings, you're ready to interpret what you've discovered.

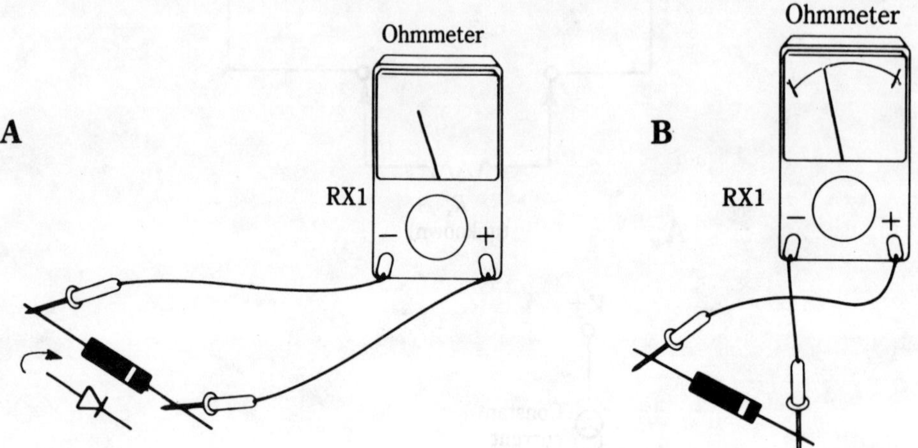

15-5 A. Ohmmeter test with reverse-bias connection; B. Forward-bias connection.

Let's assume for the sake of simplicity that the first diode under test checked out o.k. How did you know? If the diode is healthy, then one of your resistance readings will be very much higher than the other. The actual resistance readings are not terribly important; it's the ratio between the readings that counts. Consider a ratio of 5:1 or higher "normal" for older-style rectifier diodes, and 10:1 for newer rectifier diodes and small signal diodes. For example, if the low reading is 500 Ω (a typical value), then the second reading should be 5,000 Ω or more on a good diode. Diodes of very recent manufacture might read a very high ratio, or even infinite.

Small signal diodes are tested in exactly the same manner as rectifier diodes, except that the meter is set to the "RX100" scale, rather than RX1. The lower RX1 scale produces a higher current flow in the external circuit and can thus blow some small signal diodes.

Dead-shorted diodes often have a resistance near 0 Ω regardless of probe polarity (Fig. 15-6). Suppose, however, that your first and second readings are almost identical (but not 0 Ω). That diode is leaky and is almost as useless.

Shorted and open diodes test exactly as you might expect. Open diodes show a very high (or sometimes infinite) resistance, as in Fig. 15-7. On analog meters, the pointer doesn't budge off the infinity symbol, while on digital meters the display will flash "1999." A shorted diode will show either 0 Ω in both directions or a very low resistance (e.g., less than 100 Ω). Note that it is not always true that a shorted diode exhibits the same low resistance in both directions. In some cases, the resistances might be 50 and 80 Ω — that's very leaky and therefore to be considered dead-shorted for all practical purposes.

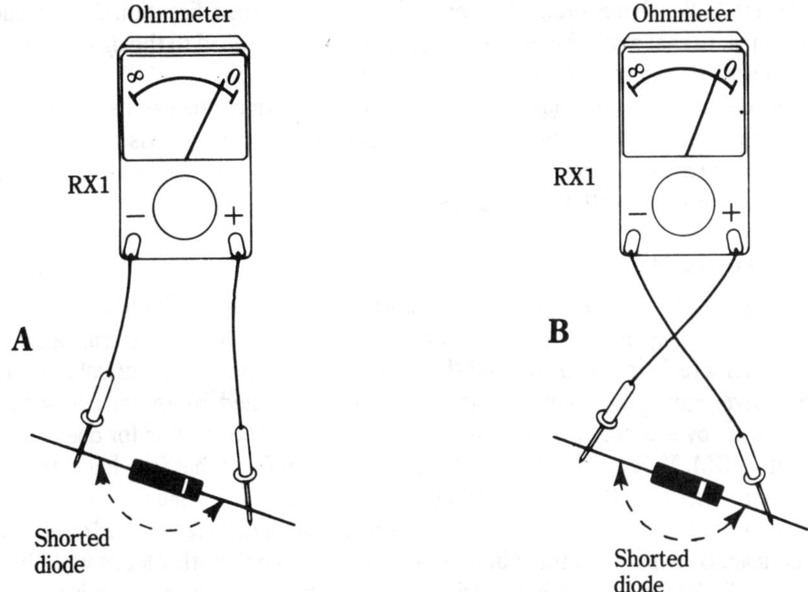

15-6   Diode testing for normal operation.

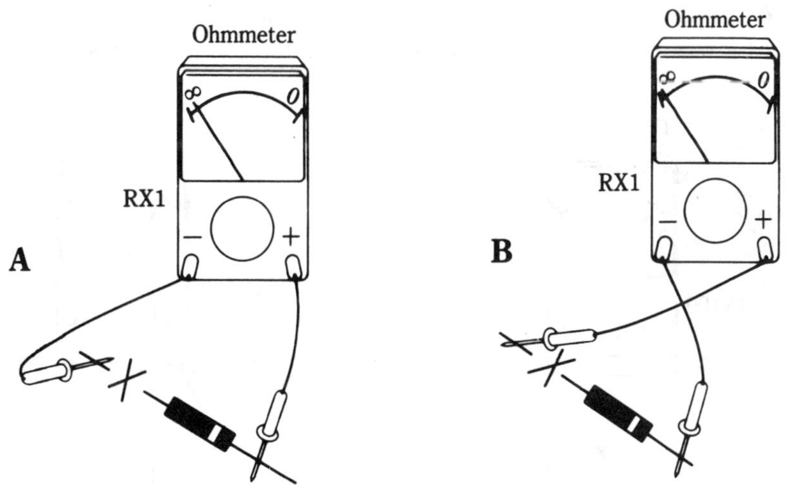

15-7   Testing an open diode.

The ohmmeter also can tell us which end of an unmarked diode is the anode or cathode. Here's where you must know the relative polarity of your ohmmeter probes. One way of determining polarity is to measure the voltage across the probes with a dc voltmeter and to note the polarity of the reading. Another method is to take a known good diode that is marked as to cathode and anode (the cathode end is usually marked somehow — with a stripe of paint, bullet shape, etc.) and connect it to the meter in a direction that produces a low resistance. In that state the ohmmeter positive lead is applied to the diode anode, and could be marked. The red ohmmeter lead is usually positive.

Once you know which probe of the ohmmeter is positive, you can use that information to tag any diode as to anode and cathode. Connect the diode across the leads to produce the low dc resistance reading—the positive lead is always connected to the anode end (and, of course, the other end is the cathode).

Testing transistors

Transistor types can be, and often are, categorized in terms of family behavior. Look in any transistor catalog and you'll see that many consecutively numbered transistor types share like characteristics. You will find that npn and pnp types are identical in performance, but have equal and opposite polarities. They are called *complementary pairs*.

Let's start by examining a small-signal pnp unit first. The method for checking transistors with VOM, VTVM, TVM, or DMM is no different from that for checking a diode. Perform your tests with the ohmmeter on the RX100 scale for small transistors and the RX1 scale for power transistors. Connect your negative ohmmeter probe to the transistor's base lead. By separately touching the collector and then emitter leads with the positive lead, you'll detect either a high resistance or a low resistance on each junction. Reversing the probe connections—placing the positive probe on the base—will show up the opposite reading (see Figs. 15-8A through 15-8D).

By now you're wondering what these resistance readings have to do with transistor testing. You've just killed two measurement birds with one stone. First, you found out if the transistor was leaky, shorted, or open between elements. This situation would have

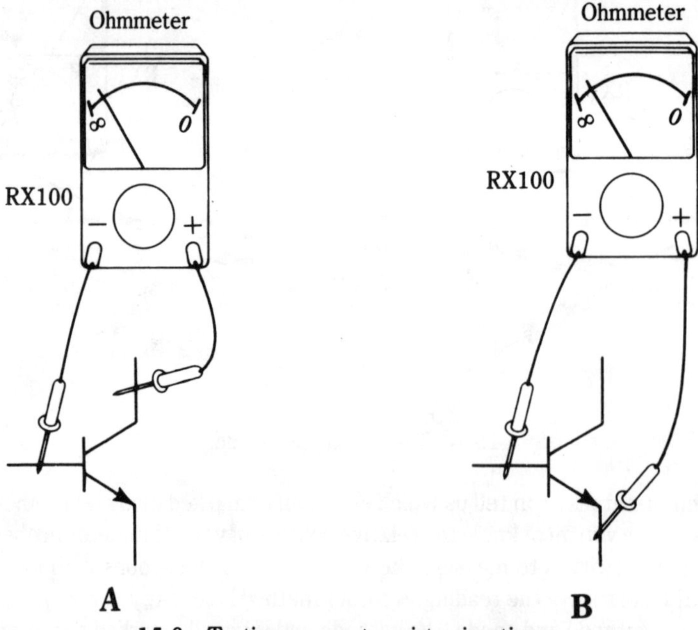

15-8   Testing an npn transistor junctions.

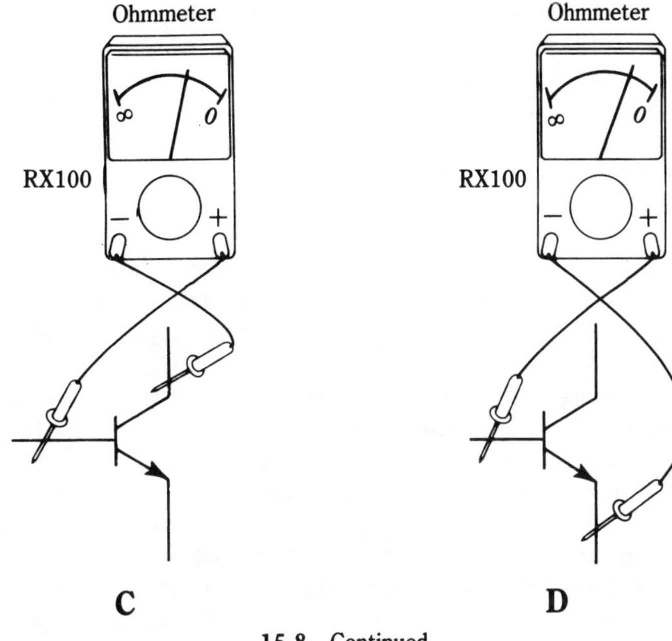

15-8 Continued

become apparent as you made your resistance measurements and found very undiodelike behavior across the junctions. Next you either confirmed or discovered the transistor's polarity; i.e., whether it was pnp or npn.

Suppose the base-emitter junction tests open. This situation is revealed as a very high resistance, whether the positive probe is connected to the base or the other way around. If the base-emitter junction tests shorted, then you'll see that this condition shows up as a very low resistance, no matter which way the ohmmeter probes are placed. Leaky transistors give you a real run for your measurement money. Indicating abnormal base-emitter resistance values, they won't always test either open or shorted — normal resistance characteristics for that particular transistor family under examination simply won't show.

Finding out whether a transistor is pnp or npn merely amounts to noting lowest resistance values as you check base-emitter junctions (refer again to Fig. 15-8). For pnp units, the lowest reading occurs when the positive probe is connected to the emitter and the negative lead is hooked to the base. For npn transistors it is just the opposite. That is, you'll get a low reading when the positive lead is connected to the base and the negative is connected to the emitter.

Explaining the unknown

Even if you didn't know that a pnp transistor was under test, you could still use the ohmmeter. The procedure isn't more difficult; you'll simply need to pay more attention to your readings. First, mount a transistor to a surface that's easy to write on — a piece of paper or cardboard will do. Label each transistor lead X, Y, and Z.

Now pick a lead at random (say X), and attach the positive ohmmeter probe to it.

Connect your negative probe to another transistor lead, say lead Y. Take a reading and write the value recorded between the leads on the paper. Now reverse your ohmmeter probes, taking another set of readings between the same two transistor leads. Again, write the value on the paper between the leads measured.

Move your ohmmeter probes to a second pair of leads (Y and Z, for instance). Repeat the resistance recording operation. Eventually, you'll want a set of values between all transistor leads.

Let's interpret the readings. One set of resistance readings will be almost identical. You've found the collector and emitter — label one lead E and the other C. (For the moment, it's entirely arbitrary which lead receives which letter.) Since you've located both emitter and collector, it stands to reason that base is your only unmarked lead. Write B next to that lead.

Assume for a moment that you've got an npn transistor labeled. If your base-emitter resistance reading was lower with the positive ohmmeter probe connected to base (base-emitter junction test), and the base-collector reading was higher with the positive ohm-meter probe base-connected (base-collector junction test), then you're looking at an npn unit. That's why the emitter and collector were arbitrarily marked — you might have to switch E and C around in order for things to make sense.

This process takes a little practice. Therefore, you might want to mount a known transistor to the paper first and experiment with this unit in order to gain practice. As expected, pnp units give opposite resistance readings with this method. Always remember to find the emitter and collector first, and then make your measurements between base-emitter and base-collector junctions.

You've gone through a lot of work sorting out good, bad, and indifferent npn and pnp transistors. Was it worth the effort? Your more immediate test result enables you to sort the "wheat" from the "chaff" in bargain lots of transistors. You also can test transistors using this method in troubleshooting situations — and get equipment back in service quicker.

Testing for gain

No transistor is worth much if its base terminal does not control collector-emitter current flow. All of the test methods given previously in this chapter will give you some gross indications of failure — an open emitter, shorted C-E junction, and so forth — but don't really tell you anything about whether or not the transistor will amplify.

Figure 15-9 shows a method used by some service technicians to determine whether or not a transistor is good. Connect the ohmmeter across the collector-emitter leads in the polarity that normally would be found in circuit. For pnp units connect the negative probe to the collector and the positive probe to the emitter; for npn units connect the positive probe to the collector and the negative probe to the emitter (as shown in Fig. 15-9). If the transistor is good, the reading will be very high, or infinity. Next, short the base lead to the collector lead. If the transistor works, then the resistance reading across C-E terminals will drop. Some people prefer using a 100K to 500K resistor instead of a short circuit (see "R" in Fig. 15-9).

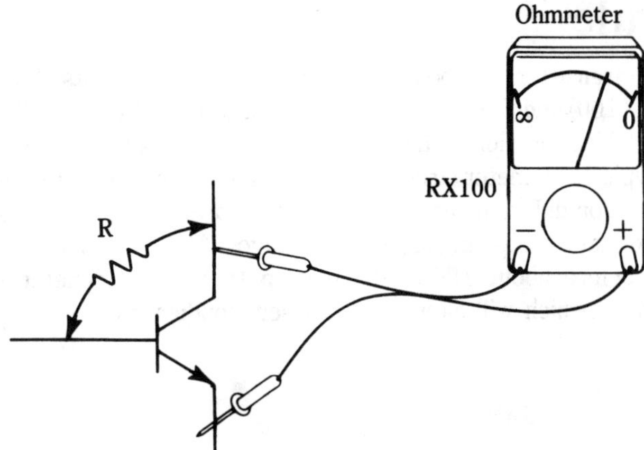

15-9   Testing an npn transistor for leakage current.

Checking leakage

Leakage in a transistor is the unwanted flow of current from collector to emitter (or vice versa). Checking a transistor for leakage is simply measuring resistance across C-E twice. Measure first with one polarity, and then switch the ohmmeter leads and measure again. The leakage reading is the higher of the two resistances. (Because of junction action, one reading is normally quite low.)

For germanium transistors the leakage will be lower than for silicon. Typically, silicon transistors will show nearly infinite resistance on the three scales normally used: RX1, RX10, and RX100. Germanium transistors might have 100K of leakage and still be good enough for use.

Using digital multimeters for testing pn junctions

The digital multimeter (DMM) has largely, but not entirely, replaced the old-fashioned VOM/VTVM. Discussed earlier was the method used in DMMs to measure resistance. Unfortunately, that method does not lend itself well to our semiconductor test method because the voltage produced across the probe tips is not sufficient to forward-bias pn semiconductor junctions. Although this feature makes it easy to make in-circuit resistance measurements without removing the semiconductors, it also prevents us from testing semiconductors. However, most recent DMMs are designed to overcome that problem. There will be a special ohms scale that will forward-bias diode junctions. These scales are marked with either words such as "High Power" or, more commonly, with the diode symbol.

Safety rule

Diodes and small transistors can be damaged by ohmmeter currents. Always start at the higher scales (RX100) and then drop down to lower scales (RX10 or RX1) only if necessary to get a readable deflection of the meter. In any event, stay on the same scale for both readings. The typical ohmmeter circuit changes currents on different ranges, so comparing readings taken on different readings means interpreting the results of two different bias levels — and that's comparing apples and mangos. Also, when using an old VTVM — i.e., one made before about 1965 — be certain that the ohmmeter battery is 1.5 Vdc, not 22.5 Vdc (which will blow out most semiconductors).

<h1 style="text-align:center">16</h1>

Selecting solid-state replacement components

EVERYONE WHO SPENDS SOME TIME AT THE WORKBENCH REPAIRING OR building electronic equipment eventually will need a transistor that he or she does not have on hand—and maybe can't easily obtain. In some cases, the type number will be found in one of the standard transistor replacement catalogs. In other cases you are on your own. The subject of transistor substitution is one that has been talked about to the point of exhaustion among communications and service technicians, yet serious problems continue to reappear. The tips given in this chapter, while most appropriate to the types of transistors normally used in audio and radio circuits, are also applicable to a wide variety of other electronic situations.

The main premise is that you are servicing radio equipment that once worked properly and then failed. This idea is terribly important. Much of what is discussed is also applicable to new construction projects and engineering laboratories, but construction project debugging is something of an arcane engineering art and is thus not suitable for general, too-broad, guidelines. Some poorly engineered equipment depends for proper operation on selected parameters of specific transistors and would not even work with all otherwise working versions of the same "2N" number devices. There are even cases on record where only those devices made by certain manufacturers will work properly in the circuit.

Ancient history note: those of us who go back to vacuum tube days recall certain very costly commercial HF receivers—the 51J4 and 51S4—that would retain their reputed "frequency meter" dial calibration accuracy only when RCA-branded tubes were used for the local oscillator. It seems that the interelectrode capacitance was crucial. There are transistor equivalents to that situation. As a result, we must limit our consideration to repair of working—hopefully properly designed—equipment.

Exact replacements

The easiest way to obtain a replacement solid-state device that will install easily and operate correctly is to order it from the original equipment manufacturer or an authorized distributor. As we are all painfully aware, however, this solution is not always either possible or practical. Some manufacturers do not sell to individuals, while others have an unrealistic minimum factory order (e.g., $100 to $500). In still other cases, the manufacturer will sell to independent shops, but wants a ridiculous amount of money for the part. Unless it is hand-selected from the general population of the same type (e.g., for some specific parameter), this pricing practice is unjustified.

Industry-standard type numbers

If the defective transistor has a standard "2N" type number, then you just obtain a replacement device having the same number and disregard the brand of the replacement. Unfortunately, some original equipment manufacturer (OEM) transistors are not marked with these standard numbers. They often have a house code number that is meaningless to anyone except the original equipment manufacturer. Sometimes, the house number is created because the transistor is specially selected from others of the same "2N" series, so only a similarly tested device will work properly in the circuit. In other cases, the house number is used because it becomes easier to control the parts inventory. In some cases it is in order to ensure themselves replacement parts business.

Crossover guides

Crossover guides would seem to be a nearly perfect source of replacement numbers. They should be used whenever possible. However, many are the gremlins that can pop up unexpectedly. Theoretically, the crossmatching has been done in advance by the use of an "infallible" computer. When you follow those recommendations, however, you will sometimes find some suggested replacements that have insufficient power or voltage ratings, too narrow a bandwidth, a different physical shape causing mounting or space problems, or wrong mounting dimensions requiring modification of the chassis. Many of these discrepancies occur because the crossovers are compiled from printed lists that sometimes contain errors. It's an open secret that the recommended substitutes are seldom tried in any kind of equipment or circuit. It is a good policy to test the reasonableness of the selection by looking at the crossover device specifications and compare them with what you know about the circuit and its requirements.

It is a good policy to return, along with a note of explanation, any crossover transistors that either don't work properly or require major reworking of the chassis or rewiring of the circuit. If everyone did so, the supplier might take the hint. The economic power of a service shop's annual semiconductor purchases makes it relatively easy to obtain a refund on bad crossovers—consumers rarely have such clout.

Another problem has nothing to do with electrical specifications, but rather with identification. Most manufacturers of radio equipment use their own "house numbers" on solid-state components. This practice causes no problems if the crossover guides list suitable replacements. In other cases it is relatively easy to guess the required transistor type. But what about the possibility that each of two manufacturers accidentally assigned the same designation to two completely dissimilar devices? It's not likely that a crossover guide will solve all such problems (although some do accommodate such ambiguities).

Remember the old rule from high school math: "Things equal to the same thing are (hopefully) equal to each other. Or, if $A = B$, and $B = C$, then $A = C$. You can use this observation to make crossover selections. Furthermore, you can use this technique at least two ways. First, you can look up the device needed to find the replacement type number. For example, suppose a 2N5xxx is found in the crossover guide as a "Z-234." You can look for other "2N" series devices also equal to ZE-234 and then use one of them.

This method is especially useful when crossing house numbers to 2N numbers — which is your second way to use the "$A = C$" theory. Suppose that your Wombat Thunderbolt VI transceiver uses a transistor with the part number "8501234." What is a "8501234?" Well, the guide calls it a ZE-234. By looking over the "2N" series columns in the crossover guide, you find a 2N5xxx is also equal to a ZE-234. Chances are good that the Wombat engineers selected the 2N5xxx and then relabeled it "8501234." It might not be the exact transistor, but it is a fair bet that it will work unless the 8501234 is a specially selected 2N5xxx. (No guarantees, however; you still have to check specs and make an educated guess.)

Up to this point, the problems of obtaining suitable replacements for transistors in mobile rigs are identical with those for base-station rigs. But there is one special problem that is more serious in mobile applications: environmental heat.

One car manufacturer became concerned about the excessive number of failures in his first all solid-state car radio models (c. 1962), and decided to investigate passenger cabin heat as the culprit. The company asked its electronics plant employees to leave their car doors unlocked for one day. During that day of 90°F weather, the engineers measured the temperatures inside many of the closed cars. They were surprised to find the average reading was 160°F on the seat and 180°F behind the dashboard.

Derating the specs

Published transistor power ratings usually are specified at room temperature, generally accepted to be 25°C (77°F). If transistors are used at higher temperatures, as in mobile applications, the maximum collector dissipation (in watts) must be reduced to prevent extra failures that could occur even when all the electrical specifications are fulfilled.

A curve typical of those used for derating is shown in Fig. 16-1. Notice that a transistor having a collector dissipation of 550 mW at 25°C can dissipate safely only 375 mW at 65°C (142°F). Therefore, a transistor that is operating below its maximum published wattage rating could be destroyed by using it in a hot car. Watch for such hazards, especially if you attempt to use "five-for-a-buck" bargain-basement replacements in which the collector dissipation rating was "optimistic."

Beyond crossmatching

Often the numbers on a bad transistor seem meaningless. There is no way to crossmatch, and you can't locate an OEM replacement from the equipment maker. The next step is to find a universal replacement from one of the many convenient sources. You must become an electronic detective and find out these things about the transistor:

1. Is it a silicon or germanium type?
2. Is it a pnp or npn?

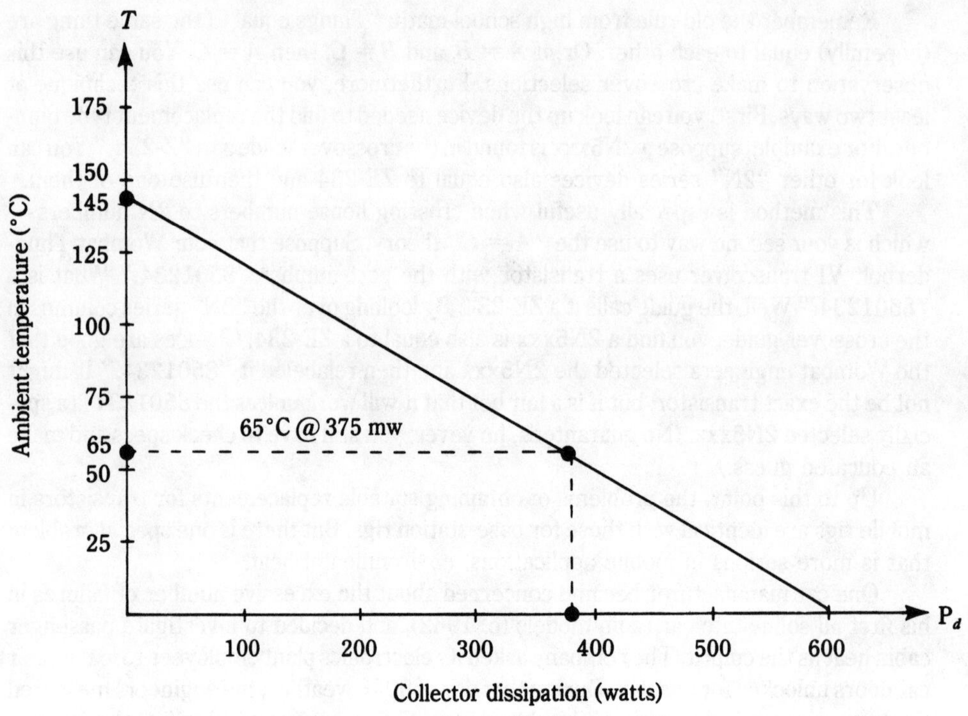

16-1   Power-derating characteristic for temperature.

3. What is the gain (alpha or beta)?
4. What frequencies must it amplify?
5. What are the collector power-dissipation requirements?
6. Are there any special mechanical mounting requirements?

After you have answered those questions, you can make a satisfactory selection from almost any brand of universal replacements.

Silicon or germanium?

Silicon transistor junctions measure higher dc resistances than germanium junctions. In fact, silicon transistors usually read "open" on all measurements except with base/emitter and base/collector forward-biasing polarity. If even one junction remains intact on a blown transistor, you can tell which material it is by comparing readings with those of known types in the same size and power category.

Forward-bias voltages for all stages other than oscillators and certain pulse circuits used in video equipment should be 0.2 to 0.3 volt for germanium transistors and 0.5 to 0.7 volt for silicon transistors. Check the schematic or another of the intact junctions (or a similarly numbered good transistor in the circuit) to see which voltage levels are found. The answer will tell you whether to look for Ge or Si transistor replacements.

pnp or npn?

When the collector voltage is more positive (or less negative) than the emitter, the transistor is an npn type. If the collector is more negative (or less positive) than the emitter, the transistor is a pnp type. Most schematics give these voltage readings. On the other hand, you can measure the collector/emitter voltages accurately enough for this purpose, right in the circuit in most cases.

If even one junction of the defective transistor is intact, you can determine the polarity by using an ohmmeter. If you obtain a normal low-resistance diode-type reading with the positive ohmmeter lead on the base and the negative lead on the collector or emitter, then it is an npn type. If you must reverse the leads to obtain a low-resistance reading, the transistor is a pnp type. This measurement must be made with an old-fashioned VOM/VTVM or a modern digital meter with a "diode" or "high-power" ohmmeter function.

Frequency response

Suppose you have installed a "universal" small-signal replacement transistor and it doesn't amplify. According to the dc voltages, however, it's drawing normal current and it doesn't heat excessively. Chances are good that the transistor selected has a bandwidth that is too narrow, and so the amplification is insufficient.

There is no one way of measuring the frequency response of transistors. Worse yet, the manufacturers can't seem to agree on the correct method of rating the frequency response. In fact, in one case three different methods are all used in the same crossover guide!

Probably that's the reason you can install a transistor rated simply as "50 MHz" and yet find it won't properly amplify a signal in a 10.7 MHz i-f amplifier in an VHF FM radio receiver. Of course, the manufacturer didn't lie, but he might inadvertently have used a rating system that doesn't fit your circuit.

As shown in Fig. 16-2, one factor is the capacitances between the junctions. Another factor is the thickness and geometry of the base region and the time it takes for the majority electrical carriers to cross the base region. If the capacitances of the replacement are too far out of tolerance, the gain will be reduced at high frequencies, and any LC tank circuits in the transistor amplifier will be mistuned.

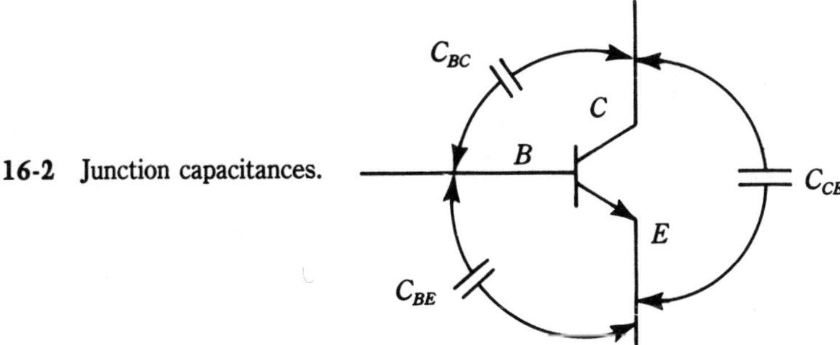

16-2   Junction capacitances.

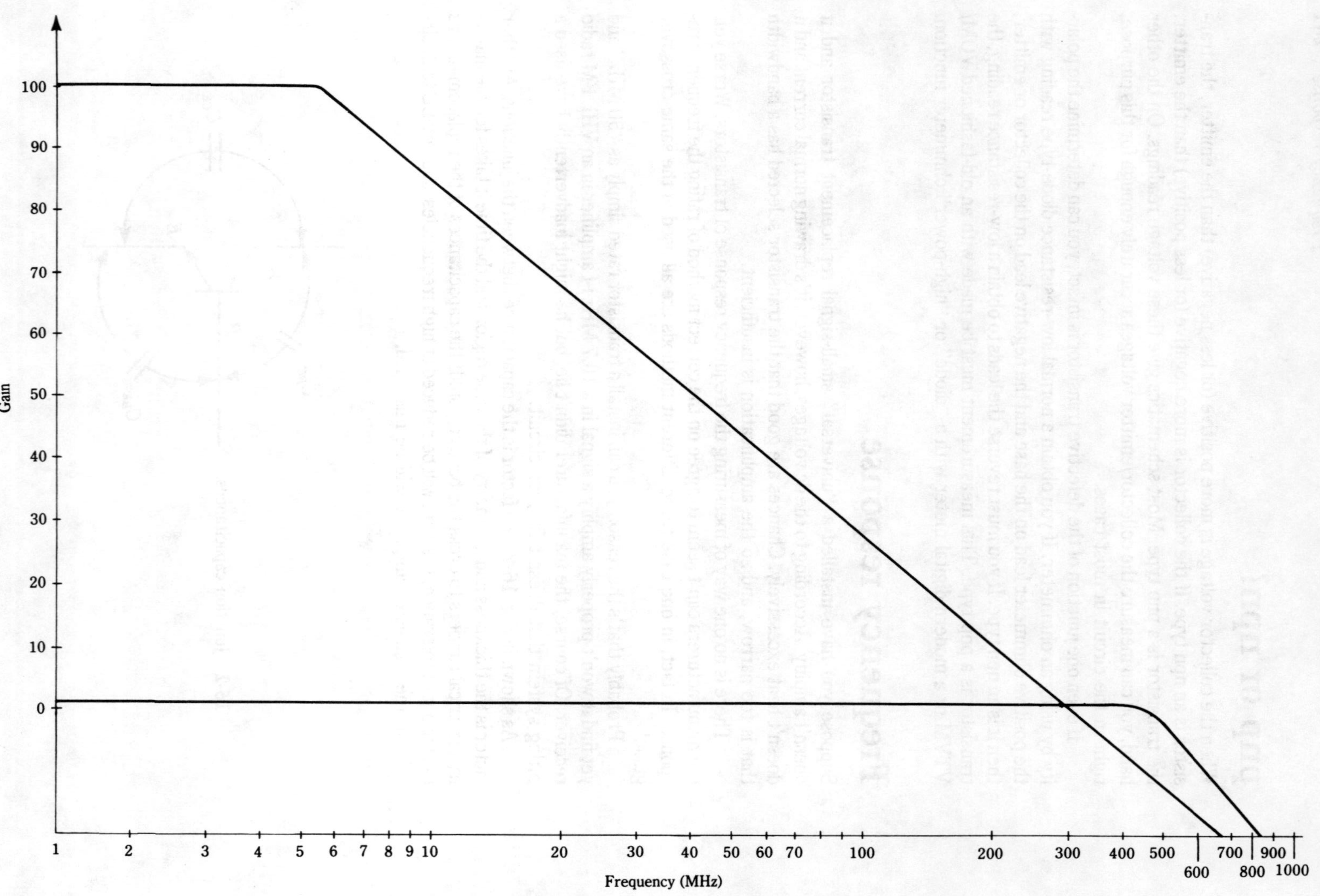

16-3 Beta and alpha frequency response characteristics.

In addition, there is the *Miller effect*, which is the effective capacitance produced by internal feedback from the output signal back to the input "amplifying" the actual capacitance. A small difference of internal capacitance can create a larger change of effective capacitance because of the Miller effect.

Alpha and beta gain

Transistor alpha gain is the ratio of the collector current to the emitter current, and can never exceed 1. Alpha gain is often measured in a common-base circuit. *Beta gain* is the ratio of collector current to base current and is usually measured in the common-emitter configuration.

The frequency cutoff point is greatly different for these two kinds of ratings, as illustrated in Fig. 16-3. A manufacturer who rates his transistors by the common-base method might correctly list them as having a far wider response than is possible by the common-emitter method. Of course, the common-base rating can provide you with a no-gain 10.7 MHz i-f stage!

Gain-bandwidth product

Another way of rating frequency response is the gain-bandwidth product method. *Gain bandwidth* is defined as the frequency at which the common-emitter gain drops to unity. For an example, let's assume a transistor with a low-frequency (1,000 Hz) beta rating of 50 and a gain-bandwidth product of 50 MHz. The gain-bandwidth product equals the beta times the common-emitter cut-off frequency, so the common-emitter cutoff frequency is found to be only 1 MHz. That's why you can act on some of the short-form specifications and still obtain a dud that won't amplify.

Analyzing maximum ratings

Care must be used in analyzing the manufacturer's maximum voltage and current ratings. Just because a transistor is listed for certain maximum collector voltage and current doesn't mean it can always operate safely at those levels. The graph in Fig. 16-4 shows that the transistor can safely stand either high current or high voltage, but not both at the same time. Maximum collector dissipation, the product of both voltage and current (watts), must not be exceeded.

Mechanical problems

Problems of physical size, connecting leads, and methods of mounting some substitute transistors are equally as vexing as those of finding a suitable electrical characteristic.

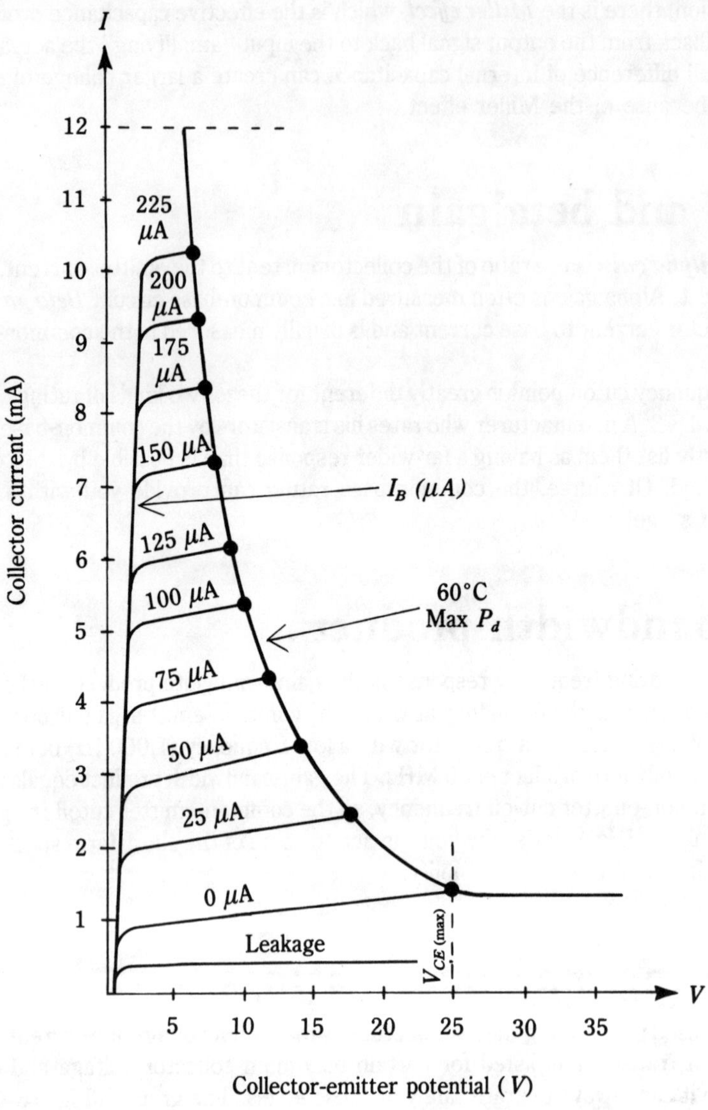

16-4 Derating curve for collector dissipation.

Conclusion

Ideally, when a replacement transistor is needed, you will have an original from the original equipment manufacturer's parts catalog. Sometimes, however, you are either unable to obtain such a transistor or the cost is prohibitive. In those cases you can usually make educated guesses as to a proper replacement type.

Appendix
Electrical safety in the electronics laboratory

ELECTRICAL APPARATUS HAS BEEN AN INTEGRAL PART OF LIFE EVER SINCE Mrs. Franklin told Old Ben to "go fly a kite." Devices operated by electricity make some jobs possible and other jobs either a whole lot easier or a whole lot more accurate. There's leaven in the electronic lump, though: electricity can be very dangerous. In fact, it can be fatally dangerous. Indeed, several of those who tried to duplicate Franklin's (and other early electrical) experiments reportedly were killed in the attempt. In this appendix you will learn the principal mechanisms of electrical hazards and some precautions to take, and who gain some insight from some nearly tragic personal experiences of my own and others who I know.

Electrical shock is only one of several different possible electrical hazards. In addition, there are other electrical safety concerns: fire, sparking (which means possible explosions), burns, microshock, and macroshock.

Fire hazards

Fire is a major home electrical hazard, as insurance company files will easily attest. Overloaded or defective electrical circuits can overheat and create a fire. Many fires every year are traced to faulty wiring or malfunctioning electrical equipment. One house down the block from my house has burned down twice in the past 20 years because of faulty electrical appliances!

Experimental circuits on the workbench often are unproven, so therefore constitute a de facto hidden fire and shock hazard. Also, things like untended soldering irons have burned down many homes and businesses over the years.

Electrical faults can also further damage the equipment that causes the problem (*secondary damage*). A short circuit that is not properly protected by either a fuse or circuit breaker might create more damage to the shorted equipment and might also damage the building wiring and electrical components. In extreme cases, a fire can result. When fuses and circuit breakers are either not used or are defeated ("penny in the fuse box" syndrome) a severe fire hazard exists, and damage to the equipment is most certainly increased.

Sparking and explosion hazards

Explosion is often less well recognized as a hazard than simple circuit overload, but nonetheless it is highly possible under the right conditions. At least three mechanisms are possible, possibly more.

Ruptures

First, an overloaded circuit or electrical component might, depending on its design, build up internal pressure (often from gas released when the device is severely overheated) and rupture spectacularly. Examples include high-voltage transformers, batteries, and capacitors.

Sparking

The second mechanism of explosion is current flow sparking in the presence of flammable liquids, gases, or vapors. If an electrical circuit is disconnected while operating or if certain faults exist, a spark might result. If that spark occurs when either flammable gases, oxygen (which isn't flammable itself, but seemingly acts that way because it promotes violent burning of other materials), or vapors (such as gasoline, certain waxes) are present, then a violent and dangerous explosion might result.

An under-the-bench "hydrogen bomb"

I once worked in a radio shop that used automobile batteries to power low-voltage dc devices being serviced on the workbench. The batteries were stored underneath the workbench on a little wooden shelf. They were charged by a trickle charger bought at an auto parts store. The system worked well as a substitute for a high-current, low-voltage ac-operated dc supply. Some amateur scientists might use this method in their own laboratory. If you do, please read and heed the lesson in this little story, for the battery nearly blinded a friend of mine — and could have killed him.

Automobile batteries are lead/acid storage batteries. The acid is sulphuric acid (H_2SO_4). Water must be added to the acid from time to time or the battery will cease working. Normal operation of these batteries produces hydrogen gas, which escapes into the atmosphere through vent holes in the water filler caps at the top of each cell of the battery. Hydrogen gas is very explosive!

In our shop, the batteries were serviced every Monday morning. My friend Hank (not his real name) stooped under the bench and disconnected the charger from the battery. That was his mistake! Immediately, a blinding flash filled the room, and our ears erupted with what sounded like a .44 Magnum "Dirty Harry" pistol going off within inches of our heads. Hank was thrown back from under the bench and, screaming, began clutching his eyes. The shop owner, my old friend the late Nelson R. Moodie, was a quick thinker and he reacted immediately. There was a garden hose in the adjacent garage (a few yards away). Nelson grabbed Hank by the shoulders and pushed him into the garage, where he turned the water hose on his eyes. Later, in the emergency room of our local hospital, Hank's clothes disintegrating from being splashed with acid, physicians told him that only Nelson's immediate reaction saved his eyesight. His face had been within inches of the battery when the spark from the charger wire exploded the hydrogen gas collected around the battery. He escaped with a few tiny flecks of gray in the white area of his eye, and no eyesight loss!

What had Hank done wrong? First, he disconnected the charger without first turning it off, so the charger current caused a spark. If a battery is to be disconnected from the circuits it powers, then they should be turned off first. Second, he failed to don the safety goggles that were provided for servicing the batteries or using machine tools. Finally, he did not use a blower fan approved for explosive atmospheres to ventilate the area and disperse the hydrogen gas. A better idea is to buy an ac-operated high-current dc power supply.

Hank was lucky, but don't you depend on "luck" for safety. Plan ahead and avoid needing "luck!"

In addition to the obvious danger of "shrapnel" from the casing of an exploding device, there is also the possibility of boiling oil or acid splattering nearby personnel. The danger from acid is obvious, but there is also a specific danger regarding the oil in some cases. Certain older capacitors and transformers were built using PCB oil as an internal coolant. PCB oil is believed to be a severe carcinogen. The importance of that statement cannot be underestimated: PCB is dangerous stuff!

Although most PCB-bearing electrical devices are now largely out of service, it is possible that some are still around, especially in older equipment. Equipment to be especially suspicious of include elderly high-power amplifiers with "oil-filled" power-supply capacitors. If you find one of these devices, it is a good idea to have it referred to a competent expert for disposal. A PCB spill can close a building for a long period of time until a proper clean-up routine is completed.

Once, when I published a PCB warning in a magazine article, a fellow wrote to me claiming the problem is greatly overstated in the popular press. That may prove to be true, but I'll leave the issue to the environmental and medical experts. They still classify PCBs as dangerous; therefore, so will I.

Static electricity

The third hazard mechanism is like the second, except that here we deal with *static electricity*. The little "bite" you get grabbing the door handle of your car after sliding across the seat or the amusing sparks seen when removing a woolen sweater in the dark is harmless under most circumstances. However, when flammable or explosive materials — especially vapors and gases — are around, a fatal explosion can occur. Many explosions in fireworks and ammunition factories are traced to static electricity.

Until about a decade ago, hospitals sometimes used flammable anesthetic gases (e.g., ether and cyclopropane) in operating rooms. Because of the very real possibility of explosion, operating rooms were designed with electrically resistive floors (5 mΩ to ground), outfitted with grounded or resistive fixtures (including carts and tables), and kept all non-explosion-proof electrical devices more than 5 feet off the floor (the gases are heavier than air). When the flammable gases were used, the room had to be rigged for explosive gases

— including conductive shoes or shoe covers for personnel, conductive liners for wastebaskets, and no synthetic undergarments. Fortunately, newer nonflammable anesthetic agents have made this hazard a thing of the past in most modern hospitals.

Electrical burns

Electrical accidents can cause third-degree burns. There are two basic mechanisms of burning. One (the most obvious) is the flash that is seen when an electrical arc occurs. The other occurs when current flows through tissue and causes the burning (basically, it's cooked). Ordinary 60 Hz ac household power can cause serious burns to human tissue, as any experienced emergency room physician can testify. However, radio frequency (RF) can also cause burning, and at least one physician (who is also a ham radio operator) told me that high-power RF burns tend to be more serious than 60 Hz power burns because RF burns go deeper into the tissue.

I recall one radio operator who (wearing Bermuda shorts) sat on a part of his antenna wire while working on it on his basement floor. Unfortunately, he had not disconnected it from the transmitter. Someone else in the family — a small child — accidentally turned on the transmitter, sending several hundred watts of radio signal into the antenna — and my friend's legs. After the cursing and screaming was over, the fellow found second-degree burns down the back of the thigh and calf of his right leg.

Microshock

Microshock is a subtle form of electrical shock, and at one time it was not widely recognized; nonetheless, microshock can prove fatal under the right circumstances. Increased use of electrical equipment in hospitals in the 1950s and 1960s led some authorities to speculate that as many as 1,200 people a year were being electrocuted accidentally in hospitals from tiny electrical currents that went unnoticed by the medical staff.

Microshock is defined as a potentially dangerous electrical shock from minute currents that are too small to affect persons with intact skin, but will affect persons if they are introduced inside the body through either a wound or medical device. Microshock is not normally a problem for amateur scientists, but you should be aware of it nonetheless.[1]

In all forms of electrical shock, a difference of electrical potential — a voltage — must exist between two points on the body. In other words, two points of contact must exist between the victim and the electrical source. That is why you sometimes see harmless "hair-raising" exhibits where a person touches an electrostatic high voltage (> 100,000 volts!) and their hair stands on end, yet they are not shocked. These potentials are essentially "monopolar" with respect to the demonstrator, so no current flow exists. Similarly, some less prudent electricians will work a circuit "hot" (i.e., without turning off the power) in seeming safety because they take care to not ground themselves, or in any other way come between the hot wire and ground, or across two hot wires. Even so, it is an extremely unsafe practice and must be discouraged at all times! So how does this situation affect microshock situations?

---

[1]Microshock can affect animals used in science experiments. Proper experimental and animal welfare protocol calls for preventing this problem.

When skin is intact, the hand-to-hand electrical resistance (i.e., opposition to the flow of electrical current) is on the order of 1,000 to 20,000 Ω. When the protective skin is breached, however, the resistance drops to a mere 20 to 100 Ω. This phenomenon is due to the fact that internal tissue is electrically similar to saltwater. Because of this lower resistance, a larger current flows when a small voltage is applied. (Ohm's law states that $I = V/R$.)

In addition to larger currents at smaller voltages, the lethal current threshold is lower in medical situations. By the late 1970s, medical and engineering experts set the maximum stray current level to which a patient can be subjected at ten microamperes (10 μA). The question is the *current density* (amperes per square centimeter, A/cm^2) in the section of the heart that contains the *sinoatrial* (SA) *node* (a natural cardiac timer or "pacemaker"). Current introduced into a blood vein has a direct electrical path to the SA node, so it can create a tremendous current density even though the absolute current level is small.

Combining lower resistance and lower lethal thresholds causes situations that normally would not be hazardous to become deadly in medical or veterinary situations. For example, using an ungrounded appliance or instrument (i.e., one that either lacks a three-wire power cord or has a broken ground wire in the three-wire cord) can cause leakage currents above the safe threshold level. Additional information about this type of hazard can be found in *Introduction To Biomedical Equipment Technology* by J.J. Carr and J.M. Brown (Prentice-Hall). Fortunately, starting in the early 1970s, medical equipment designers and hospitals became more aware of electrical problems and effectively controlled the situation by improved designs.

Macroshock

Macroshock is the type of electrical shock to which all people (not just hospital patients) are subject, and comes from direct contact with an electrical source. If you touch the 110 Vac line, while grounded, then a very painful — and possibly fatal — electrical shock will occur. This event is macroshock and does not require wounds or other breeches of the skin.

How much current is fatal?

I once worked in a hospital electronics laboratory. One day I overheard an intern claim that 110 Vac from the wall socket is not dangerous because they told him in medical school that it's not the voltage that kills, it's the current. I asked him: "Doctor, have you ever heard of Ohm's law?" According to Ohm's law, the current is the quotient of voltage and resistance, so the two are related! Besides, a little statistic that the inexperienced young doctor apparently didn't know is that 110 Vac from residential wall sockets is the most common cause of electrocution in the United States. In addition, medical studies reveal that the 50 to 60 Hz frequency used in ac power distribution almost worldwide (60 Hz is used in the United States and Canada) is the most dangerous range of frequencies.

Higher and lower ac frequencies are less dangerous (but not safe!) than 60 Hz ac. Recall that the killing factor is current density in a certain area in the right atrium of the heart, called the sinoatrial (SA) node. Any flow of current through the body that causes a high level of current to flow in that section of the heart can induce a fatal cardiac arrhythmia called *ventricular fibrillation*, which causes death within minutes if not corrected. In general, for limb-contact electrical shocks through intact skin (macroshock), the following rules of thumb are accepted:

1 to 5 mA	Threshold of perception
10 mA	Threshold of pain
100 mA	Severe muscular contraction
100 to 300 mA	Electrocution

Keep in mind that these figures are approximate, and are not to be accepted as guidelines to approximate "assumed risk." Death can occur under certain circumstances with considerably lower levels of current. For example, when you are sweating or are standing in saltwater, then risks escalate tremendously. Also, certain current-flow patterns or geometries through the body are more dangerous than others, so the levels listed might be far too low in some situations.

Mechanisms of electrical shock — some scenarios

Figure A-1 shows the schematic for the usual U.S. residential ac electrical system. Industrial electrical systems are a bit different at the service entrance, but become much like Fig. A-1 when the power is distributed throughout the building.

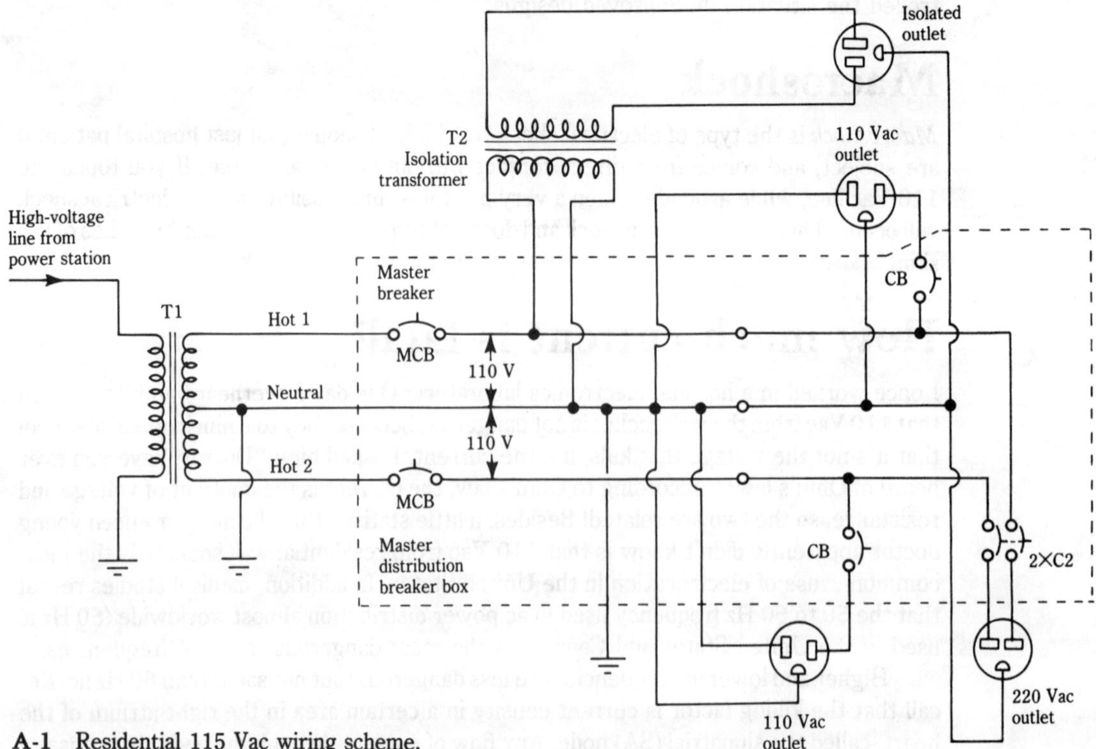

A-1   Residential 115 Vac wiring scheme.

A-2   "Pole pig" residential service transformer.

The power company distributes energy through high-voltage lines. When it arrives at a point only a short distance from the customer, it is stepped down in a "pole pig" transformer (Fig. A-2) to 220 Vac center-tapped. The center-tap (C.T.) of the transformer secondary is grounded, and therein lies the root of the problem. The two ends of the 220 Vac secondary are brought into the building as a pair of ground-referenced 110 Vac hot lines. Tapping across the two lines produces a 220 Vac outlet; tapping from the ground line (i.e., transformer C.T.) to either hot line produces a 110 Vac outlet.

In order to raise your consciousness about how shock can occur, let's take a look at certain scenarios of electrical shock that might occur. Figure A-3 shows the direct approach to fatal electrical shock. You are grounded through either bare feet or conductive shoes[2] and touch an electrically hot point. You need not be outdoors to be affected by this

---

[2]Note: Normal shoes in good condition can become conductive when wet or if previously exposed to salty or dirty water.

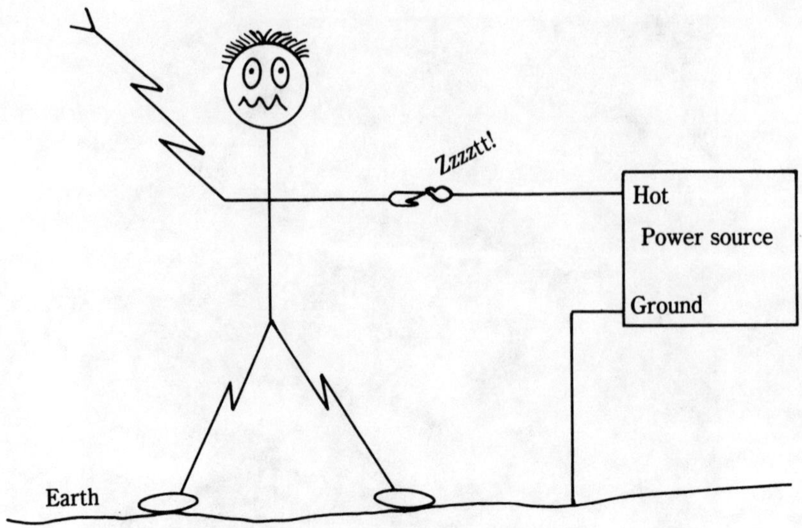

A-3   Direct contact hazard.

scenario. A concrete garage, shop, or basement floor is a reasonably good conductor, as are wet leather and some forms of rubber shoe.

Figure A-4 shows an indirect scenario that especially affects people using electronics instruments. Consider the grounded instrument probe (in this case an oscilloscope). When you grasp that probe, you might be grounded through the 'scope shield and the power cord ground conductor. If you touch a "hot" point, you will get shocked — and maybe killed.

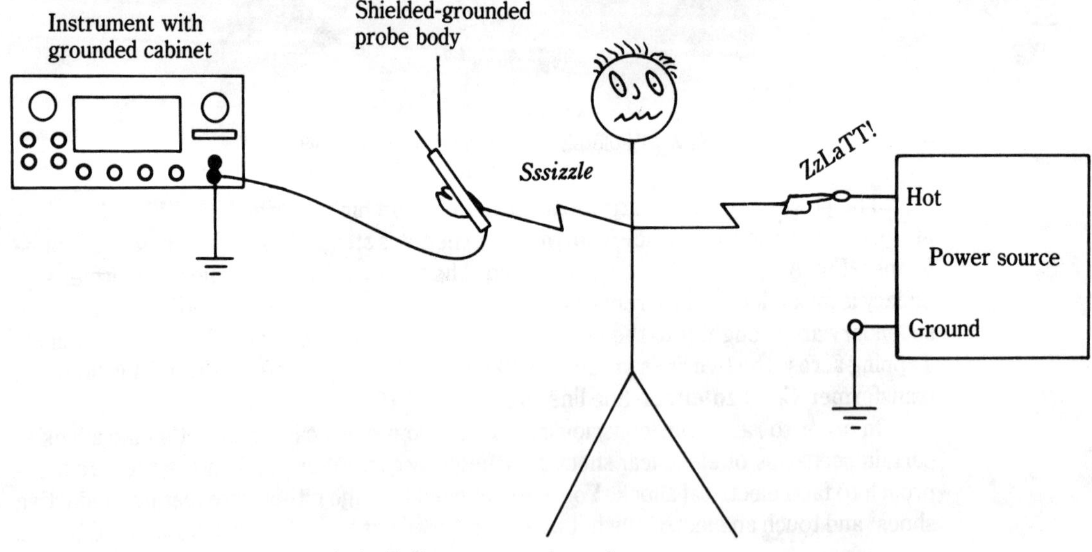

A-4   Indirect contact hazard.

A related scenario is shown in Fig. A-5. Here is an ac/dc consumer appliance, such as a low-cost radio or TV set. Note that the oscilloscope probe ground is connected to the set ground, which also happens to be one side of the ac power line. Everything is fine as long as the ac plug is oriented correctly in the wall and the wall socket is wired correctly. If, however, you plug it into the wall receptacle backward, then there will be an explosive short circuit and possible electrocution of the operator.

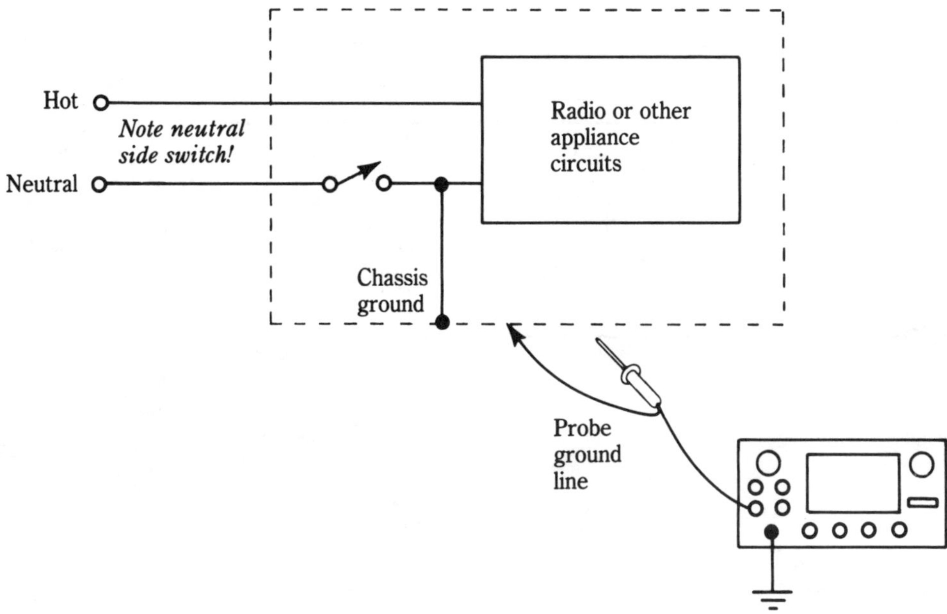

A-5   Chassis ground connection.

Another scenario is the fatal antenna or tower erection job. It is never good practice to erect an antenna near a power line! Every year we hear stories of people electrocuted because: an antenna they were working on fell across the power lines; they tried to toss a wire antenna over the power line in order to raise the antenna above the lines; or a ladder they were using fell across the power lines. Foolish! These tactics will kill you. Incidentally, this reason is why safety-approved industrial ladders are made of wood or other nonconductive material—and not of aluminum (as are consumer ladders).

Some cures for the problems

Now that we have discussed some of the mechanisms of electrical hazards, let's take a look at some cures. The problem that makes electrical apparatus so dangerous (especially outdoors or on concrete garage or basement floors) is that the electrical system in the United States is ground-referenced. The solution is to make the little "local" electrical system on your lab bench non ground-referenced. This method is used in hospital operating rooms and in some intensive care units for patient-safety reasons. It is also commonly used on radio-TV or electronic service benches, especially if ac/dc power supplies are serviced.

Figure A-6 shows the wiring for such a system. Transformer T1 is one of two forms of isolation transformer: a 1:1 transformer gives a 110 Vac isolated (non ground-referenced) ac line from a 110 Vac standard line; a 2:1 transformer does the same thing from a 220 Vac line.

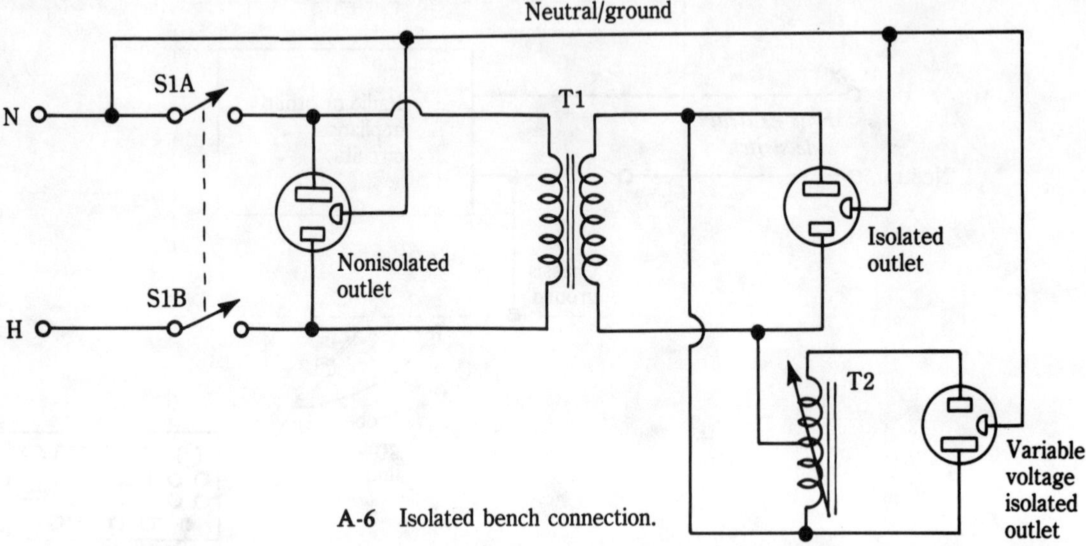

A-6   Isolated bench connection.

The second transformer in Fig. A-6, T2, is an autotransformer and is optional. It is used for varying the voltage on the ac line serving the workbench. It typically will allow you to set the output voltage from 90 Vac to 140 Vac when 110 Vac line voltage is applied.

The metal oxide varistor (MOV) is used to clip the amplitude of high-voltage power-line transients (lasting > 20 microseconds or so, > 180 volts) that could either damage or interfere with the operation of the equipment on the bench. These devices are especially needed when computers or other digital electronic devices are used, and they are the basis for the so-called "surge supressor" outlet strips sold in computer stores.

The circuit breaker or fuse is used to protect equipment on the bench, as well as the transformer. It is always placed in the hot line, or in both lines. Fuses and circuit breakers are never placed in the neutral line. The switching shown in Fig. A-6 breaks both lines. I prefer this approach on the theory that hot and neutral lines can be reversed accidentally and leave you in the position of breaking the neutral line and leaving the hot line alive when you think it is dead!

Rule to stay alive by

If you use electrical apparatus, especially outdoors or on a concrete or dirt floor, then an isolation transformer is not optional. Safety requires that it be used, and it's cheap "life insurance."

Equipotential grounding is used in many workshops to reduce the hazard of electrical shock. An *equipotential ground* is a system in which all grounded points are kept at the same exact electrical potential at or near ground (0 volts). In addition to being less hazardous, the equipotential ground system also works better where electronic recording devices and other instruments are used, so it has a double benefit.

To make an equipotential ground system (Fig. A-7), connect a heavy wire or copper braid between all normally grounded points, including the cabinets of instruments and appliances (*Warning!* Do not use ac/dc electrical devices in an equipotential ground system — in fact, don't use them at all!). This ground wire is extra to, that is redundant to, the normal power cord ground wire (the "third wire") on the equipment.

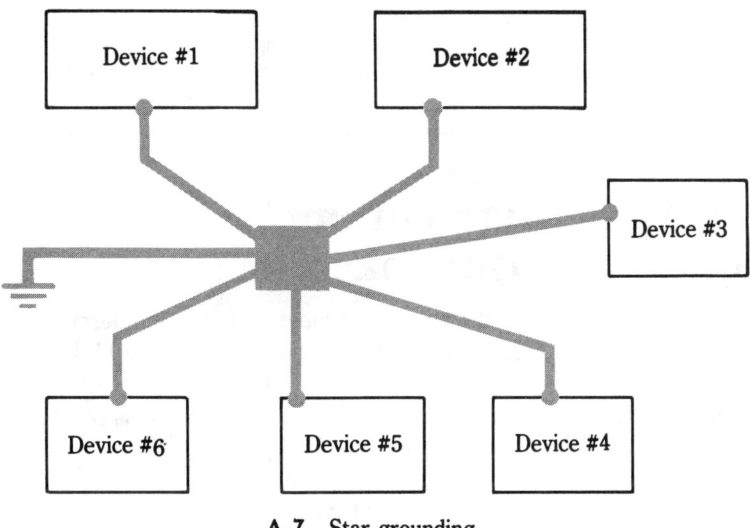

A-7   Star grounding.

The method of Fig. A-7 shows *star grounding*, also called *single-point grounding*, which is superior for scientific instrument situations. In an equipotential ground system, a short circuit to the cabinet of an instrument or other devices goes to ground even if the ground wire in the cord is broken (Fig. A-8). It blows the fuse, rather than blowing you out of existence.

Note: Power cord grounds — the third (green) wire — often break because they are under strain. That break remains hidden because the equipment continues to operate normally. However, the safety of the system is compromised when the wire is broken. That is the reason why redundant ground wires are preferred.

A second approach to safety is just the opposite of equipotential grounding. Many modern tools or appliances that are designed for use outdoors or on concrete floors are double insulated. These devices have two-wire power cords, but are clearly labelled "double insulated," or in some other manner that suggests outdoor use. Don't accept the labeling, however, unless an Underwriter's Laboratory (UL) sticker is present.

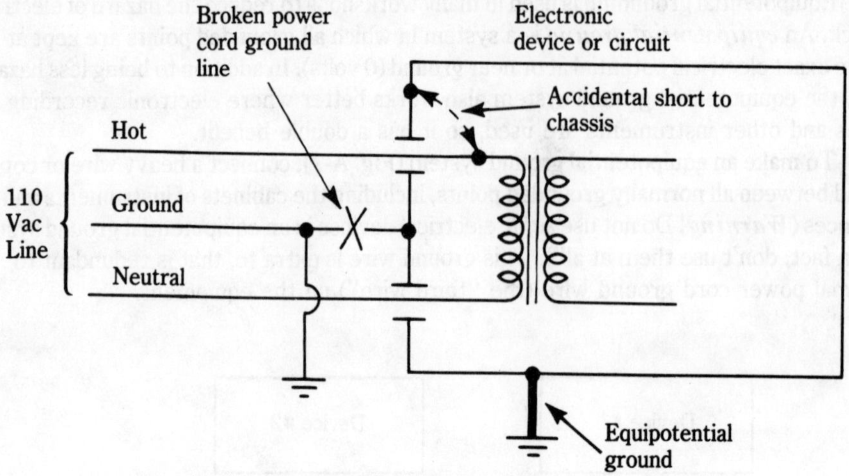

A-8   Broken ground wire hazard.

Low-voltage, high-current sources — the overlooked hazard

I once attended an engineering design review meeting for a commercial marine radio transceiver. The specification called for insulation of low-voltage (< 28 Vdc), high-current (> 10 amperes) dc power-supply terminals. One of the engineers present sneered that was something like asking him to insulate the battery terminals of his car. Implied in that contemptous snarl is that low voltage can never hurt you. There are two false premises resident in that opinion.

First, although low-voltage, high-current points rarely cause electrical shock, it is possible for dangerous shock to occur when the person has a very low electrical skin resistance (is very sweaty) or has an open wound. Although the case did not result in electrocution, an electronics technician I know recently injured himself severely when he cut himself on a $+5$ Vdc, 30 A computer power supply that he was repairing. A large amount of current flowed in his arm and caused severe pain and some physical damage.

Second, the high current is extremely dangerous if you happen to be wearing jewelry! A two-way radio shop where I once worked used 12 V lead-acid storage batteries and battery chargers for the troubleshooting bench supply for mobile service. A technician working on the battery rack dropped a wrench, and it fell onto the battery, making contact from $(-)$ to $(+)$ through his watchband. The large current turned the watchband red hot, and gave him serious second- and third-degree burns on the wrist.

Don't assume that low-voltage, high-current power supplies are harmless. Insulate them all!

Design your workshop for safety

An old safety cliché tells us that "safety is no accident." The truth behind the cliché is that good planning and good work habits result in a safer situation. Certain principles can be applied to your own workshop (or other venue) to avoid electrical hazards. Keep in mind that it's impossible for anyone to foresee all possible situations. Therefore, it is incumbent upon you to evaluate these recommendations for applicability, and to ensure that there are no other situations that require other methods or approaches to safety.

Master electrical cut-off switch   A single electrical cut-off switch in either the hot line (or both lines if 220 Vac is used) of the ac power line should be installed. This switch should be clearly labeled "Master Power Cut-off," or something similar, and everyone in the family instructed about its use. The switch should de-energize every outlet and light on the bench except the overhead lights. (How can they save your life if the room is dark?) Every member of the family, including small children, should be instructed how to use this switch — and to immediately turn off the switch in the event of a mishap.

Ground fault interrupters (GFI)   All ac power outlets in the work area should be of the ground fault interrupter (GFI) type. In most localities, the electrical code requires GFIs in wet areas (e.g., bathrooms, photo darkrooms), outdoors, or in basements and garages, so they are not optional.

Insulating floor covering   The floor of the workshop should be covered with a nonconductive carpet or mat, unless explosive or highly flammable materials are handled. In dry areas, carpeting, wood, masonite, plastic, and rubber are suitable. In wet or damp areas (including any concrete floors), however, wood, carpets, masonite, and some (porous) rubber materials will absorb moisture, so are not suitable as a floor covering.

If you work with flammable or explosive materials, consult an expert about a high-resistance grounded floor covering that will drain static electricity off harmlessly, while not putting you in a dangerous situation from ground-seeking ac power. Hospitals and some electronic assembly areas (some electronic parts are very sensitive to static voltages!) have these types of floors, so consult them on who designed their floor covering.

Isolation transformers   The workbench should be equipped with an isolation transformer that has sufficient volt-ampere (e.g., wattage)[3] capacity to power all of the devices on the bench with some to spare.

Grounded ac equipment   Select ac operated instruments and apparatus that are normally grounded. Avoid ac/dc devices! The grounded variety can be identified by the three-prong (instead of two-prong) power plugs. It is also wise to redundantly ground all the cabinets and normal ground terminals together on an equipotential ground bus.

Electrical connections   Never use bare wires or uninsulated electrical connections, even in low-voltage cases. Use proper connectors — and avoid electrical tape (the cloth kind comes loose of its own accord; the plastic kind comes loose when very hot or very cold; and they both come loose in high humidity).

Ventilation   If you work with potentially flammable or explosive materials, make sure that you ventilate the area properly. Only an expert can advise on the design of ventilation for any particular material, and in most localities the fire marshall may also have an opinion or two about the matter.

Explosion-proof fixtures   Working with explosive or flammable gases or vapors is extremely hazardous. Normal electrical switches arc internally when activated or deactivated. Normal plugs and connectors arc momentarily when brought together. They can

---

[3]Volt-amperes equals watts only in resistive circuits. If inductors, capacitors, motors, or certain other devices are in the circuit, then there is a discrepancy between V-A and watts.

cause explosions! A gasoline station near my home allegedly blew up because unvented gasoline fumes were ignited by turning on the light switch on the mechanic's drop light! If you insist on working in such an environment, then use only explosion-proof switches, connectors, and appliances.

And just where would you get an explosive environment in a residential structure? Well, unless you heat strictly with electricity, then you will more than likely use fuel oil, natural gas, or LP gas for energy. If those systems leak, the structure (especially basements) will fill with an explosive mixture of gas or fumes.

Also, take a look at other activities, including hobbies, in your building. I once took an amateur jewelry-making course. One of the people using the facility told me that he used a tank of oxyacetlyene for the soldering torch needed to melt gold and silver solders. He left the master tank valve on, with a leaky handset valve, and darn near had a serious accident. It was only coincidental that he discovered the problem and called the fire department to ventilate the room with an explosion-proof fan.

Check wiring   Under normal circumstances it might not be important whether or not the last electrician who wired the building or house accidentally switched the hot and neutral lines. However, in amateur science labs, or anytime an appliance is used outdoors, those errors (which do happen, by the way) can be fatal! Check the wiring with an outlet tester.

What to do for an electrical shock victim

The usual mechanism of death by electrical shock is usually a phenomenon called ventricular fibrillation (V. Fib.), an arrhythmic heartbeat in which the heart merely quivers, instead of beating. Unfortunately, V. Fib. is not capable of sustaining blood-pumping effectiveness, so the victim dies within a few minutes unless someone trained in cardiopulmonary resuscitation (CPR) is nearby.

Before you can aid the victim of electrical shock, you must be sure that either the victim is away from the current or the current is turned off! Otherwise, when you touch the victim in order to help him or her, you will also become a victim!

As soon as the victim is clear of the electrical current, immediately initiate CPR and send for help. CPR will not bring the victim out of V. Fib. Its function is to provide life support until properly equipped and trained medical personnel can be summoned. They will use a device called a *defibrillator* to shock the victim's heart back into correct rhythm. They will also use drugs and intravenous solutions in order to reestablish the body's balances.

None of these actions can be performed properly by untrained people. In fact, even simple CPR cannot be performed effectively by the untrained person. Everyone who works near, on, or around electrical or electronic equipment should learn CPR. In addition, teenage and adult family members also should learn CPR. After all, who is going to save you when the electrical accident occurs at home in your laboratory, ham shack, or workshop?

The local Red Cross, the Heart Association, some community colleges, and most local hospitals can direct you to certified CPR courses. It is impossible for you to learn CPR from watching "medical" or "rescue" shows on TV or in the theatre, so get trained by a knowledgable, certified instructor!

Like a hissing cobra!

Keep in mind while you read this story that I am experienced working on electrical and electronic apparatus. For more than twenty years I earned my living on the bench. Perhaps that's why I almost "bought it" one day working on some equipment.

I was working on a 110 Vac appliance in a hospital laboratory where I was employed as a technician. The equipment was live, and it was not hooked to an outlet served by the bench isolation transformer. Forgetting that it was live, I disconnected a 110 V wire and started to move it. The end of the wire hit the wrench in my other hand and found a path to ground through my elbow. My muscles tightened up as the current coarsed through my forearm, the sparking end of the wrench was "hissing like a cobra" according to a witness. I was fortunate. A friend was with me, and he yanked the cord out of the wall socket. Stupid! I knew better, but forgot the rules of good common-sense safety.

Some general safety points

There is only one way to ensure that the ac line won't shock you: disconnect it. Make it your practice to never work on equipment that has the plug inserted into the power outlet. Don't trust switches, fuses, circuit breakers, or other people. If someone were to hand you a pistol, claiming that it was unloaded, the first thing you'd do is check it yourself. The same advice also holds true for the electrical connection (which can kill you just as dead as a loaded and cocked pistol).

It is often advised that you work on electrical devices with your left hand in your pants pocket. That advice is based on the theory that the "left hand to either leg" current path is supposedly the most deadly because it passes through the heart. Even if the physiology is correct, placing one hand in your pocket leaves you awkward, and you are unable to safely work on the circuit with only one hand. It is better to use both hands and arrange it so that the work environment is safe — and use safe technique.

When working on high-voltage dc circuits, keep in mind that capacitors store electrical charge. All large-value ($> 1 \, \mu$F) capacitors must be discharged manually after the power is turned off. The capacitor must be discharged multiple times. Even when a short circuit is placed across the capacitor terminals, not all of the energy is removed the first time. Some energy is stored in the capacitor's internal dielectric material even after the main charge is discharged.

How do you go about shorting the capacitor terminals? With a screwdriver? With an alligator clip-lead? Nope! Use a shorting stick (Fig. A-8). Do not make the stick out of wood! Instead, use a dielectrically competent plastic material. Also, use more than one ground line, just in case one of them falls off while you are working, there will be one or more left to carry off the charge. Also, be sure to wear eyeglasses or safety goggles in order to prevent eye damage from the flying sparks.

The only ground available to the equipment chassis when it is working on the bench is the power cord ground, which is insufficient for safety. My workbench has a grounded 5/16-inch bolt on the back edge. When I work on electrical equipment, I ground the chassis with a heavy cable to that ground point. Don't overlook this ground.

Conclusion

There is no such thing as complete, failure-free safety anywhere. However, proper recognition of the mechanisms of danger and proper management of the risks will ensure that the environment is as safe as possible.

Index

Other Bestsellers of Related Interest

ENCYCLOPEDIA OF ELECTRONIC CIRCUITS—Vol. 1—Rudolf F. Graf

". . . schematics that encompass virtually the entire spectrum of electronics technology . . . This is a well worthwhile book to have handy."

—Modern Electronics

Discover hundreds of the most versatile electronic and integrated circuit designs, all available at the turn of a page. You'll find circuit diagrams and schematics for a wide variety of practical applications. Many entries also include clear, concise explanations of the circuit configurations and functions. 768 pages, 1,762 illustrations. **Book No. 1938, $32.95 paperback, $60.00 hardcover**

LEARNING ELECTRONICS:
Theory and Experiments with
Computer-Aided Instruction for the IBM
—R. Jesse Phagan and William Spaulding

This easy-to-use guide combines electronics theory and hands-on practice with computer-aided instruction. Designed as a self-study guide to be used in conjunction with an IBM PC or compatible, this book is perfect for the electronics student or hobbyist. Computer programs are provided to graphically show concepts and mathematical relationships. 340 pages, 271 illustrations. **Book No. 3082, $16.95 hardcover only**

BASIC ELECTRONICS THEORY
—3rd Edition—Delton T. Horn

"All the information needed for a basic understanding of almost any electronic device or circuit . . ." was how *Radio-Electronics* magazine described the previous edition of this now-classic sourcebook. This completely updated and expanded edition provides a resource tool that belongs in a prominent place on every electronics bookshelf. Packed with illustrations, schematics, projects, and experiments, it's a book you won't want to miss! 544 pages, 650 illustrations. **Book No. 3195, $22.95 paperback only**

ELECTRONIC COMPONENTS:
A Complete Reference for Project Builders
—Delton T. Horn

Get the most out of almost any electronic component. This benchtop reference catalogs characteristics, specifications, and component uses that range in complexity from basic wire and solder to transistors and ICs. And it presents insights into the theory and operation of components in typical circuit designs, the pros and cons of using devices in various situations, where and how to find parts, and criteria for making substitutions. 328 pages, 300 illustrations. **Book No. 3671, $18.95 paperback, $29.95 hardcover**

BASIC DIGITAL ELECTRONICS
—2nd Edition—Ray Ryan and Lisa A. Doyle

An easy-to-follow digital electronics reference guide, this book has long been the key to analyzing and understanding digital circuitry. Now this popular reference includes microprocessor basics, alpha-numeric codes, and information on combinational logic and memories. You will also find thorough descriptions of digital codes, conversions, and number systems; explanations of the basics of logic gates, families, and networks; and example applications. 256 pages, 166 illustrations. **Book No. 3370, $16.95 paperback only**

DESIGN & BUILD ELECTRONIC POWER SUPPLIES—Irving M. Gottlieb

Create top-quality electronics power supplies with these on-the-job tips and real-world examples. This guide brings you up to date on today's most advanced power supply circuits, components, and measurement procedures. It emphasizes practical applications over mathematical theory while focusing on the devices and techniques used with the newer types of switch-mode power supplies. 176 pages, 98 illustrations. **Book No. 3540, $17.95 paperback, $26.95 hardcover**

ELECTRONIC SIGNALS AND SYSTEMS: Television, Stereo, Satellite TV, and Automotive—Stan Prentiss

Study signal analysis as it applies to the operation and signal-generating capabilities of today's most advanced electronic devices with this handbook. It explains the composition and use of a wide variety of test instruments, transmission media, satellite systems, stereo broadcast and reception facilities, antennas, TV equipment, and even automotive electrical systems. You'll find coverage of C- and Ku-band video, satellite master TV systems, high-definition television, C-QUAM® AM stereo transmission and reception, and more. 328 pages, 186 illustrations. **Book No. 3557, $19.95 paperback, $29.95 hardcover**

MASTERING IC ELECTRONICS
—Joseph J. Carr

If you have a basic understanding of integrated circuitry but you have not yet reached an advanced level of expertise, this book is for you. It gives you all the data you need to become well versed in the theory and operation of IC electronics. This book's unique blend of principles and practice provides you with a full working knowledge of component specifications, design standards, and applications for all types of modern ICs. 416 pages, Illustrated. **Book No. 3660, $19.95 paperback, $32.95 hardcover**

SECRETS OF RF CIRCUIT DESIGN
—Joseph J. Carr

This book explains in clear, nontechnical language what RF is, how it works, and how it differs from other electromagnetic frequencies. You'll learn the basics of receiver operation, the proper use and repair components in RF circuits, and principles of radio signal propagation from low frequencies to microwave. You'll enjoy experiments that explore such problems as electromagnetic interface. 416 pages, 411 illustrations. **Book No. 3710, $19.95 paperback, $32.95 hardcover**

THE COMPLETE BOOK OF OSCILLOSCOPES—2nd Edition
—Stan Prentiss

Save hours of troubleshooting time by mastering top-of-the-line scopes with the professional advice in this book. It reviews the oscilloscope equipment and advanced testing procedures developed during the last five years. You'll appreciate this book's easy-to-read style, logical format, and original photographs taken by video communications expert Stan Prentiss in his electronics laboratory. 320 pages, 165 illustrations. **Book No. 3825, $16.95 paperback, $26.95 hardcover**